W9-DHX-416

Cell Growth, Differentiation and Senescence

The Practical Approach Series

SERIES EDITOR

B. D. HAMES
Department of Biochemistry and Molecular Biology
University of Leeds, Leeds LS2 9JT, UK

See also the Practical Approach web site at **http://www.oup.co.uk/PAS**

★ **indicates new and forthcoming titles**

Affinity Chromatography
Affinity Separations
Anaerobic Microbiology
Animal Cell Culture (2nd edition)
Animal Virus Pathogenesis
Antibodies I and II
Antibody Engineering
★ Antisense Technology
Applied Microbial Physiology
Basic Cell Culture
Behavioural Neuroscience
Bioenergetics
Biological Data Analysis
Biomechanics – Materials
Biomechanics – Structures and Systems
Biosensors
Carbohydrate Analysis (2nd edition)
Cell-Cell Interactions
The Cell Cycle
Cell Growth and Apoptosis

★ Cell Separation
Cellular Calcium
Cellular Interactions in Development
Cellular Neurobiology
★ Chromatin
★ Chromosome Structural Analysis
Clinical Immunology
Complement
★ Crystallization of Nucleic Acids and Proteins (2nd edition)
Cytokines (2nd edition)
The Cytoskeleton
Diagnostic Molecular Pathology I and II
DNA and Protein Sequence Analysis
DNA Cloning 1: Core Techniques (2nd edition)
DNA Cloning 2: Expression Systems (2nd edition)
DNA Cloning 3: Complex Genomes (2nd edition)

Cell Growth, Differentiation and Senescence

A Practical Approach

Edited by

GEORGE P. STUDZINSKI

Department of Pathology and Laboratory Medicine,
UMD—New Jersey Medical School,
Newark, N.J., USA

OXFORD

UNIVERSITY PRESS

OXFORD

UNIVERSITY PRESS

Great Clarendon Street, Oxford OX2 6DP

Oxford University Press is a department of the University of Oxford
and furthers the University's aim of excellence in research, scholarship,
and education by publishing worldwide in

Oxford New York

Athens Auckland Bangkok Bogotá Buenos Aires Calcutta
Cape Town Chennai Dar es Salaam Delhi Florence Hong Kong Istanbul
Karachi Kuala Lumpur Madrid Melbourne Mexico City Mumbai
Nairobi Paris São Paulo Singapore Taipei Tokyo Toronto Warsaw
and associated companies in Berlin Ibadan

Oxford is a registered trade mark of Oxford University Press

Published in the United States
by Oxford University Press Inc., New York

© Oxford University Press, 1999

All rights reserved. No part of this publication may be reproduced,
stored in a retrieval system, or transmitted, in any form or by any means,
without the prior permission in writing of Oxford University Press.
Within the UK, exceptions are allowed in respect of any fair dealing for the
purpose of research or private study, or criticism or review, as permitted
under the Copyright, Designs and Patents Act, 1988, or in the case
of reprographic reproduction in accordance with the terms of licenses
issued by the Copyright Licensing Agency. Enquiries concerning
reproduction outside those terms and in other countries should be
sent to the Rights Department, Oxford University Press,
at the address above.

This book is sold subject to the condition that it shall not, by way
of trade or otherwise, be lent, re-sold, hired out, or otherwise circulated
without the publisher's prior consent in any form of binding or cover
other than that in which it is published and without a similar condition
including this condition being imposed on the subsequent purchaser

Users of books in the Practical Approach Series are advised that prudent
laboratory safety procedures should be followed at all times. Oxford
University Press makes no representation, express or implied, in respect of
the accuracy of the material set forth in books in this series and cannot
accept any legal responsibility or liability for any errors or omissions
that may be made.

A catalogue record for this book is available from the British Library

Library of Congress Cataloging in Publication Data
Cell growth, differentiation, and senescense : a practical approach /
edited by George P. Studzinski.
(Practical approach series ; 215)
Includes bibliographical references and index.
1. Cells—Growth—Research Laboratory manuals.   2. Cell
differentiation Laboratory manuals.   3. Cell death Laboratory
manuals.   I. Studzinski, George P.   II. Series.
QH604.7.C447 1999   571.8–dc21   99–33469

ISBN 0 19 963 769 5 (Hbk)
     0 19 963 768 7 (Pbk)

Typeset by Footnote Graphics,
Warminster, Wilts
Printed in Great Britain by Information Press, Ltd,
Eynsham, Oxon.

Preface

This volume presents a variety of approaches to the study of mammalian cell growth, ranging from detailed presentations of the recent modifications of the established general procedures for measurement of growth and cytotoxicity, through examples of the examination of the growth signalling pathway, to the establishment of growth cessation through differentiation or senescence. The conceptual underpinnings of each approach are provided, together with the details of procedures found most useful in each author's laboratory and guidelines for interpretation of the expected results.

Current studies of growth-associated phenomena go well beyond an enumeration of cell proliferation and while it is not possible to cover every facet of this rapidly advancing field, examples are included of such advanced techniques as the assessment of cell cycle checkpoints, detection of oncogenes, and the examination of the nuclear architecture of growing cells.

The negative aspects of cell growth are perhaps of even greater importance than cell proliferation itself, since this is the focus of intense efforts to control human diseases such as cancer. In this vein, the volume presents several approaches to studies of controlled cessation of cell proliferation, through induction of differentiation or by evolving senescence. The principal focus is on human cells. It should be also noted that an important aspect of control of cell proliferation is not included here, since it is a subject of a companion publication, *Apoptosis: A Practical Approach*.

The authors and the staff of OUP have made a concerted effort to provide a truly practical compendium on how to study cell growth; it is sincerely hoped that this labour will fulfil the needs of beginning as well as experienced investigators, and help to inspire their efforts as well as to provide specific guidance.

New Jersey   G.P.S.
1999

Contents

3. Cell growth and kinetics in multicell spheroids 61

Ralph E. Durand

4. Assessment of DNA damage cell cycle checkpoints in G1 and G2 phases of mammalian cells

Patrick M. O'Connor and Joany Jackman

5. Assessment of the role of growth factors and their receptors in cell proliferation

Subal Bishayee

11. Human leukaemia cells as a model differentiation system   225

Dorothy C. Moore and George P. Studzinski

12. Induction of differentiation of human intestinal cells *in vitro*   241

Heide S. Cross and Enikö Kallay

Contents

Contributors

ALESSANDRO ALESSANDRINI
Massachusetts General Hospital - East, 149 13th Street, Charlestown, MA 02129, USA

STACEY J. BAKER
The Fels Institute for Cancer Research and Molecular Biology, Temple University School of Medicine, 3307 North Broad Street, Philadelphia, PA 19140, USA

SUBAL BISHAYEE
Department of Pathology and Laboratory Medicine, UMDNJ-New Jersey Medical School, 185 S. Orange Ave, Newark, NJ 07103, USA

STEPHEN C. COSENZA
The Fels Institute for Cancer Research and Molecular Biology, Temple University School of Medicine, 3307 North Broad Street, Philadelphia, PA 19140, USA

HEIDE S. CROSS
Institute of General and Experimental Pathology, University of Vienna Medical School, Neubau Allgemeines Krankenhaus, Waehringer Guertel 18-20, A-1090 Wien, Austria

ZBIGNIEW DARZYNKIEWICZ
Brander Cancer Research Institute, 19 Bradhurst Avenue, Hawthorne, NY 10532, USA

RALPH DURAND
Medical Biophysics Dept, British Columbia Cancer Research Centre, 601 W. 10th Ave, Vancouver, BC V5Z 1L3, Canada

W. MICHAEL FLANAGAN
Sunesis Pharmaceutical, 3696 Haven Avenue, Suite C, Redwood City, CA 94063, USA

KAREN HUBBARD
Department of Biological Sciences, City College, City University of New York, New York, NY 10016 USA

JOANY JACKMAN
Department of Biochemistry and Molecular Biology, Georgetown University School of Medicine, 3900 Reservoir Rd. NW, Washington, D.C. 21702, USA

GLORIA JUAN
Brander Cancer Research Institute, 19 Bradhurst Avenue, Hawthorne, NY 10532, USA

Contributors

ERIKÖ KALLAY
Institute of General and Experimental Pathology, University of Vienna
Medical School, Neubau Allgemeines Krankenhaus, Waehringer Guertel
18-20, A-1090 Wien, Austria

JANE B. LIAN
Department of Cell Biology and Cancer Center, University of Massachusetts
Medical School, 55 Lake Ave North, Worcester, MA 01655, USA

SANDRA MCNEIL
Department of Cell Biology and Cancer Center, University of Massachusetts
Medical School, 55 Lake Ave North, Worcester, MA 01655, USA

MARTIN MONTECINO
Departamento de Biologia Molecular, Universidad De Concepcion, Casilla
152–C, Concepcion, Chile

DOROTHY C. MOORE
Department of Pathology and Laboratory Medicine, UMDNJ–New Jersey
Medical School, 185 S. Orange Ave, Newark, NJ 07103, USA

PATRICK M. O'CONNOR
Agouron Pharmacenticals, Inc., 3565 General Atomics Court, San Diego, CA
92121, USA

HARVEY L. OZER
Department of Microbiology and Molecular Genetics, UMDNJ-New Jersey
Medical School, Newark, NJ 07103, USA

SHIRWIN POCKWINSE
Department of Cell Biology and Cancer Center, University of Massachusetts
Medical School, 55 Lake Ave North, Worcester, MA 01655, USA

E. PREMKUMAR REDDY
The Fels Institute for Cancer Research and Molecular Biology, Temple
University School of Medicine, 3307 North Broad Street, Philadelphia, PA
19140, USA

PHILIP SKEHAN
Andes Pharmaceuticals Inc., 26520 Northeast 15th Street, Redmond, Washington 98053, USA

GARY STEIN
Department of Cell Biology and Cancer Center, University of Massachusetts
Medical School, 55 Lake Ave North, Worcester, MA 01655, USA

JANET L. STEIN
Department of Cell Biology and Cancer Center, University of Massachusetts
Medical School, 55 lake Ave North, Worcester, MA 01655, USA

GEORGE P. STUDZINSKI
Department of Pathology and Laboratory Medicine, UMDNJ-New Jersey Medical School, 185 S. Orange Ave, Newark, NJ 07103, USA

THERESE THALHAMMER
Institute of General and Experimental Pathology, University of Vienna Medical School, Neubau Allgemeines Krankenhaus, Waehringer Guertel 18-20, A-1090 Wien, Austria

FRANK TRAGANOS
Brander Cancer Research Institute, 19 Bradhurst Avenue, Hawthorne, NY 10532, USA

ANDRE J. VANWIJNEN
Department of Cell Biology and Cancer Center, University of Massachusetts Medical School, 55 lake Ave North, Worcester, MA 01655, USA

ROBERT WIEDER
Department of Medicine, UMDNJ-New Jersey Medical School, 185 S. Orange Ave. Newark, NJ 07103, USA

JOHN J. WOLF
Gilead Sciences, Department of Cell Biology, 333 Lakeside Dr., Foster City, CA 94404, USA

Abbreviations

(PI)P3	phosphatidylinositol-3,4,5-triphosphate
$1,25D_3$	$1\alpha,25$-dihydroxyvitamin D_3
aa	amino acid
AEBSF	4-(2-aminoethyl)-benzenesulfonyl fluoride
AgNOR	silver affinity nucleolar organizer regions
ALB	AlamarBlue
AP	alkaline phosphatase
ATCC	American Type Culture Collection
AZRA	azure A
BCA	bicinchonic acid
BES	N,N-bis(2-hydroxyethyl)-2 aminorthane sulfonic acid
bFGF	basic fibroblast growth factor
BFP	blue fluorescent protein
BPB	bromophenol blue
BPS	between the PH domain and the SH2 domain
BrdU	bromodeoxyuridine
BS^3	3,3′-bis(sulfosuccinimido)suberate
BSA	bovine serum albumin
cdk	cyclin-dependent kinase
CFE	colony-forming efficiency
cki, or Cki	cyclin-dependent kinase inhibitor
CMV	cytomegalovirus
cpm	counts per minute
CREB	cAMP response element binding protein
CS	calf serum
CSF	colony-stimulating factor
CSK	cytoskeletal buffer
CTR	chromotrope 2R
DAG	diacylglycerol
DAPI	4,6-diamidino-2-phenyl indole
DEPC	diethylpyrocarbonate
DMEM	Dulbecco's modified Eagle's medium
DMSO	dimethyl sulfoxide
DNase	deoxyribonuclease
DOPE	L-α-dioleoylphosphatidylethanolamine
DTT	dithiothreitol
EBV	Epstein–Barr virus
ECL	enhanced chemiluminescence
EDTA	ethylenediaminetetra-acetic acid, disodium salt
EGF	epidermal growth factor

EGFP	enhanced green fluorescent protein
EGFR	EGF receptor
EGTA	ethyleneglycol-bis(β-aminoethyl ether)-N,N,N',N''-tetra-acetic acid
eIF	eukaryotic initiation factor
ELISA	enzyme-linked immunosorbent assay
EMSA	electrophoretic mobility shift assay
Eph	erythropoietin-producing hepatocellular carcinoma-produced growth factor
ERK	extracellular signal regulated kinases
FBS	fetal bovine serum
FDA	fluoroscein diacetate
FGFR	fibroblast growth factor receptor
GAGs	glycosaminoglycans
GF	growth factor
GFP	green fluorescent protein
GNEF	guanine nucleotide exchange factor
GR	glucocorticoid receptor
Grb	growth factor receptor-binding protein
GST	glutathione-S-transferase
H&E	haematoxylin and eosin
HA	haemagglutinin
Hepes	N-2-hydroxyethylpiperazine-N'-2'-ethanesulfonic acid
HF	human fibroblasts
hGH	human growth hormone
HPV	human papillomavirus
HRPO	horseradish peroxidase
IGFR	insulin-like growth factor-1 receptor
IHC	immunohistochemistry
IL	interleukin
IR	insulin receptor
IUdr	iodouridine
i.v.	intravenous
KL	Kit ligand
LB	Luria Broth
LMPCR	ligation-mediated polymerase chain reaction
LTR	long-term recovery and long terminal repeat
M-CSF	macrophage colony-stimulating factor
MAP	mitogen-activated protein
MBP	myelin basic protein
MEK	Map/Erk kinase
MES	2-(N-morpholino) ethanesulfonic acid
MHC	major histocompatibility complex
MLV	murine leukaemia virus

MMCT	microcell-mediated cell fusion
MMTV	mouse mammary tumour virus
MNase	micrococcal nuclease
Mnk	MAP kinase-interacting serine/threonine kinase
MoAb	anticlonal antibody
MPF	maturation promoting factor
MTT	3-(4,5-dimethylthiazol-2-yl)-2,5-diphenyltetrazolium bromide or thiazolyl blue
NBT	nitroblue tetrazolium
NM-IF	nuclear matrix-intermediate filament
NRG	neuroregulins
NSE	non-specific esterase stain
NTP	nucleotide triphosphate
ON	oligonucleotide
ORG	orange G
PAGE	polyacrylamide gel electrophoresis
PBS	phosphate-buffered saline
PBSA	PBS with bovine serum albumin
PC	phosphatidylcholine
PCNA	proliferating cell nuclear antigen
PCR	polymerase chain reaction
PD	population doubling
PDGF	platelet-derived growth factor
PDK	phosphatidylinositide-dependent kinase
PGO	phosphate–glucose oxidase
PH	pleckstrin homology
PI	propidium iodide
PI3K	phosphatidylinositol 3′-kinase
Pipes	1,4 piperazinediethane sulfonic acid
PKA	protein kinase A
PKC	protein kinase C
PMA	phorbol-12-myristate-13-acetate (= TPA)
PMSF	phenylmethylsufonyl fluoride
pNP	*p*-nitrophenol
pNPP	*p*-nitrophenylphosphate
pRB	retinoblastoma protein
PSA	prostate-specific antigen
PTB	phosphotyrosine-binding domain
RBC	red blood cell
RSK	ribosomal S6 kinase
RSV	Rous sarcoma virus
SCF	stem cell factor
SDS	sodium dodecyl sulfate
SH2	src homology region 2

SI	sucrase–isomaltase
SRB	sulforhodamine B
SRE	serum response element
SRF	serum response factor
SSV	simian sarcoma virus
STC	spermine–tetrahydrochloride
TBE	Tris–borate
TBO	toluidine blue O
TBS	Tris-buffered saline
Tc	doubling time
TCA	trichloroacetic acid
TCF	ternary complex factors
TCPK	L-1-tosylamido-2-phenylethylchloromethyl ketone
TE	Tris-EDTA buffer
TGF	transforming growth factor
TNN	thionin
TPA	12-*O*-tetradecanoyl phorbol-13-acetate (= PMA)
VDR	vitamin D receptor
VVGF	vaccinia virus growth factor
WC	whole cell

1

Selection of methods for measuring proliferation

ROBERT WIEDER

1. Introduction

1.1 Defining parameters of proliferation

In the process of defining the phenotype of a cell, one of the most frequent parameters scientists measure is the rate of proliferation. Often, the first decision that confronts us is that of choosing 'the right assay'. The selection of an assay is critical to the outcome and depends on a variety of factors. These include the experimental conditions, the cell cycle parameters affected by the variable under investigation, and the source and state of the cells or the tissue to be assayed. To select an appropriate assay, we have to be aware of the events that constitute the cell cycle.

Figure 1a depicts the phases of the mammalian cell cycle. After undergoing cell division, some cells that have become terminally differentiated, such as neural tissue, enter a quiescent phase, termed G0, where they remain for their entire existence. Cells in other organs, such as liver, for example, are able to re-enter the cell cycle after a partial removal of the organ stimulates regeneration. Other cells, such as those from the crypts of the small intestine, bone marrow progenitors, tumour cells, or cells growing in tissue culture, enter a growth phase called G1 following cell division. The amount of DNA or ploidy of a cell in G1 is $2n$. These cells remain in G1 at a place called the restriction point and do not begin to replicate their DNA unless they overcome some physiological barriers. The cells first repair any DNA damage prior to initiation of DNA replication. If the damage is too extensive, the cells decide to commit suicide using a process termed programmed cell death, or apoptosis. Passage through G1, as well as other phases of the cell cycle, and the decision to progress past the restriction point are governed by protein complexes containing cyclins, cyclin-dependent kinases (cdk), their inhibitors, and proliferating cell nuclear antigen (PCNA), an activator of DNA polymerase that also participates in excision repair of accumulated DNA damage. Extracellular signalling modulates the levels of cyclins in G1 and the activity of the cyclin–cdk complexes. The cyclin–cdk complexes phosphorylate the retinoblastoma

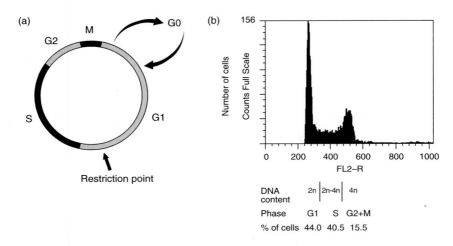

Figure 1. (a) Phases of the cell cycle. After mitosis (M), cells enter a growth phase termed G1. Some cells enter a quiescent phase after mitosis, called G0, from which some cells can emerge under certain conditions and re-enter the cell cycle through G1. After G1, cells replicate their DNA in the synthetic, or S, phase. This is followed by another growth phase, G2, prior to entering mitosis. (b) Cell cycle distribution in rapidly growing cells that were labelled with propidium iodide as determined by flow cytometry. The *y*-axis represents the cell numbers and the *x*-axis represents the PI fluorescence intensity that translates to the quantity of DNA per cell. The leftmost peak represents G1/G0 cells with 2*n* amounts of DNA, the rightmost peak represents cells in G2 and M phases with 4*n* amount of DNA, and the cells in between these peaks are in the S phase undergoing active DNA synthesis that have amounts of DNA between 2*n* and 4*n*.

protein, pRb. This event overcomes the restriction point and results in the initiation of DNA synthesis. The mechanisms of cell cycle control are eloquently reviewed by Sherr (1) and by Juan *et al.* in Chapter 7. Chapter 7 also presents technical details for flow cytometric analysis of the cell cycle-specific cellular content of cyclins, cdks, and cdk inhibitors, and a unique flow cytometric analysis of the retinoblastoma protein phosphorylation status.

The G1 phase is followed by the S phase, during which the cellular DNA is duplicated. The ploidy of cells during S phase is between 2*n* and 4*n*. Following DNA duplication, cells enter a second growth phase called G2. During this time the cells duplicate their infrastructure, including the amount of ribosomes, ribosomal RNA, cellular proteins, and other functional elements in preparation for cell division. The amount of DNA in G2 is 4*n*. Following G2, the cells undergo mitosis, completing the cycle. *Figure 1b* demonstrates the cell cycle distribution of propidium iodide (PI)-labelled SK-Hep1 hepatoma cells undergoing optimal proliferation in a semi-confluent tissue culture plate and the percentage of cells in the various phases of the cycle.

Cell proliferation is reflected in a number of associated processes, either directly or indirectly, that can be quantitated. The most direct measure of proliferation is the rate of doubling of the number of cells. This can also be reflected by the average doubling time of cells either in culture or *in vivo*. Proliferation is also reflected by the fraction of cells in a population *in vitro* or *in vivo* undergoing cell division. However, cells can also undergo DNA synthesis without undergoing cell division. While this is not proliferation, assays that measure DNA synthesis will register as strongly positive under these conditions. Proliferation can be a measure of the cells undergoing cell cycle progression minus the population of cells simultaneously undergoing cell death. It can also be reflected in the length of time it takes cells to traverse

Table 1. Assays for parameters of proliferation

	Property measured	**Property derived**	**Assay**
I. Cells in tissue culture	1. Cell number	a. Rate of division exclusive of cell death b. Percent viable cells c. Doubling time	cell count in trypan blue
	2. DNA content	a. Fraction of cells in specific phases of cell cycle b. Fraction of cells undergoing DNA synthesis c. Ploidy d. Fraction of subdiploid apoptotic cells	PI labelling and flow cytometry
	3. DNA synthesis	a. Relative rate of DNA synthesis b. Fraction of cells undergoing DNA synthesis	a. [^{3}H]Thymidine uptake, scintillation counts of DNA on filters b. [^{3}H]Thymidine, BrdU autoradiography, IHC[a]
	4. Metabolic rate	a. Relative metabolic rate b. Relative viability	a. MTT assay
	5. Mitosis	Fraction of cells undergoing cell division	a. Colchicine, colcemid, or nocodazole b. Mitotic figure count c. Flow cytometry
	6. Collective cell volume	Relative rate of proliferation	a. RNA content b. Protein content
II. *In vivo* cells	1. DNA content	Cell cycle distribution	Flow cytometry
	2. DNA synthesis	Fraction of cells undergoing DNA synthesis	*In vivo* [^{3}H]thymidine, BrdU, IUdr uptake, autoradiography, IHC
III. *Ex vivo* cells	1. Mitotic figures	M phase—related to fraction of cells dividing	Histochemistry
	2. DNA content	Cell cycle distribution and fraction in S phase	Flow cytometry
	3. Ki67, PCNA, AgNOR	Related to fraction of cells proliferating	Immunohistochemistry

[a]IHC = immunohistochemistry.

various phases of the cell cycle and the effects of various interventions on each phase. Various aspects of the cell cycle are affected by interventions, and only by selecting the appropriate assay can the complete effect of the intervention be measured.

Table 1 outlines the parameters of proliferation measured by available assays and refers to the relationship of the results to proliferation.

2. Cells in tissue culture

Measuring the growth of cells in tissue culture allows for the greatest choice in the selection of an assay because of the ease of handling. Unlike the case of *ex vivo* tissue samples or *in vivo* organs, all of the available methods can be applied to cells in tissue culture. The following are a sampling of some available methods and the processes they reflect.

2.1 Counting cell numbers

2.1.1 Determining the rate of proliferation and assessing viability

The most direct measure of proliferation is a change in cell number with time. While seemingly a simple measurement, the change in cell number is affected by a number of factors. The number of cells after a given time is a measurement reflecting cell proliferation, minus the fraction of cells that died or stopped proliferating. The rate of proliferation is affected by a number of factors, including the choice of medium, the concentration, source and batch of serum, supplements, growth factors, the density of the cells, and, in the case of non-immortalized cells, whether the cells have undergone a limiting number of divisions and are undergoing senescence. *Protocol 1* outlines the steps for counting cells in tissue culture. The measurement of cell number should take place in the presence of a dye, such as trypan blue, excluded only by viable cells. By counting both blue and white cells, an estimate of cell viability can also be obtained. Percent viability = white cells/(blue cells + white cells) $\times$ 100%.

Protocol 1. Counting cell numbers

Equipment and reagents
- Ca^{2+}, Mg^{2+}-free phosphate-buffered saline (PBS)
- 0.05% trypsin/0.5 mM EDTA
- 10% fetal calf serum
- Haemocytometer

A. *Adherent cells*

(The quantities apply to 100 mm diameter tissue culture plates; adjust numbers proportionately.)

1. Incubate 5×10^5 cells per plate in culture medium in 9–12 tissue culture plates at 37°C in 5%CO_2.

4

2. On days 1, 3, and 5 (and 7 for slowly growing cells), harvest triplicate dishes by first aspirating the medium.

3. Rinse cells with 3–5 ml Ca^{2+}, Mg^{2+}-free phosphate-buffered saline (PBS) and aspirate.

4. Incubate cells with 1 ml 0.05% trypsin/0.5 mM EDTA (1 ×), enough to cover all cells, for 1–2 min at 37 °C.

5. Tap the edge of the dish with horizontal force until the cells detach. Return the dish to the incubator for an additional 30 sec to 1 min if the cells have not detached completely. The efficiency varies with the trypsin batch and the cells.

   [PBS/1 mM EDTA may be used instead of trypsin in experiments that require the intact maintenance of membrane-bound proteins that protrude from the cell surface, such as cellular receptors. A 10–15 min incubation at 37 °C is required to cause cell detachment and disrupt cell–cell adhesion.]

6. Triturate the detached cells with a 2 ml pipette several times to disrupt cell clumps and effect a single cell suspension.

7. Add 4 ml medium containing 10% fetal calf serum to inactivate the trypsin and triturate the cells several times to disrupt clumps. Transfer to a 15 ml conical tube.

B. *Cells in suspension*

1. Incubate 5×10^5 cells in triplicate 25 mm^2 culture flasks in 5 ml medium and supplements at 37 °C 5% CO_2 with the cap loosened.

2. Mix the suspension prior to sampling by trituration with a pipette and sample 20 µl of cells on days 1, 3, 5, and 7.

C. *Counting the cells*

1. Mix 20 µl cells that have been well mixed prior to sampling with an equal volume of trypan blue 0.4%

2. Apply to a haemocytometer by pipetting from the edge of the coverslip and permitting diffusion by capillary action. Count cells from all samples within a short fixed time after mixing with trypan to obtain a constant and reproducible measure of viability. (Even healthy cells eventually take up trypan if incubated long enough.)

3. The haemocytometer has counting chambers on two sides. Count both sides and average the numbers.

4. There are nine grids in a counting chamber. Count cells in as many grids as needed to count at least 100 cells/chamber. Count cells in symmetric grids to avoid over- or undercounting due to sedimentation of cells during diffusion.

Protocol 1. *Continued*

5. The concentration of cells in the original suspension in cells/ml is:

$$\frac{\text{no. cells counted}}{\text{number of grids counted}} \times 10^4 \times 2$$

Where 10^4 is the conversion of cells/0.1 mm^3 [volume of a grid is 1 mm $\times$ 1 mm $\times$ 0.1 mm (height)] and 2 is the trypan blue dilution.

2.1.2 Determining the doubling time

The growth curve of cells in a tissue culture dish is S-shaped when plotted on linear co-ordinates and linear when plotted as the log of cell number vs. time on a linear scale. Take, for example, a hypothetical experiment in which 5 $\times$ 10^4 MDA-MB-435 human breast cancer cells were incubated in 100 mm tissue culture dishes in Dulbecco's modified Eagle's medium (DMEM)/10% fetal calf serum (FCS) supplemented with 2 mM glutamine with and without the presence of 1 ng/ml transforming growth factor β (TGF-β) and total viable cells were counted 1, 3, 5, and 7 days later. Let us assume the following cell counts were obtained.

Table 1.2

Day	– TGF-β	+ TGF-β
1	5.0×10^4	4.9×10^4
3	1.4×10^5	8.0×10^4
5	3.8×10^5	1.25×10^5
7	1.0×10^6	2.0×10^5

The arithmetic and semi-log curves of the plotted data are shown in *Figure 2a* and *2b*, respectively. No increases in cell numbers are generally noted the

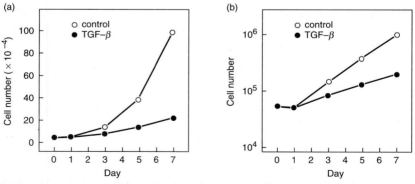

Figure 2. Graphs representing growth curves of control cells (○) and cells treated with TGF-β (●). The *y*-axis represents cell number and the *x*-axis represents the number of days in culture. Graph (a) represents the linear relationship of growth while graph (b) is the same data presented as a semi-log graph.

first day after incubation. To determine the doubling time of the cells in the two arms of the above experiment, we first begin with the general formula that describes the growth of cells:

$$A = A_0 2^n$$

where A is the number of cells at any time, A_0 is the number of cells at an initial point, and n is the number of cell divisions that have taken place. The value n can also be described as:

$$n = T/T_c$$

where T is the time elapsed and T_c is the doubling time of the cell. Thus the proliferation formula can be written as:

$$A = A_0 2^{T/T_c}$$

Solving for T_c, by taking the natural log of both sides, we obtain the formula:

$$\ln A/A_0 = T/T_c \ln 2$$

$$2.3 \log A/A_0 = T/T_c \times 0.69$$

$$T_c = \frac{(T \times 0.69/2.3)}{\log (A/A_0)} = \frac{0.3T}{\log (A/A_0)}$$

In the experiment described, the time T between day 7 and day 1, in which the data are linear on the semi-log plot, is 6 days, or 144 hours. In the case of the cells grown without TGF-β , the number of cells at the end of this time is 1×10^6 while the number of cells on day 1, A_0, is 5×10^4. Thus the equation reads:

$$T_c = \frac{(0.3 \times 144 \text{ h})}{\log 10^6/5 \times 10^4} = \frac{43.2 \text{ h}}{\log 20} = \frac{43.2 \text{ h}}{1.3} = 33.2 \text{ h}$$

When the cells were grown in the presence of TGF-β, the final number of cells was only 2×10^5 while the original number was similar to that of the control cells at 4.9×10^4. Thus the equation becomes:

$$T_c = \frac{(0.3 \times 144 \text{ h})}{(\log 2 \times 10^5/4.9 \times 10^4)} = \frac{43.2 \text{ h}}{\log 4.08} = \frac{43.2 \text{ h}}{0.61} = 70.8 \text{ h}$$

The total cell cycle time should be determined between two time points where the shape of the growth curve on the semi-log plot is a straight line. This exponential part of the growth curve represents the time when, in principle, nearly all of the cells are dividing, whereas in the initial lag phase or the terminal plateau phase, a considerably smaller fraction of cells are dividing. Proliferation data are published as semi-log plots. Error bars should be supplied and p values should be reported between significant differences in two curves at a given time point using Student's *t*-test. These calculations can be obtained from a statistics manual.

2.2 Measuring DNA content

2.2.1 Direct measurement of DNA content

Direct measurement of the DNA content of a population of cells gives a rough estimate of cell number. While this method is not used very often in tissue culture any longer, it is more useful when obtaining tissue samples *ex vivo*. It is rarely used to assess proliferation. Diploid cells contain about 3×10^9 bases per haploid genome or about 6×10^9 bases. At 660 daltons per phosphonucleotide, this relates to approximately 6.6 pg of DNA per mammalian cell. This calculation is often off by a small factor in benign cells owing to the presence of a small fraction of cells in the S and G2+M phases, but is even more inaccurate in cells from tumour tissue where more cells are cycling and a significant fraction of the cells may also be aneuploid. The method of extracting DNA and measuring its content is outlined in *Protocol 2*.

Protocol 2.   Extraction and measurement of DNA (from ref. 2)

Reagents

- TBE: Tris–borate (0.045 M)/EDTA (1 mM)
- TE: 0.01 M Tris–HCl, pH 8.0, 5 mM EDTA
- Resuspension buffer: 0.01 M Tris–HCl, pH 8.0, 10 mM EDTA, 20 μg/ml pancreatic RNAse, and 0.5% SDS
- Proteinase K
- Phenol
- Chloroform
- 70% ethanol

Method

1. Wash the cells growing in DME/10% FCS in monolayers with TBE solution.

2. Add 0.5 ml TBE, scrape the cells into a centrifuge tube, and wash twice with TBE.

3. Resuspend in TE to an approximate concentration of 5×10^7 cells/ml.

4. Transfer to an Erlenmeyer flask and add 10 ml of resuspension buffer and incubate for 1 h at 37 °C.

5. Add proteinase K to 100 μg/ml and, using a glass rod, mix until viscous.

6. Extract DNA with phenol and chloroform, precipitate with 70% ethanol, and resuspend in 0.01 M Tris/5 mM EDTA.

7. Take OD (optical density) readings at 260 nm and 280 nm. A_{260}/A_{280} should be about 2 in a protein-free preparation.[a]

[a]$A_{260} = elc$, where the extinction coefficient, e, of DNA is 7×10^3. The value for l, the path length of light in the cuvette, is usually 1 cm and the concentration, c, of DNA is in moles/l. Therefore, at an OD of 1.0, the concentration of DNA is 50 μg/ml or 1.52×10^{-4} M. Do not make concentrations of DNA lower than 0.1 and higher than 1.8 A_{260} units/ml, or they will be less accurate.

2.2.2 Flow cytometry

The best and most accurate measurement of DNA content for the purpose of assessing proliferation is that obtained by cytofluorometry. This method has revolutionized the measurement of cell cycle progression and has enabled scientists to ask detailed, significant, and mechanistic questions regarding molecular events in all phases of the cell cycle. Flow cytometry and its application to the understanding of proliferative controls will also be addressed in Chapter 7. In this chapter, I shall discuss how flow cytometry correlates with cell proliferation and how it identifies the fraction of cells in specific phases of the cell cycle and cells that have undergone cell death and DNA fragmentation and loss. After various interventions, cells are harvested by trypsinization or from suspension cultures, labelled with propidium iodide (PI) or another DNA-binding dye, as in *Protocol 3*, and assayed by flow cytometry. The technical aspects of using a flow cytometer and programmes to analyse the data can be obtained from a specialist manual.

Protocol 3. Labelling cells with propidium iodide for flow cytometry

- Detach adherent cells with trypsin, as described in *Protocol 1*, taking steps to ensure a good quality single cell suspension. Following trypsin treatment and trituration, resuspend in PBS rather than medium. For cells growing in suspension, centrifuge for 10 min at 2000 r.p.m. (600 $\times$ g) and resuspend in PBS.

To introduce DNA intercalating dyes into cells, the membrane has to be permeabilized in one of two ways. Cells either have to be fixed with ethanol or the dye has to be dissolved in a detergent when added to live cells. If cells are stained and assayed within half an hour, they may be stained directly, without fixing. If cells are fixed, they may be stored prior to assay.

Equipment and reagents

- propidium iodide (PI)
- sodium citrate
- 1% v/v Triton X-100
- PBS
- RNase A (DNase-free)
- flow cytometer

Direct staining

1. Place approximately 1×10^6 cells into a tube in PBS at 4°C.
2. Prepare a solution of 500 μg/ml propidium iodide, 10 mg/ml sodium citrate, and 1% v/v Triton X-100, and filter through a 0.8 μm filter (This is 10$\times$ PI solution).
3. Add 1/10 volume 10$\times$ PI solution and acquire cells in a flow cytometer within half an hour.

Protocol 3. *Continued*

Fixing cells

1. Centrifuge cells and aspirate off the PBS.
2. Add 2 ml 70% ethanol in water or PBS at 4°C for 1 h to fix cells. They may be stored for several days at 4°C at this point.
3. Prior to analysis, centrifuge at 2000 r.p.m. (600 × g) for 10 min, aspirate off the ethanol, and wash twice with 2 ml PBS.
4. Resuspend the cells in 1 ml PBS, add 10 μl RNase A (DNase-free), and incubate 37°C for 1 h.
5. Centrifuge cells at 2000 r.p.m. (600 × g) for 10 min and aspirate supernatant.
6. Resuspend the cells in 500 μl of PI (50 μg/ml) in PBS and acquire the samples in the flow cytometer.

Figure 1b demonstrates the cytofluorometric distribution of cells by their DNA content. The cell fit programme (CELLQuest Version 2.0 program, Becton Dickinson Immunocytometry Systems, San Jose, CA) calculated the fraction of cells with $2n$, $2n$ to $4n$, and $4n$ amounts of DNA, corresponding to G1, S, and G2+M phases of the cell cycle. While cell counting or quantitation of total DNA content in a tissue culture dish or a sample of tissue are direct or indirect measurements of the number of cells in the unit sample, and serial measurements can be correlated with rates of proliferation and doubling time, the determination of the distribution of the amount of DNA in a population of cells is a relative measure of proliferation. Since the purpose of determining the rate of proliferation is to assess the effects of an intervention, relative measures of proliferation between control and experimental populations are valuable, regardless of the methods. Flow cytometric analysis of cell cycle distribution provides a potentially far greater amount of information than other methods. The effects of an intervention are apparent on all phases of a cell cycle. Treatments that stimulate proliferation increase the fraction of cells in the S phase of the cell cycle as well as of cells in the G2+M phases, compared with controls. Agents that inhibit proliferation and block progression from G1 to S cause an increase in the fraction of cells in G1 with $2n$ ploidy, while agents that block exit from G2 cause an increase in the fraction of cells in G2 with $4n$ ploidy, both at the expense of the S phase.

The rate of progression through all phases can be determined through the help of compounds that arrest cells at specific points. However, to maintain the specificity of cell cycle arresting agents, careful titration needs to be carried out for each different cell line under study, to ensure that the lowest effective concentration of the inhibitor is used and that no effects are imposed on other phases of the cell cycle.

(a) Determining the fraction of cells in a population undergoing active cycling

Nearly 100% of cells in a transformed cell line growing in log phase in tissue culture are cycling. However, in a population of non-immortalized cells that has undergone 40 or more divisions, for example, considerably less than 100% of the cells are cycling. To determine what fraction of cells are in G0, exponentially growing cells should be arrested in G2+M with nocodazole {methyl[5-(2-thienyl-carbomyl)-1*H*-benzimidazole-2-yl]-carbamate}, a reversible mitotic blocking agent, and the fraction of cells with 2*n* DNA should be determined after the cycling cells have all left G1.

The appropriate concentration of nocodazole should be determined for each cell line. To do this, a log phase population of cells should be incubated with varying concentrations of nocodazole, for example 0, 0.05, 1, 1.5, 2, 2.5, and 4 µg/ml for 16 h, and the DNA content should be determined flow cytometrically. The lowest concentration of nocodazole that induces the greatest emptying of the G1 phase should be selected for further studies. The percentage of cells in G1 at this and higher concentrations of nocodazole is the percentage of cells in G0. Alternatively, if the concentration of nocodazole has already been established for a cell type, that concentration alone can be used for determining the fraction of cells not actively cycling.

(b) Determining the length of the G1 phase

There are two methods of determining the length of G1 by flow cytometric analysis. The first one calculates the half-time of emptying of cells from G1. This is achieved by incubating cells on six tissue culture plates, or tissue culture flasks for cells that grow in suspension, and culturing them for two or three days until they are in log phase. Nocodazole is added at the concentration determined before, and the cells are sampled and the DNA is labelled with PI before adding nocodazole and at 1, 2, 3, 4, and 5 h after treatment. The percentage of cells in G1 is plotted against time, *t*, of collection using the formula

$$G1 = G1_{initial}e^{-kt}$$

$$\ln G1/G1_{initial} = -kt$$

which shows a first-order decrease with time, denoting the probability of a cell exiting G1. To determine the half-time for G1 decrease, the formula becomes

$$\ln G1_{half}/G1_{initial} = \ln 0.5 = -kt_{half}$$

$$t_{half} = \frac{0.693}{k}$$

where *k* is the slope of the line. The calculated t_{half} is the time it takes half the cells to exit G1. The time spent in G1 is then obtained by doubling this number.

The second method of determining the length of G1 is by calculating it as a fraction of the doubling time. First, determine the doubling time as described earlier in this chapter by serial cell counting. Then, calculate the length of the G1 phase, T_{G1}, by the formula

$$\frac{T_{G1}}{T_c} = \frac{\ln(F_{G1}+1)}{\ln 2}$$

where T_c is the doubling and F_{G1} is the fraction of cells in G1 as obtained by flow cytometric analysis of DNA content. This second method can only be used in cells in which 100% of the population is cycling. If the population contains both cycling and non-cycling cells, a change in the duration of G1 by this calculation can mean a change in the cell cycle phase duration, or it can mean a change in the proportion of cycling to non-cycling cells by the intervention whose effects are being assayed (3).

(c) Determining the length of the S phase

A similar concept of arresting cells prior to entry into the S phase and subsequent monitoring of the rate of progression through this phase can be used. Exponentially growing cells should be incubated overnight in hydroxyurea at a concentration in the range of 0.1 mM, determined previously to inhibit entry into the S phase. The hydroxyurea is washed out with PBS and the cells are incubated in fresh culture medium. Cells are sampled at hourly intervals for DNA content determination by flow cytometry. The relative DNA content is plotted against time. Progress through the S phase is characterized by an initial rapid increase in DNA content in the S phase channel, followed by a progressive decrease. The S phase duration is estimated as the difference between the time at the half-maximal peak height of the decreasing slope and the time at the half-maximal peak height of the increasing slope (3).

(d) Determining the length of the G2+M phase

To determine the effect of an intervention on the length of G2+M phases of the cell cycle, DNA synthesis is reversibly inhibited and the emptying of the G2+M phases is monitored by flow cytometry, as above. Aphidicolin, a DNA polymerase inhibitor, can be used to prevent cells from entering the S phase or from leaving the S phase once DNA synthesis begins (4). As before, cells in log phase are incubated with the cell cycle inhibitor, this time 5 μg/ml aphidicolin overnight. Cells are washed with PBS, incubated with fresh medium, fixed, and stained with PI at half-hour intervals for five or six time points. The exit of a population of cells from G2+M differs from the exit from G1. While the curve of the G1 exit is linear on a semi-log plot, indicating that each cell has some condition-dictated probability of exiting G1, the initial decline of asynchronously growing cells in G2+M plotted against time is straight on a linear plot, suggesting that the time is fixed for each cell under any given condition once it completes DNA synthesis (3). The half-time for exiting G2+M is determined by a curve fit through the initial linear portion and identifying the time at which half of the maximal decrease of cells in G2+M took place. Doubling this number yields the average time in G2+M. Adding up the derived times in G1, S, and G2+M should be close to the doubling time calculated by cell counting experiments.

2.3 Measuring the rate of mitosis

While a number of agents increase the proliferative rate, promote entry into the DNA synthetic phase, and modify the duration of specific phases, the one true direct measure of cellular proliferation, besides counting changes in actual cell number with time, is a determination of the fraction of cells undergoing mitosis. This value is affected by a number of factors when fixing and staining cells and visually determining fractions of cells undergoing mitosis. These include the range of time following cell harvest prior to fixation. During a period of delay, cells that have entered the mitotic phase will complete cell division and result in an undercount. Mitosis consists of four phases: prophase, metaphase, anaphase, and telophase, with the metaphase and telophase being the most obvious to recognize. The efficiency of the count is investigator specific, with experienced investigators able to identify a higher fraction of true mitotic cells than inexperienced ones. Addition of a hypotonic shock step to the slide preparation accentuates metaphase cells and results in easier identification.

A novel method of separation of the G2+M phase cells into characteristic G2 and M phases was developed by Giaretti *et al.* (5). The method measures light scattering in BrdU and PI double-labelled cells that were fixed in ethanol, their histones extracted with HCl, and their DNA thermally denatured. The procedure enhances differences in chromatin structure in different phases of the cell cycle, permitting identification by 90° scatter. By this procedure, mitotic cells have a much lower scatter than G2 phase cells, while G1 post-mitotic cells have lower scatter than G1 cells ready to enter the S phase. The procedure is outlined in *Protocol 4.*

Protocol 4. Flow cytometric differentiation between G2 and M and between post-mitotic and pre-synthetic G1 (from ref. 5)

Equipment and reagents
- BrdU
- PBS
- RNase A
- 0.1 M HCl
- PBST: 5 ml ice-cold PBS and 0.4% Tween-20
- FITC-conjugated goat anti-mouse IgG
- 0.5% BSA
- PI
- dual laser cell sorter

13

Protocol 4. *Continued*

Method

1. Pulse cells growing in tissue culture either in suspension or adhered to dishes with 10 μg/ml BrdU for 15 min at 37 °C.

2. Harvest cells (trypsinize adherent cells, as in *Protocol 1*), centrifuge at 2000. r.p.m. (600 × g) for 10 min at 4 °C, resuspend in ice-cold PBS, and recentrifuge at 2000 r.p.m. (600 × g) for 10 min at 4 °C.

3. Resuspend in 3 ml ice-cold PBS and add 7 ml ice-cold absolute ethanol to fix cells.

4. Centrifuge 2–3 × 10^6 cells at 2000 r.p.m. (600 × g) for 10 min and resuspend in 2 ml PBS with 1 mg/ml RNAse A at 37 °C for 20 min.

5. Centrifuge at 2000 r.p.m. (600 × g) for 10 min, resuspend cells in ice-cold 0.1 M HCl for 10 min.

6. Centrifuge at 2000 r.p.m. (600 × g) for 10 min, resuspend cells in ice-cold distilled water. Repeat this step once.

7. Heat to 95 °C for 10–50 min.

8. Quench on ice, add 5 ml ice-cold PBS with 0.5% Tween-20 (PBST).

9. Centrifuge at 2000 r.p.m. (600 × g) for 10 min, resuspend cells in 0.4 ml anti-BrdU mouse monoclonal antibody at 1:100 dilution in PBST and 0.5% BSA. Incubate at room temperature for 30 min.

10. Wash cells twice with PBST.

11. Stain with in 0.4 ml FITC-conjugated goat anti-mouse IgG at 1:100 dilution in PBST and 0.5% BSA for 20 min at room temperature.

12. Wash the cells again and resuspend in PBS containing 10 μg/ml PI.

13. Analyse using a dual laser cell sorter measuring red (PI) and green (FITC) fluorescence, forward, and 90° scatter simultaneously. Tune the incident light to emit 500 mW power at 488 nm. Measure the green emission fluorescence at 530 and the red emission fluorescence at 630 nm using appropriate filters. Measure the scattering of the incident beam at 90° using a beam splitter which reflects about 10% of the light and a band-pass filter at 488 nm. Analyse all four signals using a multichannel analyser.

14. Display bivariate distributions of combinations of any of the two parameters as 'contour plots'. Plots of DNA content by PI versus 90° scatter are able to differentiate between G2 and M and between early and late G1 phase cells.

2.4 Measuring DNA synthesis

2.4.1 Determining the relative rate of DNA synthesis by [³H]thymidine uptake

The rate of DNA synthesis is a reflection of proliferation under many conditions. However, cells can duplicate their DNA without undergoing cell division or they can duplicate their DNA and divide without a concomitant net increase in the total cell population because of ongoing cell death. Cells arrested in G1 under density-induced contact inhibition or cytokine-induced arrest undergo excision repair of DNA damage that can also translate to replication-independent nucleotide incorporation. Therefore, to measure proliferation as a reflection of DNA synthesis, the experimental conditions must be carefully controlled. These, as well as all proliferation studies are most reflective of the actual cellular proliferation rate when performed in the exponential growth phase of cells in tissue culture. These cells are least likely to be affected by artefacts of density and a suboptimal culture environment, and are most sensitive to interventions whose rate-limiting effects are to be assayed. The effects of interventions that promote proliferation, on the other hand, need to be assayed either earlier than the log phase or in a more dense culture, where growth has slowed somewhat.

To measure the proliferative rates by [³H]thymidine uptake, cells are cultured in microtitre wells, thymidine is added, and the uptake by DNA is measured, after lysing and washing on a filter, by scintillation counting. The method is outlined in *Protocol 5*. Bromodeoxyuridine (BrdU) can be incorporated instead of [³H]thymidine and the incorporation can be assayed with antibodies to BrdU in a non-radioactive assay.

Protocol 5. [³H]Thymidine uptake filter assay

Equipment and reagents
- 96-well, flat bottom, tissue culture plates
- [³H]thymidine
- scintillation counter

Method

1. Incubate $5 \times 10^3 - 1 \times 10^4$ cells per well in 96-well, flat bottom, tissue culture plates in 100 μl tissue culture medium and supplements appropriate for that cell type or the experiment, at 37°C in 5% CO_2. The starting cells should be from log phase, nearly 100% viable cultures whose passage number is recorded.

2. The next day, add 100 μl additional medium containing the variable concentrations of growth factors, or the substance whose effect on proliferation is being tested, in triplicate wells, along with the appropriate control wells. Incubate the cells at 37°C in 5% CO_2 for an

---

15

Protocol 5. *Continued*

additional one to four days, depending on the experimental design. Check the wells under the microscope to determine the extent of confluence of the control cells the day of labelling.

3. After the several day incubation period, dilute [³H]thymidine (1 mCi/ml) in medium at 1:50 and add 30 μl diluted [³H]thymidine (0.6 μCi) to each well. Incubate at 37°C for 4–5 h.

4. After incubation, scrape the cells with a cell masher and aspirate along with the medium on to a Whatman filter. Wash the wells with distilled water extensively and aspirate through the Whatman filter to hypotonically lyse the cells and wash through the unincorporated tritiated thymidine.

5. Place each filter containing the bound DNA with incorporated thymidine from one well into a scintillation vial, add liquid scintillant, and count the rate of decomposition of tritium in a scintillation counter.

The multiwell plate assay allows for easy handling of a large number of experimental variables with minimal processing. The data are graphed as counts/minute vs. the variable tested or the type of intervention. The graphs represent relative values. As the fraction of cells in S phase and the percentage of cells able to cycle can both be modulated, it is difficult to make any deterministic statements about the effects of the intervention on these values. To determine these effects, flow cytometry, or [³H]thymidine auto-radiography and incorporation into individual cells, must be carried out. For tissue culture assessments, it is far easier to obtain cell cycle data by flow cytometry than by autoradiography. For determination of *in vivo* doubling time and the fraction of cells undergoing DNA synthesis, [³H]thymidine or BrdU incorporation and autoradiography are still very useful techniques.

2.4.2 Determining the fraction of cells undergoing DNA synthesis by autoradiography and immunohistochemistry of incorporated [³H]thymidine and BrdU

The detailed methods of preparing cells and carrying out the autoradiography are presented in a straightforward and easy to follow manner elsewhere (6).

In contrast to flow cytometry, autoradiography can determine the average doubling time of a cell population. Aside from this advantage, most of the data that can be obtained by autoradiography in cells in tissue culture can also be determined by flow cytometric analysis of DNA content in a less tedious, less labour-intensive, and more precise manner. Also, while flow cytometry provides an instantaneous profile of the cycle distribution of cells at any one time, [³H]thymidine incorporation provides a less instantaneous view of the fraction of cells that enter the DNA synthetic phase over the period of time the cells are exposed to radioactive thymidine. Nevertheless, it is a valuable, time-tested tool that still has significant applications. The incubation of [^{3}H]thymidine (7) or of BrdU (8, 9) can be used interchangeably (10) with the latter being less labour intensive and generating data that are easier to reproduce.

The duration of pulsing of the cells with [^{3}H]thymidine or BrdU can be as low as 15 min, however, when sufficiently high levels of tritiated thymidine such as 0.2 λCi/ml are used. The method provides a freeze-frame profile of cells in the S phase at a certain time point in the proliferation curve of a cell population and the effects of an intervention on this phase of the cell cycle. The effects at different times after the intervention can be assessed by separate autoradiographs. The recurrent theme of setting up the appropriate conditions to measure the effects of a particular intervention continues to emerge. When measuring a cell cycle inhibitor, the culture conditions require log phase growth. However, when assaying the effects of a mitogen, less than optimal growth conditions must be created. These can be early, lag phase culture, confluence, serum deprivation, or arrest in various phases of the cell cycle by specific physiological inhibitors. Autoradiography can be used to determine the fraction of cells in S phase, the average doubling time of a cell population, the fraction of non-cycling cells, the duration of specific phases of the cell cycle, and the effects of various interventions on any of these variables. I will briefly outline the methodology for determining the length of the S phase and the average doubling time of a population using auto-radiography or BrdU incorporation with an anti-BrdU antibody on microscope slides.

To determine the fraction of cells in the S phase at any time point, cells are pulsed with tritiated thymidine for 15 min, and the fraction of labelled mitoses are scored on autoradiograms several days later, as described elsewhere (6). Cells are counted at 100 × magnification and cells with granules are scored as a fraction of total cells. The increase in labelled mitoses is followed by a decrease and a subsequent increase as the cells continue to progress through the cell cycle (*Figure 3*). The time between the time point where 50% of the mitoses have autoradiographic granules while the percentage of labelled mitotic figures is increasing and when 50% of the mitotic figures are labelled when the percentage is decreasing, is equal to the length of the S phase, or T_S. The time between the point where 50% of the mitotic figures are labelled in the first rise in mitoses and the time when 50% of the mitotic figures are labelled in the second rise in the frequency of mitotic figures, is the average doubling time of the cell population, or T_c.

To determine how many cells should be counted, one can estimate the variance by the formula:

$$\text{variance} = p(1-p)/c$$

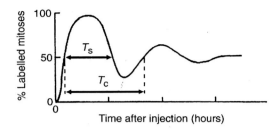

Figure 3. The percentage of labelled mitoses in a population pulsed with tritiated thymidine is plotted against the time after injection. The time period labelled T_s is the length of the S phase, while the time period labelled T_c is the duration of the total cell cycle.

where p is the probability of label per cell, or the labelling index (LI), and c is the number of cells counted (11). According to this formula the number of cells that should be scored depends on the labelling index. Populations of cells growing in log phase have nearly 100% of their cells take up tritiated thymidine or bromodeoxyuridine when incubated with label over several hours, and a large fraction when incubated for 30 min, and, thus, scoring 100 cells should provide a sufficiently precise estimate of the fraction of cells in the S phase.

2.5 Measuring active metabolism as a reflection of viable cell number: the MTT assay

All assays for measuring proliferation have advantages and disadvantages and are more appropriate and easier under some circumstances than others. An assay was developed for the purpose of measuring the effects of various cytokines at different concentrations in a large number of experiments simultaneously using a simple, rapid, colorimetric assay that can be carried out in a multiwell plate, requires no washing or harvesting of the cells, and can be read using a scanning multiwell spectrophotometer, also known as an ELISA reader (12). The effects of radiation and cytotoxic agents, as well as effects on cell survival in general, can also be quantitated using this assay (13, 14). The assay takes advantage of the ability of live cells, but not dead ones, to reduce a colourless tetrazolium salt, [3-(4,5-dimethylthiazol-2-yl)-2,5-diphenyltetrazolium bromide (MTT, thiazolyl blue)]. The tetrazolium ring is cleaved by mitochondrial dehydrogenases of living cells forming a blue formazan salt with a peak absorbance at 570 nm. A reading at 690 nm can be used as a reference wavelength where neither MTT nor formazan absorb (15), to eliminate error from scratches on the plastic or turbidity. Often, the reference reading may be eliminated without sacrificing the validity of the results. Prior to the absorbance reading, the MTT formazan crystals are solubilized with a 10% SDS/0.01 N HCl solution. Other solubilization agents of the formazan salt, such as 10% Triton X-100, 10% Nonidet P-40, or DMSO, can also be used with similar results. A solution of 0.04 N HCl in isopropyl alcohol was used in the initial development of this assay (12). This solvent requires a shorter time period to achieve solubility of the formazan salt, but the mixing with the medium is more difficult and results in a more turbid mixture, and can result in protein precipitation from the serum (16). Parallel experiments measuring the interleukin-2 (IL-2) content in dose–response experiments of an IL-2-dependent murine natural killer cell line, NKC3, using an MTT assay and [^{3}H]thymidine deoxyribose incorporation, yielded virtually identical values within experimental error by the two assays (16), and, in fact, the MTT assay was much more sensitive and provided a higher precision in detecting IL-2 and IL-4 in lymphocyte stimulation assays (17). The method, its advantages and pitfalls, as well as other methods relying on the ability of live cells to carry out metabolic processes, are described in detail in Chapter 2.

2.6 Measuring the collective cell volume

2.6.1 RNA content

The amount of total RNA in a cell population is a relative indication of proliferation, although the amount of RNA in G2 is about twice that in G1. However, in a well-distributed, rapidly proliferating population, the value of total RNA can be a measure of relative cell number. Most of the RNA in the cell is ribosomal RNA. In one rapid assay for isolating total cellular RNA, the cells in log phase of growth are washed twice with ice-cold Ca^{2+}-, Mg^{2+}-free PBS on ice. Ice-cold 0.01 M EDTA, pH 8.0, 0.5% SDS (2 ml) is added to the plate and the cells are scraped into a tube. The plate is then rinsed with 0.1 M citrate–EDTA buffer and the contents are added to the tube. Phenol is then added to precipitate the DNA and the tube is centrifuged at 3000 × g for 10 min. The top layer is transferred to another tube and the RNA is ethanol precipitated and redissolved in Tris–HCl/ EDTA, pH 8.0. The RNA concentration is then determined by measuring the optical density at 260 nm (18). The extinction coefficient of RNA is about 20% higher than that of double-stranded DNA so that an RNA concentration of about 40 μg/ml has an A_{260} of 1.

2.6.2 Protein content

A relative determination of cell number can be made by measuring the collective protein content on a tissue culture dish. The easiest and most reproducible method is that of Bradford (19) that makes use of an absorbance wavelength change in an acidic solution of Coomasie blue from 465 nm to 595 nm, when the dye binds to basic and aromatic amino acid residues. To determine total protein concentration on a plate, cells are either scraped into a lysis buffer containing 20 mM Tris–HCl, pH 7.4, 1 mM EGTA, 50 μM

NaVO$_4$, 50 mM NaF, 0.01 U/ml aprotinin, 1 µg/ml leupeptin, and 1 mM phenylmethylsulfonyl fluoride (PMSF), or detached by trypsinization, washed, and resuspended in lysis buffer. Other lysis buffers containing protease inhibitors can be used that also contain detergent, without substantial differences in protein measurements. The cells are then disrupted by sonication and an acid containing Coomasie blue solution is added. The absorbance at 595 nm is determined in diluted samples and graphed against a linear albumin concentration curve as a standard. The concentration of protein varies somewhat with the cell cycle but insufficiently to affect the average protein measurement of a proliferating population.

3. Determining proliferation *in vivo*

Proliferation of cells *in vivo* is governed by many factors, including the state of differentiation of the cells and nature of the tissue, the number of prior divisions, blood supply, nutritional status, reason for proliferation: whether it is part of normal physiology such as organogenesis, continuous epithelial or haematopoietic cell generation, or whether it is part of a neoplastic process. In neoplastic growth, as a consequence of these factors, cells in different parts of a tumour do not grow at uniform rates. In addition, tumours grow according to a Gompertzian model (Norton–Simon hypothesis) (20). This model posits that while doubling times of tumours can range from one week to one year, tumours do not grow exponentially for most of their existence. Instead, as tumours grow larger, their doubling time continues to increase, eventually reaching immeasurably large values. At this stage, the growth rate of the tumour reaches an asymptotic plateau, both as a consequence of increased doubling time and a greater propensity for cell death. Thus, measuring the proliferative rate of representative cell samples from a tumour at a specific time is not characteristic of that tumour at other times, or perhaps of other portions of the tumour. Keeping these issues in mind, investigators still wish to correlate cellular proliferation with various histological types or interventions, and thus devised methods to determine the proliferative rate of biopsy specimens.

3.1 Measuring DNA synthesis

3.1.1 Measuring the fraction of cells undergoing DNA synthesis by autoradiography of [³H]thymidine uptake or immunohistochemistry of BrdU incorporation

Measurement of DNA synthesis *in vivo* in animal experiments involves the injection of nucleotide analogues labelled with [³H]thymidine (0.2–1.0 mCi/kg body weight) or BrdU (200 mg bolus) (*Protocol 6*) that can be incorporated during the DNA synthetic phase in the tissue of interest. After 15–30 min, a biopsy is obtained or the animal is sacrificed, the tissue is sectioned and mounted on slides, and autoradiograms or immunohistochemical staining of the slides is carried out to determine the fraction of cells with label.

In deciding how many cells to score and how many animals to use to compare the labelling index of an intervention to that of a control animal, or to that of another intervention, several factors have to be taken into consideration. These include the expected labelling index, the minimum difference in the labelling index expected to be determined, and the homogeneity of labelling in the tissue of the animals to be assessed. The variance is defined by the formula:

$$\text{variance} = \frac{\sigma^2\left[(c-1)+p(1-p)\right]}{nc}$$

where p is the mean probability of label amongst all of the animals in a group, σ^2 is the variance of p amongst the animals, n is the number of animals, and c is the number of cells scored per animal (11). Generally, the lower the labelling index, the more cells need to be counted per animal. These should be at least 300 if the labelling index is around 1%, or as low as 150 cells if the labelling index is above 5%. In either case, far greater decreases in variance can be obtained by increasing the number of animals than by counting more cells per animal. Comparative studies should include 10 animals per point for obtaining maximal statistical significance in the experimental differences recorded. The data should be reported as the labelling indices ± the variance, rather than fold increases, because the latter may be a potentially misleading statistic whose variance can be artificially much larger than the variances of the labelling indices (11).

3.1.2 Measuring the labelling index, duration of S phase, and potential doubling time with single BrdU injection *in vivo*

Begg *et al.* (21) describe a method whereby the DNA synthesis time (length of S phase) of cells taken from an *in vivo* tumour sample several hours after an *in vivo* administration of BrdU can be estimated using a simple calculation from variables obtained by flow cytometric analysis of BrdU-associated fluorescence and DNA content. A mouse or a rat bearing a tumour is injected with 100 mg/kg BrdU, intraperitoneally. For patients, an intravenous dose of BrdU (200 mg/m^2) is administered under an Institutional Review Board (IRB)-approved protocol after obtaining signed informed consent (22, 23). Several hours later, after biopsy of the tumour or when the entire tumour is excised, a single cell suspension is obtained (*Protocol 6*) and the cells are labelled with a monoclonal antibody to BrdU and a fluoresceinated anti-mouse IgG secondary antibody giving a green fluorescence to the labelled cells, and the DNA in the cell is labelled with propidium iodide, giving off a red fluorescence. Simultaneous measurements of red fluorescence and green fluorescence are analysed by flow cytometry.

Protocol 6. *In vivo* BrdU injection labelling index determination (from ref. 22)

Determinations of labelling index and average doubling times for tumour cells in human patients is still an experimental intervention. This method is usually performed only as part of a clinical protocol requiring approval by an Institutional Review Board and is governed by all laws and regulations dealing with the protection of human subjects.

Reagents

- BrdU
- H&E
- 0.2% collagenase type II
- anti-BrdU antibody
- 0.02% DNase I in Ca^{2+}-, Mg^{2+}-free Hanks medium
- FITC-labelled secondary antibody

Method

1. Dissolve 200 mg BrdU in 20 ml 0.9% saline immediately prior to injection.

2. Administer BrdU in the 20 ml volume as an i.v. push.

3. Perform a biopsy, or excision of the tumour, 4–9 hours after injection of BrdU.

4. Cut 2 μm sections, stain with H&E, and examine adjacent cuts for histological identification. Estimate the fraction of tissue made up of tumour cells vs. stroma or necrosis.

5. Prepare a single cell suspension (24):

   (a) Dissociate by mechanical means first.

   (b) Incubate tumour fragments in 0.2% collagenase type II and 0.02% DNase I in Ca^{2+}-, Mg^{2+}-free Hanks medium. Then proceed to step (e). Or,

   (c) Fix solid pieces in 70% ethanol (25).

   (d) Incubate in 0.4% pepsin in 0.1 M HCl at 37 °C to release nuclei.

   (e) Denature DNA in 2 M HCl for 15–30 min at room temperature.

   (f) Stain with anti-BrdU antibody at 1:25 dilution for 1 h at room temperature.

   (g) Incubate with 2 ml FITC-labelled secondary antibody for 30 min at room temperature.

   (h) Add 10 μg/ml PI for 10 min in 10 ml PBS.

   (i) Analyse cytofluorometrically. Excite at 488 nm (100 mW) with a tungsten laser. Collect green fluorescence (BrdU) between 510 and 560 nm and red fluorescence (PI) at > 620 nm.

(j) Plot (i) DNA content vs. relative cell number (DNA for proliferation) and (ii) bivariate cytogram of DNA content vs. BrdU uptake [relative green (BrdU) vs. red (PI) on a linear scale].

(k) Labelling index (LI) is defined as the total number of cells in the population that are synthesizing DNA and can be measured by setting a region around those cells that show significant BrdU incorporation. The lower limit of this region should be determined by a control incubation of the cells without addition of the anti-BrdU antibody.

The relative movement (RM) of cells through the S phase can be quantitated using the following formula:

$$RM = \frac{(F_L - F_{G1})}{(F_{G2M} - F_{G1})}$$

where the mean red fluorescence of cells that are positive for green fluorescence is termed F_L, and the mean red fluorescence of G1 and G2+M cells, are termed F_{G1} and F_{G2M}, respectively. Immediately after labelling, F_L will be approximately one-half of the way between the G1 and G2+M peaks. As all of the cells labelled with BrdU leave the S phase and enter G2+M, the F_L will approach the fluorescence of G2 and RM will become 1.0. The time between cells entering the S phase and the time they reach G2 is T_S. The formula for calculating T_S is:

$$T_S = \frac{0.5}{(RM - 0.5)} \times t$$

where t is the time after the BrdU injection that the sample was taken. This formula is an approximation and assumes that the increase in RM is linear with time.

The potential doubling time, T_{pot}, can be estimated from this calculation using the formula:

$$T_{pot} = \frac{\lambda T_S}{LI}$$

where LI, the labelling index, is estimated as the proportion of the whole population with BrdU uptake as determined by green fluorescence, and λ is a correction factor for non-linear distribution of cells through the cell cycle, always within 30% of 1.0. The sampling times should be at least 3 h after the BrdU administration for rodent tumours and 6 h in human tumours to minimize errors due to the estimates for the initial RM (21).

These calculations do not account for the non-dividing cells and the rate of cell death within a tumour. To obtain an estimate of the cell death rate using these approximations, an estimate of the tumour doubling rate, T_D, needs to be obtained using tumour size measurements with time. To estimate the cell death fraction (ϕ) the following formula can be used:

$$\phi = 1 - \frac{T_{pot}}{T_D}$$

3.1.3 Measuring the labelling index, duration of S phase, and doubling time using a BrdU/IUdr double label method *in vivo*

While there are many methodologies for labelling cells *in vivo* with different radionuclides, with BrdU, IUdr, or other labels able to incorporate into DNA, the double label method of Yanik *et al.* (26) provides the type of clear, useful, and easy to obtain information that can be correlated with clinical outcomes. Many other methods provide a labelling index, but using a time-delayed double label with BrdU and IUdr provides data for calculating both a labelling index and the duration of the S phase. These experiments also permit the calculation of the average doubling time of cells in a population. The calculation can, however, fall victim to the estimate of 1.0 for the growth fraction of tumours, which is clearly an overestimate. There are few, good *in vivo* methods for obtaining an adequate estimate of a tumour's growth fraction. Thus, depending on what part of a tumour the biopsy is obtained from, the calculation of the doubling time of tumour cells can vary with the availability of blood supply, degree of necrosis, nutritional status of the tumour cells, and infiltration by tumour-infiltrating lymphocytes, to name a few examples. The method is described in *Protocol 7*.

Protocol 7. *In vivo* BrdU/IUdr double-label labelling index/doubling time determination (from ref. 26)

Determinations of labelling index and average doubling times for tumour cells in human patients is still an experimental intervention. This method is usually performed only as part of a clinical protocol requiring approval by an Institutional Review Board and is governed by all laws and regulations dealing with the protection of human subjects.

Equipment and reagnets

- Iudr, BrdU (Accurate Chemical & Scientific Corp,Westbury, NY)
- Bouin's solution (picric acid, formalin, acetic acid and water)
- PBS
- 1.5% horse serum
- biotinylated anti-rat IgG
- avidin–DH (Vectastain Elite ABC Kit, Vector Laboratories, Burlingame, CA)
- biotinylated horseradish peroxidase H
- rabbit anti-mouse IgG
- alkaline phosphatase–anti-alkaline phosphatase antibody complex (Universal DAKO APAAP kit, Dako Corp., Santa Barbara, CA)
- 0.5 M TBS (Tris-buffered saline)
- 1 M Levamisol
- Fast blue BBsalt
- microtome

24

Method

1. Inject patients with IUdr (100 mg/m^2) infused intravenously over 1 h.

2. Following a 1 h rest, infuse BrdU (100 mg/m^2) intravenously over 1 h.

3. At the completion of the BrdU infusion, obtain a biopsy specimen of the tumour, fix in Bouin's solution for 3 h, dehydrate in ethanol and embed in plastic with glycol methacrylate, and cut 2 μm sections using a microtome.

4. Rehydrate the slides with distilled water for 10 min.

5. Incubate the slides with 3% H_2O_2 for 30 min.

6. Incubate the slides with pronase 1 mg/ml for 45 min. Rinse with 0.15 M PBS after each incubation.

7. Incubate with 4 N HCl for 20 min and rinse with 0.5 M PBS.

8. Sequentially stain slides with two monoclonal antibodies, one that recognizes BrdU only and one that recognizes both BrdU and IUdr.

Anti-BrdU antibody staining

(a) Stain first with the rat anti-BrdU antibody at a 1:3000 to 1:5000 dilution (or the dilution recommended by the package insert) in 0.5 M PBS, 1.5% horse serum for 60 min at room temperature. Rinse with 0.5 M PBS.

(b) Subsequently stain with a biotinylated anti-rat IgG diluted to 1:200–500 in 0.5 M PBS, 1.5% horse serum, for 30 min. Rinse with 0.5 M PBS.

(c) Develop slides by incubating with a mixture of avidin–DH and biotinylated horseradish peroxidase H complex diluted 1:50 with 0.5 M PBS (ABC reagent from the Vectastain Elite ABC Kit). Rinse the samples with 0.5 M PBS.

(d) Stain the cells with 0.025% 3,3'-diaminobenzidine tetrahydrochloride [diluted in a solution of 100 ml 0.5 M Tris-buffered saline (TBS), 0.01 ml 30% H_2O_2] for 5 min. Rinse with distilled water and place in 0.5 M TBS for 30 min.

Anti-BrdU/IUdR antibody staining

(e) Subsequently, stain the slides with a second, mouse monoclonal antibody recognizing both BrdU and IUdr, at a dilution suggested in the vendor's data sheet with 0.5 M TBS, 0.25% Tween-20 for 60 min at room temperature, as above. Rinse the sections with 0.5 M TBS.

(f) Incubate the slides with a rabbit anti-mouse IgG at 1:20 dilution in 0.5 M TBS for 30 min. Rinse the sections with 0.5 M TBS.

Protocol 7. *Continued*

(g) Cover the slide with a 1:40 dilution of mouse alkaline phosphatase–anti-alkaline phosphatase antibody complex in 0.5 M TBS for 30 min. Rinse with 0.5 M TBS.

(h) Develop with a fast blue substrate solution prepared as follows:
- dissolve 20 mg naphthol AS-MX phosphate in 2 ml *N,N*-dimethyl formamide
- add to 100 ml 0.1 M Tris buffer, pH 8.2
- add 0.1 ml Levamisol 1 M (to inactivate endogenous alkaline phosphatases)
- add 100 mg Fast blue BBsalt
- mix well and filter

Immerse the samples in the fast blue solution for 15 minutes, rinse in distilled water, mount with Fluoromount and allow to dry overnight.

9. Count cells under a light microscope. Cells will appear as unlabelled, brown only (BrdU only), blue and brown (double stained with both antibodies, contain both BrdU and IUdr), and blue only (IUdr only).

10. Calculate the labelling index (LI) according to the formula:

$$LI = \frac{(\text{BrdU only cells} + \text{double-labelled cells})}{\text{total cells counted}} \times 100$$

To calculate the duration of S phase (T_S) the following formula is used:

$$T_S = \frac{(\text{BrdU only cells} + \text{double-labelled cells})}{\text{IUdr only cells}} \times t$$

where *t* is the time interval between the start of the infusion of the first label and the start of the second label infusion. The doubling time of the average tumour cells is calculated from the equation:

$$T_c = (LI/T_S) \times \text{growth fraction}$$

The growth fraction of tumours is assumed to be 1.0, which is an underestimate.

4. Cells and tissue samples obtained from human patients or animals

When intervention *in vivo* is not possible prior to obtaining a biopsy, a measure of the proliferative rate of cells in a tissue can be roughly estimated using a variety of techniques.

4.1 Determining the mitotic index

As described earlier for tissue culture cells, the mitotic index is a true reflection of the fraction of cells undergoing cell division, provided there is no mitotic delay or arrest. Only a small portion of cells in *ex vivo* tissue samples contain discernible mitotic indices when observed on fixed biopsies. As discussed earlier, the method may undercount true mitoses because cells entering the phase may complete cell division by the time processing takes place and G2 cells may not enter it. Prophase and telophase cells may not be counted as reliably as metaphase and anaphase cells. Pathologists regard the appearance of any mitotic figures as a sign of a high grade malignant phenotype, but quantitation is not commonly done. The incidence of mitotic figures has been used to try to correlate proliferative rate with survival, without success.

4.2 Measuring cellular DNA content

The most straightforward way of measuring the proliferative fraction of tumour cells from a patient or an experimental animal is determining the fraction of cells in the S phase using flow cytometric analysis of DNA content per cell. Single cell suspensions are prepared by mincing fresh biopsy samples with scalpels or scissors, incubating the tissue with 0.2% collagenase type II and 0.02% DNase I in Ca^{2+}-, Mg^{2+}-free Hanks medium, or fixing the tissue in 70% ethanol and releasing the nuclei in 0.4 mg/ml pepsin in 0.1 M HCl for 1 h at 37 °C, pelleting the cells or nuclei, and resuspending in PI, 10 µg/ml in PBS.

4.3 Nuclear antigens associated with proliferation: PCNA, Ki67, and AgNOR

Very often, fixed or frozen tissue is all that is available for determination of proliferative status. By identifying antigens associated with proliferation in tissue whose architecture is preserved, the proliferative status of specific cells of interest is identified, rather than that of all of the cells in the tissue. Several nuclear antigens that are assayable by immunohistochemistry have been correlated with proliferative status in cells. Among these are the proliferative cell nuclear antigen (PCNA), the antigen recognized by the Ki67 antibody (27) and AgNORs, proteins from nucleolar organizer regions with an affinity for silver (28). This is a relatively non-specific indicator of proliferation.

AgNORs are nucleolar organizer regions representing loops of DNA transcribed to ribosomal RNA associated with proteins that have an affinity to silver. The characteristics of AgNOR labelling are based on several visual characteristics, such as clusters or dots per cell, percentage of cells with one cluster, mean AgNOR size, total AgNOR area, and its percentage of the nuclear area. Based on a comparison with the BrdU labelling index, the number of clusters or dots are related to the percentage of cells in S phase and the AgNOR area may be associated with the duplication time, with the mean size of AgNORs being smaller in rapidly proliferating cells (29). The specific significance of AgNORs is not understood and many groups have tried to correlate AgNORs with tumour aggressiveness and rapidity of proliferation, with variable success rates that depended on the tumour type. The most useful correlations have been in differentiating malignant from benign tissue samples. A method for staining for AgNORs in paraffin-embedded tissue samples is described in *Protocol 8*.

The antigen recognized by the antibody Ki67 is a nuclear matrix non-histone protein that is expressed in all phases of the cell cycle except G0 and early G1. The amount of antigen increases with cell cycle progression and reaches a peak in G2+M. A number of studies confirmed a correlation between Ki67 and other markers of proliferation: tritiated thymidine uptake, PCNA, S phase and LI by BrdU, and flow cytometry in breast cancers and lymphomas (30–33). However, Ki67 positive rates were generally higher than the LI by BrdU uptake (30, 31) with the formula Ki67 $= 1.6 \times (BrdU)+1.22$ functioning as the relationship between the two values, because of the cell cycle phase expression characteristics (30). Staining with Ki67 requires fresh, snap-frozen tissues and may not be done in paraffin-embedded samples. *Protocol 8* describes a method for immunostaining fresh tissue samples for Ki67.

Proliferating cell nuclear antigen (PCNA) is a nuclear protein that is a member of the G1 cyclin complexes and is a cofactor to DNA polymerase, having a function in DNA synthesis and an independent function in DNA repair. The synthesis of PCNA initiates in G1, increases during the remainder of G1 and the beginning of S phases, then begins to decline through the rest of the cycle. PCNA has been used as a proliferation marker in many studies but its value as a predictor of labelling index and tumour grade is not certain. *Protocol 8* also describes a method of staining for PCNA from archival tissue.

Protocol 8.   Determining the expression of AgNORs, Ki67, or PCNA from *ex vivo* samples

Equipment and reagents

- xylene
- ethanol (96, 90, and 80%)
- PBS
- gelatin
- 1% aqueous formic acid
- 50% aqueous silver nitrate
- 0.1% v/v H_2O_2
- sodium azide
- mouse monoclonal antibody to Ki67 (Dako Corp, Santa Barbara, CA)
- peroxidase-conjugated rabbit anti-mouse IgG (Dako)

- 40% human AB serum
- peroxidase-conjugated pig anti-rabbit IgG antiserum
- diaminobenzidine tetrahydrochloride (DAB) solution: 10 mg DAB in 50 μl H_2O_2, 3 ml, 0.1 M citric acid, 6.9 ml sodium hydrogen phosphate (0.2 M), pH 6.4 (prepare fresh)
- Gill's haematoxylin
- primary anti-PCNA antibody (Dako)
- 5% pig serum
- biotinylated horse anti-mouse IgG (Vector Labs)

AgNOR staining of paraffin-embedded tissue section (from ref. 34)

1. Cut paraffin blocks into 3 μm slices and mount on glass slides. Dissolve the paraffin by two successive 5 min baths in xylene, two successive 3 min baths in 96% ethanol, a 3 min 90% ethanol, and a subsequent 3 min 80% ethanol bath. Wash the slides with PBS four times.

2. Prepare the AgNOR staining solution:
   (a) Dissolve 2 g gelatin in 100 ml of 1% aqueous formic acid.
   (b) Mix the gelatin solution with a 50% aqueous silver nitrate solution at 1:2 v/v for the final working solution.

3. Incubate the slides with the AgNOR staining solution for 60 min at room temperature.

4. Wash off the silver colloid with deionized water.

5. Counterstain with neutral red, wash with deionized water.

6. Dehydrate the slides with xylene and visualize the AgNOR spots under the microscope. Count the number of AgNOR spots per cell and determine the mean.

Immunohistochemical staining for Ki67 (from ref. 35)

1. Snap-freeze the biopsy tissue. Cut 3 μm slices and mount on glass slides. Air-dry overnight.

2. Permeabilize the cells with acetone at 4 °C for 10 min. Air-dry for 1 h.

3. Block endogenous peroxidase by incubating the slides with 0.1% v/v H_2O_2 and 1 μg/ml sodium azide in PBS for 15 min at room temperature. Wash with PBS for 5 min, twice.

4. Incubate with mouse monoclonal antibody to Ki67 diluted 1:30 in PBS for 1 h at room temperature. Wash with PBS for 5 min, twice.

5. Incubate with peroxidase-conjugated rabbit anti-mouse IgG diluted 1:40 with 40% human AB serum in PBS. Wash with PBS for 5 min, twice.

6. Incubate with peroxidase-conjugated pig anti-rabbit IgG antiserum diluted 1:30 in 40% human AB serum in PBS for 30 min at room temperature. Wash with PBS for 5 min, twice.

7. Apply the DAB solution (freshly prepared) for 6 min. Wash twice with tap water.

8. Counterstain sections in 2% Gill's haematoxylin for 7 min.

9. Dehydrate with xylene and air-dry.

Immunohistochemical staining for PCNA

1. Prepare paraffin-embedded slides, dissolve the paraffin, and rehydrate with ethanol, as above in the AgNOR protocol.

Protocol 8. *Continued*

2. Block endogenous peroxidase by incubating the slides with 0.1% v/v H_2O_2 in 10% v/v methanol in PBS, pH 7.4, for 30 min at room temperature. Wash with TBS.
3. Incubate with primary anti-PCNA antibody diluted 1:25 in TBS 5% pig serum. Wash with TBS.
4. Incubate with the second antibody, a biotinylated horse anti-mouse IgG diluted 1:200 in 5% pig serum for 30 min at room temperature. Wash with TBS.
5. Apply a horseradish peroxidase–streptavidin complex diluted 1:50 with TBS (Vector) for 45 min.
6. Develop with DAB solution, as above.

Wilson *et al.* (35) compared five different methods of determining the proliferative rate of tumour tissue *ex vivo* obtained from 125 patients with colon cancer 4–5 h after *in vivo* administration of iododeoxyuridine (IUdr). The methods of labelling index determination were flow cytometric and immunohistochemical determinations of IUdr using a monoclonal antibody, immunohistochemical determination of Ki67, and PCNA, the latter in both methanol- and formalin-fixed tissue. There was significant variation and poor or absent correlations of the values obtained in the labelling indices between the methods. Even when the markers measured the same stage of the cell cycle, i.e. IUdr uptake as measured by flow cytometry or immuno-histochemistry, the assays yielded data with little correspondence. Paired analysis of data obtained using different assays showed a poor correlation between the different methods of estimating the labelling index. There was, however a positive correlation between IUdr uptake by both assays and Ki67 positivity. The IUdr uptake by flow cytometry and immunohistochemistry were significantly correlated but the labelling index determined by IUdr uptake was consistently higher when determined by immunohistochemistry than by flow cytometry in both aneuploid and diploid cells. The LIs in aneuploid cells as determined by IUdr uptake, by both flow cytometry and immunohistochemistry, were significantly higher than in diploid cells, while the ploidy did not affect Ki67 or PCNA positive rates. While there was no correlation between labelling indices by any method and stage or aggres-siveness of disease in colorectal cancers (35), there was a correlation in other tumour types such as thyroid tumours, thymomas, male breast cancer, bladder carcinomas, pharyngeal carcinoma, and myeloma.

References

1. Sherr, C.J. (1994) *Cell* **79**, 551.
2. Blin, N. and Stafford DW. (1976) *Nucl. Acid Res.* **3**, 2303.

3. Sladek, T.L. and Jacobberger, W.J. (1993) *J. Virol.* **66**, 1059.
4. Pedrali-Noy, G., Spadari, S., Miller-Faures, A., Miller, A.O.A., Kruppa, J., and Koch, G. (1980) *Nucl. Acids Res.* **8**, 377.
5. Giaretti, W., Nusse, M., Bruno, S., Di Vinci, A., and Geido, E. (1989) *Exp. Cell Res.* **182**, 290.
6. Baserga, R. and Malamud, D. (1969) *Autoradiography*. Harper and Row, New York.
7. Steel, G.G. (1977) *Growth kinetics of tumors*. Clarendon Press, Oxford.
8. Pera, F., Mattias, P., and Detzer, K. (1977) *Cell Tissue Kinet.* **10**, 255.
9. Trent, J.M., Gerner, E., Broderick, R., and Crossen, P.E. (1986) *Cancer Genet. Cytogenet.* **19**, 43.
10. Cawood, A.H. and Savage, J.R.K. (1983) *Cell Tissue Kinet.* **16**, 51.
11. Morris, R.W. (1993) *Environmental Health Perspectives* **101** (suppl. 5), 73.
12. Mosmann, T. (1983) *J. Immunol. Meth.* **65**, 55.
13. Carmichael, J., DeGraff, W.G., Gazdar, A.F., Minna, J.D., and Mitchell, J.B. (1987) *Cancer Res.* **47**, 936.
14. Carmichael, J., DeGraff, W.G., Gazdar, A.F., Minna, J.D., and Mitchell, J.B. (1987) *Cancer Res.* **47**, 943.
15. Denizot, F. and Lang, R. (1986) *J. Immunol. Meth.* **89**, 271.
16. Tada, H., Shiho, O., Kuroshima, K-i., Koyama, M., and Tsukamoto, K. (1986) *J. Immunol. Meth.* **93**, 157.
17. Gieni, R.S., Li, Y., and HayGlass, K.T. (1995) *J. Immunol. Meth.* **187**, 85.
18. Stallcup, M.R. and Washington, L.D. (1983) *J. Biol. Chem.* **258**, 2802.
19. Bradford, M.M. (1976) *Anal. Biochem.* **72**, 248.
20. Norton, L. (1982) In *Clinical interpretation and practice of cancer chemotherapy* (ed. E.M. Greenspan), pp. 53–70. Raven Press, New York.
21. Begg, A.C., McNally, N.J., Shrieve, D.C., and Karcher, H. (1985) *Cytometry* **6**, 620.
22. Wilson, G.D., McNally, N.J., Dische, S., Saunders, M.I., Des Roches, C., Lewis, A.A., and Bennett. (1988) *Br. J. Cancer* **58**, 423.
23. Bourhis, J., Wilson, G., Wibault, P., Bosq, J., Chavaudra, N., Janot, F., Luboinski, B., Eschwege, F., and Malaise, E.P. (1993) *Int. J. Radiation Oncol. Biol. Phys.* **26**, 793.
24. Wilson, G.D., McNally, N.J., Dunphy, E., Karcher, H., and Pfragner, R. (1985) *Cytometry* **6**, 641.
25. Schutte, B., Reynders, M.M., van Assche, C.L., Hupperets, P.S., Bosman, F.T., and Blijham, G.H. (1987) *Cytometry* **8**, 372.
26. Yanik, G., Yousuf, N., Miller, M.A., Swerdlow, S.H., Lampkin, B., and Raza, A. (1992) *J. Histochem. Cytochem.* **40**, 723.
27. Gerdes, J., Schwab, U., Lemke, H., and Stein, H. (1983) *Int. J. Cancer* **31**, 13.
28. Crocker, J., McCartney, J.C., and Smith, P.J. (1988) *J. Pathol.* **154**, 151.
29. Lorand-Metze, I., Carvalho, M.A., and Metze, K. (1998) *Cytometry* **32**, 51.
30. Sasaki, K., Matsumura, K., Tsuji, T., Shinozaki, F., and Takahashi, M. (1988) *Cancer* **62**, 989.
31. Kamel, O.W., Franklin, W.A., Ringus, J.C., and Meyer, J.S. (1989) *Am. J. Pathol.* **134**, 107.
32. Sahin, A.A., Ro, J.Y., El-Naggar, A.K., Wilson, P.L., Teague, K., Blick, M., and Ayala, A.G. (1991) *Am. J. Clin. Pathol.* **96**, 512.

33. Pich, A., Ponti, R., Valente, G., Chiusa, L., Geuna, M., Novero, D., and Palestro, G. (1994) *J. Clin. Pathol.* **47**, 18.
34. Crocker, J. and Nar, P. (1987) *J. Pathol.* **151**, 111.
35. Wilson, M.S., Anderson, E., Bell, J.C., Pearson, J.M., Haboubi, N.Y., James, R.D., and Schofield, P.F. (1994) *Surgical Oncol.* **3**, 263.

2

Cell growth and cytotoxicity assays

PHILIP SKEHAN

1. Introduction

This chapter describes non-radioactive assays commonly used to measure changes in cell growth and cytotoxicity induced in a cell population by treatment with agents such as drugs, hormones, nutrients, and irradiation. The focus is specifically on mammalian cell cultures, but the methods can be used with a wide variety of cells, tissues, and microorganisms. Because of the ease and speed with which large numbers of samples can be processed by optical plate readers, most of the protocols which follow are designed specifically for 96-well plates, but are easily adapted to other culture vessels.

2. Growth and cytotoxicity assays

There are four types of non-radioactive cell growth and cytotoxicity assays: (1) cell or colony counts; and assays of (2) macromolecular dye binding, (3) metabolic impairment, and (4) membrane integrity. No single method is universally appropriate for all situations. Each has limitations, and all are subject to potentially serious artefacts under certain circumstances.

The ideal assay is safe, simple, inexpensive, rapid, sensitive, quantitative, objective, reliable, and foolproof. It would also be desirable if the assay required no special handling or disposal methods, was rigorously proportional to cell number or biomass under a wide range of environmental and biological conditions, was non-destructive so that samples could either be monitored repeatedly to generate kinetic curves or could be collected at the end of an assay for examination by other methods, and could be used equally effectively with adherent cultures, single cell suspensions, floating multicellular aggregates, and mixed adherent–non-adherent populations.

Cell and colony counts are time consuming, tedious, and sensitive to minor variations in methodology. Cell counts enumerate morphologically intact cells but do not distinguish between living and dead cells. Colony counts often require subjective judgments about what constitutes a colony, and are subject to a wide variety of troublesome artefacts that greatly complicate their interpretation.

Dye-binding assays probably come closest to fulfilling the ideal requirements for growth and cytotoxicity assays. They are simple, rapid, reliable, sensitive, and quantitative, but do require access to a 96-well plate reader or spectrometer. They also fail to distinguish between living and dead cells *per se*, although with proper experimental design they will distinguish between net cell killing, net growth inhibition, and net growth stimulation.

Metabolic impairment assays measure the decay of enzyme activity or metabolite concentration following toxic insult. They are generally more complex and artefact prone than dye-binding assays. Their validity requires that the same precise conditions be met that are required by a bimolecular chemical reaction analysis. Deviation from these conditions can lead to extremely serious errors that invalidate the assay. Metabolic impairment assays are nevertheless popular because they distinguish between normal and reduced levels of cellular metabolism, which is a surrogate index of metabolic viability, though not necessarily an accurate predictor of proliferative capability.

Membrane integrity assays measure the ability of cells to exclude impermeant extracellular molecules. They can be either colorimetric or fluorescent, and require the same instrumentation as dye-binding assays. They tend to be less artefact prone than metabolic impairment assays, and have the same capacity to estimate 'viability', which in this case is the ability to distinguish between the normal and impaired exclusion of extracellular molecules. Membrane integrity assays are complicated by the fact that living cells slowly accumulate probe molecules; hence protocols for these assays must be very carefully optimized. Like metabolic impairment assays, they provide a surrogate index of viability that is not always an accurate predictor of proliferative capacity.

3. Quantifying cell growth

3.1 Experimental design

Growth is the change in a quantity over time (1). It is a rate process that requires at least two separate measurements. In growth end-point assays, measurements are made at the start and end of an experiment. In kinetic assays, intermediate measurements are also recorded to produce a time series of values.

Five different types of samples are required to determine the biological effect of an experimental treatment upon a cell population: (i) medium blanks—growth medium with no cells or drugs; (ii) drug blanks—growth medium with drug but no cells; (iii) time zero values—cells plus medium at the start of an experiment; and end of assay values for (iv) control cells plus medium and (v) test cells plus medium.

The population size of samples is calculated by subtracting medium blanks from time zero and control measurements, and drug blanks from test samples receiving drugs (1). This produces three final population measurements: the time zero population size (Z), the control population size (C) at the end of the assay, and the test population size (T) at the end of the assay. The net effect of a treatment is determined in the following way (1):

- net growth stimulation has occurred if T is greater than C
- net growth inhibition has occurred if T is greater than Z but less than C
- total growth inhibition has occurred if $T = Z$
- net cell killing has occurred if T is less than Z

3.2 Growth rate and doubling time

Cellular growth processes are usually described by one of two interrelated equations, the specific growth rate (R), and the population doubling time (T_d):

$$R = -1 + (S_2/S_1)^{T_u/T_{obs}} \qquad \text{change per day} \qquad (1)$$
$$T_d = 16.636/\ln(1+R) = 16.636/\ln[(S_2/S_1)^{T_u/T_{obs}}] \quad \text{in hours} \qquad (2)$$

The specific growth rate describes the per capita change in the size of a population over a specified observation period (T_{obs}). S_1 and S_2 represent population size values at the beginning and end of the observation period. These are calculated from T, C, and Z values as described in Section 3.1. For studies of mammalian cell cultures, the most convenient unit period of time (Tu) is usually one 1 day of exactly 24 h. If T_{obs} is not an exact multiple of 24 h, fractional days should be used. An R of 1 means that during the observation period there has been the equivalent of one additional cell per day added to the population for each cell that was present at time zero. In other words, R is a per capita rate of population change. The doubling time is a reciprocal function of the natural logarithm of R. As presented here it specifies, in hours, the length of time needed for the population to double. T_d is better at describing very slow growth, while R is better at describing very fast growth.

R is a net rate of growth; it is the difference between the birth or production rate (B) and the death or loss rate (L), i.e. $R = B-L$. Because cellular birth and death rates can only be measured with special techniques such as time lapse cinematography, growth studies generally measure only the net growth rate. The assays described in this chapter all measure net rate, and are not capable of determining the extent to which changes in birth and loss rates contribute to the overall growth process.

3.3 Potency

The effects of different drugs on a cell line, or of one drug on different types of cells, can be compared using indices of either drug potency or drug efficacy (1, 2). Commonly used indices of drug potency include:

(a) GI_{50}: the concentration producing 50% net growth inhibition, i.e. $(T-Z)/(C-Z) = 0.50$.

(b) *TGI*: total growth inhibition, which occurs at the concentration for which $T=Z$.

(c) *LC*$_{50}$: the concentration that produces 50% net lethality, which occurs when $(T{-}Z)/Z = -0.50$.

The major disadvantage to the use of potency indices is that they are interpolations rather than actual measured values. They require that an assumption be made about the shape of the dose–response curve between two measured points; this assumption can be a source of significant error.

3.4 Efficacy

To compare visually a drug's efficacy against different types of cells or microbes, data are calculated as $(T{-}Z)/(C{-}Z)$ for all non-negative values $(T \geqslant Z)$ of cellular response, and as $(T{-}Z)/Z$ for all negative values $(Z > T)$. When these calculated response function values (R_f) are plotted against drug concentration (usually on a log scale), they produce curves for which the control value of R_f for all cell lines is 1.0, total growth inhibition occurs at $R_f = 0.0$ for all cell lines, and the total lysis of all cell lines occurs at $R_f = -1.0$. The greatest differential or selective efficacy occurs at the single test drug concentration that generates the greatest variability between cell lines, as measured by mean absolute deviation or standard deviation. The mean absolute deviation is the preferred calculation because of its lesser sensitivity to the biasing influence of outlier data points. The difference between samples in R_f at this concentration is a measure of drug efficacy. Efficacy has the advantage of using actual measured values rather than interpolated approximations.

4. Cell cultures

4.1 Seeding density

Seeding density depends on cell size, growth rate, and assay duration. It must be determined individually for each cell type (2). In a 2–3 day assay, seeding densities are typically in the range of 5–25 thousand cells per well (kcpw) in 96-well microtitre plates. Time zero values must be sufficiently above background to allow net cell killing to be accurately quantitated, while end of assay values must remain within the linear range of the assay, typically 1.5–2.0 OD units.

4.2 Dissociation and recovery

Chemical and enzymatic dissociation are traumatic processes that overtly kill some cells while sensitizing survivors to other toxic insults. This sensitization can produce false-positive artefacts with cells subsequently exposed to growth inhibitors or cytotoxins. Dissociated cells require a recovery period before an experiment is started. One day is usually sufficient (2).

4.3 Drug solubilization

- Make stock solutions of polar compounds in water, buffer, or medium, then dilute in complete growth medium to the final test concentration (2).
- Dissolve non-polar compounds in a solvent such as dimethyl sulfoxide (DMSO) or ethanol (EtOH) and filter sterilize (0.22 μm pores). A 1:1 mixture of DMSO and EtOH is also a good solvent: it evaporates more slowly than EtOH and chemically sterilizes most test materials. DMSO is toxic to cells at concentrations above 0.1–1.0%.
- Conduct preliminary experiments to determine its toxicity threshold for each individual cell line. Ethanol is usually growth stimulatory in the 1–2% range.

4.4 Assay duration

The assay duration is determined by two factors: (i) the length of time cells need to respond to an experimental treatment; and (ii) the length of time that cells can grow before nutrient depletion sets in. Depletion typically develops within 3–4 days after plating unless cultures are re-fed (3). Once depletion begins, a progressive deterioration of cellular health and viability develops rapidly, becoming a major artefact in data interpretation. Depletion can be calibrated by comparing the day by day growth kinetics of cultures that receive no feeding with cultures that are fed daily (3). The two curves begin to diverge when depletion sets in. This normally sets the upper limit of assay duration if cultures are not fed. If experiments must be continued for longer periods, then medium must be replaced. Above densities of near-confluence, most types of cells require daily feeding. With cytotoxicity assays, a 36–48 hour assay period following a 1 day recovery period is usually adequate to detect a drug's effect, while avoiding the need to re-feed in mid-experiment.

4.5 Control wells on every plate

Every test plate requires its own control wells. Because temperature and humidity are not uniform in tissue culture incubators, plates located at different positions tend to experience different local microenvironments which differentially affect cell growth. Control values from one plate are often different from and cannot reliably be used for test samples on other plates.

5. Dye-binding assays

This section describes assays for culture protein, biomass, double-stranded polynucleic acid, and DNA. The protein and biomass assays are colorimetric, and measure the binding of coloured dyes to cell cultures fixed with trichloroacetic acid (TCA). The polynucleic acid and DNA assays are fluorescent, and measure the amount of fluorochrome intercalation into RNA or DNA in freeze–thaw permeabilized cells.

5.1 Optimizing and validating dye-binding assays

Dye-binding assays are bimolecular chemical reactions that use a change in the amount of one reactant (the dye) to measure the amount of a second (the cellular target). Valid binding assays must meet several requirements which serve collectively to guarantee that a single cell in one test sample produces the same signal as an identical cell in another. If these conditions are not met, then two samples cannot be quantitatively compared. The necessary conditions are established by optimizing a protocol for each experimental system in each laboratory using the cells and reagents of that laboratory. They should not be adopted in a cookbook manner from the literature. The dye-binding assays described in this chapter have the desirable characteristic that their optimized protocols, once established, are identical for a wide variety of cell types (2, 4, 5). This is not true with most other assays.

5.1.1 Supramaximal dye concentration

The concentration of a dye must be sufficiently high (supramaximal) to saturate all target-binding sites. A supramaximal concentration is any dye concentration sufficiently high that its further increase causes no further increase in staining intensity (4, 5). If the condition is not met, dye-binding does not accurately measure the amount of a macromolecular target.

5.1.2 Supramaximal dye volume

To saturate target binding sites, dye concentration and the volume of staining solution must both be supramaximal for assay validity (5). A supramaximal volume is one on the plateau region of the curve relating staining volume to measured signal for a concentration of target molecules higher than will ever be encountered experimentally.

5.1.3 Supramaximal staining time

A third requirement for the validity of dye-binding assays is that the staining time must be supramaximal (4, 5). A supramaximal staining time is one in which the signal being measured has reached a stable plateau value that does not increase further with time. It should be determined using a culture density higher than will ever be encountered experimentally.

5.1.4 Washing away unbound dye

Several of the assays require that unbound dye be washed away with an appropriate buffer at the end of a staining period. Washing can cause the loss of cell-associated dye. Rinse procedures must be carefully optimized to determine the minimum number of rinses necessary to remove excess unbound dye without desorbing any significant amount of cell-associated stain (4, 5). Rinses should be performed rapidly. Remove rinse solutions from cultures as quickly as possible; never leave them sitting in wells. Failure to reach a stable staining intensity of cultures with successive washes invalidates the assay.

5.2 Protein and biomass stains

Many colorimetric biological stains bind electrostatically to macromolecular fixed ions of opposite electrical charge. The binding and desorption of these dyes can be controlled by varying pH. Dilute acids and bases, such as 1% acetic acid and 10 mM unbuffered Tris base, can be used to bind and solubilize the dyes (1).

Under mildly acidic conditions, anionic dyes bind to basic amino residues, serving as cellular protein stains. They can be extracted and solubilized by weak base. Protein stains suitable for growth and cytotoxicity assays (*Table 1*) include sulforhodamine B (SRB), orange G (ORG), bromphenol blue (BPB), and chromotrope 2R (CTR).

Under mildly basic conditions, cationic dyes bind to negative fixed charges of proteins, RNA, DNA, and glycosaminoglycans, serving as stains for culture biomass. They can be quantitatively extracted from cells and solubilized for optical density measurement by weak acid. Biomass stains suitable for growth and cytotoxicity assays (*Table 1*) include thionin (TNN), azure A (AZRA), and toluidine blue O (TBO). Phenosafranin and safranin O are also acceptable as biomass stains.

5.2.1 Cell fixation

Protein and biomass are both measured colorimetrically on fixed cell cultures. Trichloroacetic acid (TCA) and perchloric acid (PCA) are the fixatives of

Table 1. Optimal staining protocols for selected dyes

	Protein stains			Biomass stains		
	ORG	BPB	CTR	TNN	AZRA	TBO
Concentration (%w/v)	1.5	1.0	0.25	0.3	0.25	0.2
Staining solution	AcOH	AcOH	AcOH	Tris	Tris	Tris
Minimum stain time (min)	5	20	30	10	30	10
Number of washes	3	4	3	4	4	3
Solubilizing solution	Tris	Tris	Tris	AcOH	AcOH	AcOH
Optimal wavelength (nm)	478	590	508	594	628	626
Resolution (kcpw)	2–3	3–4	4–5	>5	>10	>10

kcpw = thousands of cells per well
Solubilizing solutions: 1% acetic acid (AcOH); 10 mM unbuffered Tris base.
Protein stains: ORG (orange G), BPB (bromphenyl blue), CTR (chromotrope 2R).
Biomass stains: TNN (thionin), AZRA (azure A), TBO (toluidine blue O).

choice (1). They fix cells instantly, permeabilize their membranes to allow rapid dye penetration, extract small molecular weight metabolites, greatly strengthen cell adhesion, and harden cells to withstand rough handling. Formaldehyde is less effective as a fixative. Glutaraldehyde, because of its bifunctionality, can potentially interfere with the assay, and should not be used. Organic fixatives such as ethanol and methanol cause significant cell lysis, and are not appropriate for these assays.

Protocol 1 describes the *in situ* TCA fixation of adherent cell cultures (1). Once fixed, cells are extremely resistant to damage and dislodgement. Solutions can be poured directly on to fixed plates from a beaker or even a running water tap with no risk of cell dislodgement. The only delicate part of the fixation process is the initial addition of the fixative; it must be added gently without producing any fluid shearing forces that would dislodge cells before they become fixed.

Protocol 1. *In situ* TCA fixation of adherent cells in 96-well microtitre plates (1, 4)

Method

1. Grow cells with 200 μl of medium per well. To fix, gently layer 50 μl of cold 50% TCA on top of the 200 μl already in the well to produce a final TCA concentration of 10%.

2. Incubate plates for 30 min to complete fixation.

3. Wash plates five times with distilled or tap water.

4. Air-dry at room temperature; dried plates can be stored indefinitely before further processing.

For suspensions of single cells and very small aggregates (about 5 cells or less), use the modified fixation procedure of *Protocol 2*. For larger aggregates, which are not attached to the plastic by this fixation method, use the propidium iodide assay of *Protocol 4*.

Protocol 2. *In situ* TCA fixation of single cell suspensions in 96-well microtitre plates (4)

Method

1. Grow cells with 200 μl of medium per well. Lay plates on a stable bench top and allow cells to settle to the bottom of the wells. Gently layer 50 μl of cold 80% TCA on top of the 200 μl already in the well to produce a final TCA concentration of 16%; cells must be in contact with tissue culture plastic at the instant that the TCA front reaches them in order to be attached.

---

2. Leave the plates undisturbed on the bench top for 30 min to complete fixation.

3. Wash five times with distilled water, as with adherent cultures.

4. Air-dry at room temperature; dried plates can be stored indefinitely before further processing.

5.2.2 Sulforhodamine B (SRB) colorimetric assay of cell protein

Sulforhodamine B (SRB) is a bright pink aminoxanthene dye with two sulfonic groups. It is one of a family of related dyes, such as naphthol yellow S and Coomassie brilliant blue, that are used widely as protein stains in histochemistry and molecular biology. Under mildly acidic conditions, SRB binds to basic amino acid residues of TCA-fixed proteins. It provides a sensitive index of culture cell protein that is linear with cell number over a cell density range of more than 100-fold. Of more than 20 protein and biomass stains investigated for possible use in microtitre assays, SRB provided the best combination of staining intensity and signal-to-noise ratio (4). It provides a stable end-point that does not have to be measured within any fixed period of time. Once stained and air-dried, plates can be kept indefinitely without deterioration before solubilization and reading. SRB has proven particularly useful in very large-scale drug screening (2).

Protocol 3.   Sulforhodamine B (SRB) assay for culture cell protein in 96-well microtitre plates (4)

Reagents

- 0.4% (w/v) SRB
- 1% acetic acid

- 10 mM unbuffered Tris base (pH 10.5)

Method

1. Harvest plates by fixing with TCA as described in *Protocols 1* or *2*. Be sure to collect Z, C, and T samples as described in Section 3.1.

2. Stain TCA-fixed cells for at least 30 min with 0.4% (w/v) SRB in 1% acetic acid.

3. Remove SRB then quickly wash plates four times with 1% acetic acid to remove unbound dye.

   (a) Wash quickly to avoid desorbing bound dye molecules; pour acetic acid on to the plates from a large beaker; remove by turning the plate upside down and flicking to shake out residual fluid.

   (b) Do not aspirate individual wells; this takes too much time and will desorb protein-bound dye molecules, causing artefacts in your data.

   (c) Complete all four washes of one plate before going on to the next.

Protocol 3. *Continued*

4. Air-dry until no standing moisture is visible (overnight is recommended); plates can be stored indefinitely before proceeding.

5. Solubilize bound SRB by adding a fixed volume (usually 100–200 μl) of 10 mM unbuffered Tris base (pH 10.5) to each well and shaking for at least 5 min on a shaker platform.

6. Read the optical density in a 96-well plate reader; the optimal wavelength is 564 nm.[a]

[a] The assay is very sensitive; confluent cultures often give ODs that are above the linear range of the assay (about 1.8 OD units). It is frequently desirable to use a suboptimal wavelength so that higher cell densities will remain within the range of linearity. Wavelengths in the 490–530 nm range work well for this purpose.

5.2.3 Other protein stains

Orange G, bromphenol blue, and chromotrope 2R have chemistries identical to SRB, and stain TCA-fixed protein nearly as well (4). Their optimized assays follow *Protocol 3* modified as shown in *Table 1*. Coomassie brilliant blue and naphthol yellow S are much less satisfactory.

5.2.4 Cellular biomass assays

Thionin, azure A, and toluidine blue O are biomass stains that approach the protein stains in sensitivity (4). Protocols for their use are identical to those for the protein stains, except that the acidic and basic solutions are reversed. Use the modifications of *Protocol 3* indicated in *Table 1*.

5.2.5 Propidium iodide fluorescent assay of double-stranded polynucleic acid

Propidium iodide (PI) is a general biomass stain that binds to RNA, DNA, protein, and glycosaminoglycans (GAGs). Its fluorescence increases by as much as 100-fold when it intercalates into double-stranded sequences of RNA and DNA (6). This allows double-stranded polynucleic acids to be measured in the presence of free, unbound PI, provided there are no significant amounts of proteins or GAGs present. An important exception involves a narrow excitation peak centred at 531 nm with a band width of approximately 2 nm (5). At this excitation wavelength, fluorescence is independent of cell protein over a broad range of emission wavelengths centred at 604 nm. Using narrow bandwidth 531/604 nm excitation/emission filters, culture double-stranded polynucleic acid content (RNA + DNA) can be measured in the presence of complete growth medium and serum, providing the basis for an exceptionally simple assay that requires no washing steps at all (5). Cultures are permeabilized by freeze-thawing, an aliquot of propidium iodide is added to each well, and after a short incubation period PI fluorescence is read in a fluorescent plate reader at wavelengths of 530/590 or 530/620 (*Protocol 4*).

These wavelength combinations reflect RNA more than DNA. Because RNA usually changes earlier than protein in cellular growth shifts, the PI assay tends to detect cellular responses earlier than protein, biomass, and DNA assays. It is usually the method of choice where short assays are desirable. It is also the method of choice for extreme assay simplicity.

The excitation filters of fluorescent plate readers are broad band, and do allow passage of some fluorescence from protein-bound PI. This produces a low level of background fluorescence from serum in the growth medium. The sensitivity of the PI plate reader assay is 1–3 thousand cells per well in 5% fetal calf serum, comparable with the SRB assay. With 10% serum, resolution is reduced by a factor of 2. These values are obtained with ordinary 96-well plates; somewhat better resolution can be achieved with more expensive low-fluorescence plates.

The PI assay was originally designed for use with human lymphoma cells, which aggregate extensively and cannot be measured with most other assays (5). It works equally well with single cell suspensions, adherent cultures, and bacteria; it is probably the simplest of all plate reader-based growth and cytotoxicity assays.

With all of the cell lines that we have examined, a single freeze–thaw cycle has been sufficient to permeabilize cells to PI, even though it does not always kill cells. Detergent permeabilization must not be used with this assay: detergents have complex and unpredictable effects, and can lead either to strong quenching or high background levels of fluorescence. High salt, EDTA, and proteinase K treatment do not significantly influence the sensitivity of the PI assay (5).

Unlike many fluorescent assays, PI fluorescence is not sensitive to pH under the conditions of *Protocol 4*. Thus, ordinary media like RPMI can be used without worrying about the pH shifts caused by cell growth or plate removal from a CO_2 incubator. Fluorescence is steady for several hours, although polynucleic acids sometimes degrade over longer periods of time.

Protocol 4. Propidium iodide (PI) assay of cellular RNA and DNA in 96-well plates (5)

Equipment and reagents
- 200 μg/ml PI solution
- spectrofluorimeter or a fluorescent plate reader

Method

1. Harvest plates by freezing at –30 °C for at least 2 h; frozen plates can be stored indefinitely. Be sure to collect Z, C, and T samples as described in Section 3.1.

2. Thaw at 50 °C for 15 min.

Protocol 4. *Continued*

3. Add 50 μl/well of a 200 μg/ml PI stock in distilled water to the 200 μl of growth medium in each well; the final PI concentration is 40 μg/ml.

4. Incubate in the dark at room temperature for 60 min.

5. Read fluorescence at 531/604 nm excitation/emission wavelengths in a spectrofluorimeter or at 530/590–620 in a fluorescent plate reader.

    (a) Excitation wavelength is critical, emission is not.

    (b) In a spectrofluorimeter, readings are independent of culture protein and serum to at least 10% fetal calf serum; in a plate reader there is a slight background fluorescence from serum and cell protein that cuts sensitivity in half when serum is raised from 5 to 10%.

Note. Wrap PI solutions in foil and protect from light.

5.2.6 Hoechst 33258 fluorescence assay of DNA

Hoechst 33258 is a UV-excited blue bisbenzimidazole dye which selectively intercalates into A–T-rich regions of DNA, undergoing a fluorescence enhancement in the process. Unlike PI, the Hoechst dye is specific for DNA rather than for macromolecules or polynucleic acids generally. It provides a fluorescence signal that is linearly proportional to DNA content over a wide range of DNA values. The Hoechst 33258 assay is nearly as simple as the PI assay, the only additional step being an initial removal of medium from culture wells (7). This step is required to allow the addition of an EDTA-high NaCl buffer which partly dissociates DNA, allowing better dye access and therefore brighter fluorescence. The assay is comparable to the PI assay in sensitivity. It is well suited for use with firmly adherent cultures, but is less suitable for single cell suspensions or adherent lines that produce viable subpopulations of floating cells.

Protocol 5. Hoechst 33258 fluorescence assay of cellular DNA in 96-well microtitre plates (7)

Reagents
- TNE
- Hoechst 33258 dye

Method

11. Collect Z, C, and T samples as described in Section 3.1.

12. Remove medium from plates.

13. Freeze plates at –80 °C; store frozen until ready for further processing.

14. Thaw plates.

15. Add 100 μl distilled water to each well and incubate for 1 h at room temperature.
16. Refreeze at –80 °C for 90 min.
17. Thaw to room temperature.
18. Add 100 μl of TNE buffer (described below) containing 20 μg/ml of Hoechst 33258 dye; after dilution, final Hoechst 33258 concentration is 10 μg/ml.
19. Incubate in the dark for 90 min.
10. Read in a fluorescent plate reader at 350/460 nm excitation/emission wavelengths.

Note. Wrap Hoechst 33258 solutions in foil and protect from light.
TNE buffer contains 10 mM Tris, 1 mM EDTA, 2 M NaCl, pH 7.4.

6. Metabolic impairment assays

6.1 Neutral red cell viability assay

Neutral red is a vital dye that accumulates in the lysosomes of living, un-injured cells. Dead and severely traumatized cells lose their ability to accumulate and retain neutral red, providing the basis for a simple colorimetric viability assay used widely in cellular toxicology (8).

With vital dye assays, great care must be taken in selecting the length of the dye incubation period. The rate of dye uptake differs widely from one cell type to another, and the uptake itself is sometimes preceded by a lag period of unpredictable length. The dye incubation period of 3 h listed in step 5 of *Protocol 6* is only a rough guide; the staining duration should be individually optimized for each cell line. Because the dye is simultaneously taken up and lost by cells, the staining time must be sufficiently long to allow intracellular neutral red levels to reach a stable equilibrium value. This usually requires a few tens of minutes. Cells lose neutral red rapidly once the dye is removed. It is advisable to conduct efflux studies to determine just how rapidly the loss occurs, and to then ensure that the wash period is sufficiently short to avoid leakage artefacts.

Protocol 6. Neutral red cell viability assay (8)

Reagents

- 4% paraformaldehyde
- 1% calcium chloride
- solubilization fluid: 1 ml glacial acetic acid in 100 ml of 50% ethanol

Method

11. Prepare a 0.4% stock solution of neutral red in distilled water. This may be stored in the dark at 4 °C for several months.

Protocol 6. *Continued*

12. Within 24 h of use, dilute the neutral red stock solution 1:80 in growth medium to produce a final concentration of 50 μg/ml; optimize this concentration individually for each cell line.

13. Pre-warm the neutral red solution to 37 °C, and centrifuge at 1500 r.p.m. (~500 g) for 5 min immediately before use to remove undissolved dye crystals.

14. Collect *Z*, *C*, and *T* samples as described in Section 3.1.

15. Remove medium from the test plates and replace with 0.2 ml of neutral red solution per well.

16. Incubate for ~3 h at 37 °C; individually optimize the incubation time for each cell line.

17. Remove the neutral red solution, and rinse the wells once with 0.2 ml per well of 4% formaldehyde containing 1% calcium chloride. This washes away residual neutral red and fixes the cells to prevent their detachment during dye solubilization. The rinsing process must be completed within 1–2 min to avoid loss of intracellular dye.

18. Add 0.2 ml of solubilization fluid to each well.

19. Solubilize the dye for 15 min on a microtitre plate shaker.

10. Read the optical density at 540 nm.

Note. Wrap neutral red solutions in foil and protect from light.

6.2 MTT redox assay

Tetrazolium reactions have been used widely to histolocalize dehydrogenase enzyme activity. The tetrazoliums are oxidation–reduction (redox) indicators that do not react with the dehydrogenases *per se*, but rather with their reaction products, NADH and NADPH, and also with any small molecules that possess reducing capacity (9). The reaction reduces water-soluble tetrazolium salts to a highly coloured and insoluble formazan that precipitates out of solution in the immediate vicinity of the reaction. A yellow coloured tetrazolium, 3-(4,5-dimethylthiazole-2-yl)-2,5-diphenyl tetrazolium bromide (MTT), is converted by the reduction reaction to a purple formazan, and has been adapted for use in microtitre plates as a cellular viability assay (10).

The MTT assay has the same validity requirements as a Lineweaver–Burke analysis: (i) initial reaction velocity must be measured over (ii) a linear range with (iii) a supramaximal concentration of MTT and (iv) a protocol that is insensitive to microenvironmental fluctuations in parameters such as pH and temperature. The assay only measures cell number accurately when the ability of cells to reduce MTT is rigidly constitutive; if the ability is non-constitutive, differences in microenvironment or past histories of two otherwise identical cells will produce different levels of enzyme expression, invalidating their quantitative comparison.

The MTT assay has a number of serious problems (9). The intensity of MTT reduction varies considerably from one cell line to another, and frequently declines with increasing culture age. The reaction tends to be non-linear with cell number, and both the reaction rate and the spectral absorbance curve vary with pH. The MTT reaction rate varies with cellular and medium glucose levels; it is only linear for short periods of time, and the period of linearity varies with the cell line employed. There are frequently long lag periods between the addition of MTT and the onset of its metabolism, making it virtually impossible to accurately measure initial reaction velocity. The supramaximal concentration of MTT varies widely with cell line.

These several difficulties collectively can make apparent IC_{50}s (drug concentration producing 50% growth inhibition) vary by as much as 20-fold depending on MTT concentration, culture age, population density, and length of assay. In using the MTT reaction as a viability assay, great care must be exercised to optimize it fully for each individual cell line and to work strictly within its limits of documented validity.

The MTT assay described in *Protocol 7* is an unusually well-optimized method that is distinctive in its use of a pH 10.5 buffer that increases sensitivity and reduces or eliminates many of the artefacts normally associated with the assay (11).

Protocol 7. MTT redox assay in 96-well microtitre plates (11)

Reagents

- Sorenson's glycine buffer: 0.1 M glycine, 0.1 M NaCl, adjusted to pH 10.5 with 0.1 M NaOH
- DMSO

Method

1. Prepare a 5 × stock (~ 2 μg/ml) of MTT in phosphate-buffered saline (PBS). The final concentration is typically about 0.4 μg/ml, but should be individually optimized for each cell line.

2. Collect *Z, C,* and *T* samples as described in Section 3.1.

3. Add 50 μl of the MTT stock to the 200 μl of medium in each well.

4. Incubate cultures at 37 °C for a period that is typically 30 min to 4 h in length but must be individually determined for each cell line.

5. Aspirate off medium from each well. For non-adherent cells, leave about 20 μl of medium in the wells to avoid cell loss, then centrifuge plates at 200 *g* for 5 min.

Protocol 7. *Continued*

6. Add 25 μl of Sorenson's glycine buffer.

7. Add 200 μl DMSO to each well and incubate for 10 min while vibrating on a plate shaker to solubilize the formazan crystals.

8. Read the optical density at 570 nm. This wavelength is critical; do not change it.

6.3 AlamarBlue (ALB) oxidation–reduction assay

AlamarBlue (ALB) is a proprietary redox indicator sold by AccuMed International. In the presence of cellular metabolism the colour of ALB changes from a fully oxidized, non-fluorescent blue to a fully reduced, fluorescent red. Redox reactions can be driven either enzymatically, or, as with the MTT assay, by a variety of metabolites with reducing capacity (9). The cellular mechanisms of ALB reduction have not been established, but mitochondria are implicated, and in some fungi there is a suggestion that surface oxidases may be involved as well (12). On the basis of oxidation–reduction potentials, it is expected that ALB will be reduced by a variety of enzymes and small molecules, including the cytochrome system, FMN, FAD, NAD, and NADP (13).

The reduction of ALB can be monitored spectrophotometrically, fluorometrically, or visually. Fluorometric measurements are preferable with some experimental systems, spectrophotometric methods with others; either is satisfactory in most (14–16). Colorimetric changes provide the basis for a visual assay that can be used in microbial susceptibility tests (17) or under field conditions where instrumentation is not available (16).

The ALB assay is simple, rapid, inexpensive, and requires no lysis, extraction, or washing of samples. The assay has been used successfully with bacteria, fungi, protozoa, and both adherent and non-adherent mammalian cells to study metabolism, proliferation, cytotoxicity, and cell adhesion. It generally correlates well with other assay methods, although occasional exceptions have been found (12, 14). A disadvantage of the assay is that the ALB signal from a culture is not stable upon storage. The absorbance of oxidized (blue) ALB is stable for at least 3 days, but the absorbance of reduced (red) ALB increases from one day to the next (13). With fluorescence measurements, the reverse is true. The assay is also temperature sensitive (13). For these reasons, ALB should be measured shortly after an assay is completed, and care should be taken to ensure that all cultures are at the same temperature.

The AlamarBlue assay measures the integral of a culture's chemical reducing capacity over time, not cell growth or cell number. The assay can measure the metabolic activity of non-proliferating cultures (18). ALB molecules that have been reduced by cellular metabolism do not reoxidize when cell death occurs, so that the colour persists stably even if cells are removed (14, 19). Thus if an ALB-containing culture experiences a period of growth prior to experimentally induced cell death, the assay will report only the growth-associated metabolism, not the cell death, and not the cell number at the end of the assay.

With proliferating cultures, the assay is proportional to cell number only under the special conditions of serial cell dilutions and very short ALB incubation times initiated immediately after plating. This proportionality breaks down progressively with increasing time after plating and duration of dye incubation, and the ALB signal becomes increasingly a reflection of metabolic history and less an indicator of cell number.

Because ALB reduction is irreversible, short dye incubation times are critical for distinguishing between net cell growth inhibition and net cell killing. Except with serial dilutions, the curve relating ALB signal to cell number is usually either sigmoidal or concavely approaches an asymptotic upper limit. Regions of quasi-linearity can often be identified, but their breadth varies with both the type of cell and the details of the assay protocol.

The ALB assay needs to be individually optimized for each experimental system. For mammalian cells, optimal dilutions of ALB in culture wells are typically in the range of 1/10 to 1/25 while optimal incubation times range from 1 to 6 h; the most common optima appear to be a 1/25 dilution and 3 h incubation (20). ALB is toxic to cells at high concentrations and exposure times (14), but non-toxic protocols have been successfully developed for some cell lines (19, 21). Under conditions of rigorously documented non-toxicity, repeated measurements can be made on a single sample continuously incubated with ALB, and cells can be harvested at the end of an assay for further analysis by other methods.

An important disadvantage of the ALB assay in optical density mode is the fact that the absorbance spectra of the oxidized and reduced forms overlap substantially. This requires that measurements be made at two different wavelengths in order to correct for the spectral overlap. Failure to make this correction can result in serious numerical errors.

ALB sensitivity is similar to that of the MTT but less than that of the PI, SRB, and Hoechst assays. In terms of its underlying chemistry the ALB assay is similar to the MTT assay. The ALB assay is faster, simpler, and less artefact prone than the MTT assay, and is generally a preferable assay.

Protocol 8.   AlamarBlue metabolic reduction assay (13, 22, 23)

Method

11. A concentrated stock solution of AlamarBlue is available from Accumed International, Inc. Keep ALB solutions foil wrapped in the dark.

Protocol 8. *Continued*

12. Collect *Z*, *C*, and *T* samples as described in Section 3.1.

13. At the end of an experimental incubation period, add 1 vol of ALB stock solution per 25 vols (4% v/v) of growth medium in each well (8 μl ALB for 200 μl of growth medium).

14. Incubate plates at 37 °C for 3 h to allow metabolic dye reduction.

15. Equilibrate plates to room temperature for 30 min in the dark.

16. Measure the relative fluorescence at 530–560 nm excitation and 590 nm emission wavelengths. Fluorescence is temperature sensitive; either equilibrate plates at room temperature before reading or measure plates in a warm room at the culture incubation temperature. For better sensitivity, measure the fluorescence in bottom-reading rather than top-reading mode. The ratio of test to control fluorescence values at 590 nm measures the effect of a treatment on cell growth or metabolism.

17. For spectrophotometric assays, correct for the spectral overlap of the oxidized and reduced forms of ALB by measuring each sample at two different wavelengths, between, approximately, 540 and 630 nm. One of these must be a low wavelength (LW) and the other a high wavelength (HW); for example, 570 and 600 nm, respectively.

18. A correction factor (RO) for the absorbance of oxidized ALB must be calculated:

    (a) measure the absorbance (AM) of growth medium alone (no ALB)

    (b) measure the absorbances of oxidized (blue) ALB in growth medium at the low and high wavelengths

    (c) subtract AM from each of the measured ALB absorbances to produce, respectively, AOLW and AOHW, the absorbance of oxidized (blue) ALB at the low and high wavelengths

    (d) calculate the correction factor RO of oxidized ALB: RO = AOLW/AOHW

19. Measure the absorbance values (ALW and AHW) of a test sample at each wavelength.

10. Calculate the percentage of reduced ALB (ARLW) in a sample as: ARLW = 100 × [ALW−(AHW × RO)].

11. Calculate the percentage difference in reduction (PDR) between treated and control cells: PDR = 100 × (test ARLW/ARLW for positive growth control).

Note. Final ALB concentration and incubation time with the dye should be individually optimized for each experimental system.

6.4 Cellular ATP assay

ATP is the primary donor of chemical high energy in living cells. It is present in all metabolically active cells, drives a wide range of biochemical reactions, and is rapidly degraded by ATPases in dead and dying cells. These properties make ATP levels a good index of viable biomass in a wide range of biological systems. The ATP-driven luciferin–luciferase system that produces firefly bioluminescence is the basis for an exquisitely sensitive assay that can detect as little as 10^{-12} moles of ATP, the equivalent of about 50–100 mammalian cells (24, 25). The method correlates well with a variety of other assays, including cell number, biomass, colony-forming units, dye exclusion, and thymidine incorporation. A particular advantage of the ATP assay is that, like PI, it can be used with adherent cultures, single cell suspensions, floating multicellular aggregates, mixed adherent-floater populations, and three-dimensional tissues.

The major drawback of the ATP assay is that it uses a specialized and expensive type of detection system, the luminometer, which is not available to many laboratories. Although 96-well plate luminometers have recently become available, most existing instruments are tube luminometers that read a single tube or cuvette at a time. For ATP measurement, an automatic injector is highly desirable for introducing the luciferin–luciferase mixture into a sample. A variety of companies, including most luminometer manu-facturers, sell luciferin–luciferase reagent kits designed for ATP assays. *Protocol 9* uses the reagent from Sigma Chemical Company; reagents from other companies are similar, although slight protocol modifications may be required as suggested by the manufacturer.

Protocol 9.   ATP assay (14)

Method

1. Collect *Z, C,* and *T* samples as described in Section 3.1.

2. At the end of an experimental incubation period, add an equal volume of 2% trichloroacetic acid (TCA) to the medium present in each culture well.

3. Incubate for 15 min at room temperature to extract ATP. The ATP in the extract is stable for several hours in TCA; do not neutralize until just before samples will be read in the luminometer.

4. Remove a 100 μl extract from each extract and neutralize with an equal volume of 0.1 M Tris–HCl buffer (pH 9.0).

5. If an automatic injector is not available, transfer a 100 μl sample of the neutralized extract to a luminometer cuvette containing 100 μl of Sigma luciferin–luciferase reagent. If an automatic injector is available,

---

Protocol 9. *Continued*

place the extract in a clean cuvette, place in the luminometer, then add luciferin–luciferase mixture.

6. Measure the luminescence over a 30 sec interval.

Note. The final TCA concentration must be at least 1% to extract cellular ATP. Modify the protocol as suggested by the manufacturer for luciferin–luciferase reagents other than Sigma's. Coloured compounds at high concentration will sometimes quench bioluminescence. Quenching can be corrected for by running ATP calibration curves in the presence of each drug concentration that produces quenching.

7. Membrane integrity assays

7.1 Fluorescein diacetate (FDA)

Fluorescein diacetate (FDA) is an electrically neutral, non-fluorescent molecule. Viable cells accumulate FDA intracellularly and hydrolyse it to an anionic reaction product, fluorescein, which is highly fluorescent. Fluorescein is retained intracellularly by living cells for short periods of time, causing them to become temporarily fluorescent. Dead cells lack the ability to accumulate and hydrolyse FDA, and are therefore non-fluorescent. Both FDA and fluorescein are non-toxic; they provide the basis for a vital assay of cell metabolic viability that allows test cultures to be subsequently reused for other purposes (26).

Because FDA is hydrolysed by serum, producing high levels of background fluorescence, assays must be conducted in serum-free solution; this is usually acceptable for short assays, but can be highly traumatic with long incubation times. Fluorescein is pH sensitive, and undergoes a fluorescence enhancement with increasing alkalinity. Keep the pH rigidly constant by using a well-buffered solution such as PBS as a solvent and rinsing solution. Serum-free non-CO_2-buffered media such as PDRG (27) or GIBCO's CO_2-independent growth medium are also suitable, and are more physiological than simple buffers such as PBS. Alternatively, serum-free RPMI and MEM can be used if they are supplemented with a CO_2-independent buffer such as 25 mM Hepes or β-glycerophosphate.

Like all viability assays, the FDA to fluorescein conversion does not, strictly speaking, measure viability *per se*. The conversion is a function of cell volume, cell physiology, and level of hydrolytic enzyme activity as well. However, the assay correlates well with, and is linearly proportional to, cell number under a variety of conditions (28).

Protocol 10. Fluorescein diacetate cell viability assay in 96-well microtitre plates (28)

Reagents
- PBS
- FDA

Method

1. Prepare a 10 μg/ml stock solution of FDA in DMSO; store at –20 °C.
2. Collect *Z, C,* and *T* samples as described in Section 3.1; centrifuge at 200 *g* for 5 min.
3. Remove medium by inverting plates and flicking gently.
4. Wash once with PBS or a serum-free, pH stable, non-CO_2-buffered medium.
5. Add 200 μl of pre-warmed FDA (10 μg/ml) in the same solution used in step 4 (optimize the exact FDA concentration for each individual cell line).
6. Incubate at 37 °C for 60 min (optimize the exact incubation time for each individual cell line, with particular attention to a possible lag phase in fluorescence production).
7. Centrifuge plates at 200 *g* for 5 min; remove solutions as in step 3.
8. Add 200 μl of pre-warmed PBS or serum-free medium to each well.
9. Read the fluorescence immediately at 485/538 nm excitation/emission wavelengths.

Note. Wrap FDA solutions in foil and protect from light.

---

8. Survivorship assays

In cultures treated with cytotoxic drugs, some cells die and lyse, others die but do not lyse, and still others are mortally injured but do not die or lyse within the time limit of the experiment. Survivors often go through an extended period of proliferative quiescence as part of their response to environmental stress or trauma, and their subsequent outgrowth is often a very slow process requiring weeks or months.

The short assays described in *Protocols 3–10* commonly fail to detect small subpopulations of surviving cells, and do not identify populations of momentarily surviving cells that are destined to die later as the result of experimental insult. Survivorship assays provide a means for identifying such populations. They represent an extended second stage assay that is conducted following completion of a short-term primary assay such as those described in *Protocols 3–10*.

There are two classes of survivorship assays: (1) recovery, and (2) colony formation. Recovery assays follow bulk cultures for extended periods of time to determine whether they eventually exhibit cellular recovery or delayed mortality. Colony-forming assays convert populations to single cell suspensions, then determine the proportion of cells that give rise to progressively growing colonies. Colony-forming efficiency (CFE) is the percentage of single cells plated that gives rise to multicellular colonies. CFE can be measured by plating cells on a solid substratum such as tissue culture plastic, or by seeding them in a non-adherent medium such as soft agar.

8.1 Adherent versus non-adherent cells

Mammalian cells of solid tissue origin usually require some degree of adhesive contact to other cells or a physical surface. When deprived of these contacts, their proliferation slows and their death rate increases, causing serious artefacts in survivorship assays that restrict adhesion and cell to cell contact (29, 30). Soft agar CFE assays are best suited to non-adherent cells that grow naturally as single cell suspensions. Adherent cells tend to exhibit elevated mortality rates in soft agar, a tendency that can be potentiated by cytotoxic insult. The adhesion of cells to a plastic substratum at least partly satisfies their anchorage requirement, and supports better proliferation and viability than growth in soft agar. Long-term recovery and plastic CFE are generally preferable to soft agar assays with adherent cell populations. Recovery assays are in turn preferable to plastic CFE assays because they do not require cytotoxic chemical or enzymatic dissociation, and better maintain cell to cell contacts during the highly stressful period of early recovery.

8.2 Long-term recovery (LTR) assay

The simplest survivorship assay is the long-term recovery (LTR), which is based on the fact that living cells secrete metabolic acids that reduce the pH of a culture's growth medium (31). In media containing phenol red as the pH dye indicator (most media), this metabolism is evidenced by a change in colour from red to either orange or yellow.

Before starting an LTR assay, perform a primary assay using the methods of *Protocols 3–10* to construct a dose–response curve for a drug of interest. From this, select three drug concentrations which produce effects that are: (1) half-maximal, (2) just maximal, and (3) supramaximal.

Conduct the LTR assay in T25 flasks rather than 96-well plates. Seed the cells in 5–10 ml of medium, but otherwise use the same protocol (cells/cm^2, assay duration, etc.) that was employed in the primary assay. At the end of the drug incubation period, replace the test solution with fresh drug-free medium, equilibrate the atmosphere in the flask with that of the incubator chamber, close the flask top tightly, and incubate at 37°C for as long as desired. *Protocol 11* uses 60 days as a standard recovery period, but the assay can be abbreviated or extended indefinitely. Replicate flasks can be collected at intermediate times to explore the possibility that delayed cell killing has occurred.

To use the LTR assay qualitatively, examine the colour of flasks visually 2–3 times each week. Use cell-free flasks treated with drugs then incubated in drug-free medium as a reference. The survivorship of metabolically viable cells is established when test cultures become less red and more orange or yellow than the reference flasks. Verify microscopically that the survivorship is by the experimental cell population rather than by contaminating micro-organisms. Data are expressed as growth delay: the number of days following drug removal at which cultures are judged to have clearly exhibited metabolic recovery.

To determine whether metabolically recovered cells are also proliferatively viable, dissociate and replate the cells. Collect a time zero sample, then incubate the replicate flasks for several days. At the end of a normal culture growth period, count the cells or conduct one of the assays in *Protocols 3–5*. An increase above time zero value establishes progressive growth and proliferative viability.

Metabolic recovery can be documented quantitatively by measuring the change in culture pH. The absorbance maximum for phenol red in cell-free growth medium in the pH range 7.4–7.6 is approximately 560 nm. As the medium shifts from red towards orange or yellow, the optical density at 560 nm decreases. If a CO_2-independent medium is used (Section 7.1), pH is stable in air so that aliquots can be collected and read at 560 nm in a microtitre plate reader or spectrophotometer. With CO_2-dependent media, prepare a set of tightly capped reference flasks containing bicarbonate-free medium adjusted to various pH values; use these to visually estimate the pH of experimental flasks.

Protocol 11.   Long-term recovery assay for T25 flasks (31)

Method

1. Perform a short-term dose–response analysis in 96-well microtitre plates (*Protocols 3–10*).

2. Repeat the experiment in T25 flasks exactly as in microtitre plates except:

   (a) plate cells in 5–10 ml of growth medium

   (b) scale up the innoculum size to correct for different vessel sizes (i.e. use the same number of cells per cm^2 in the T25 flasks that you used in the microtitre plates).

3. Treat flasks with three drugs determined by data from step 1: use concentrations that are: (1) half maximal; (2) just maximal; and (3) supramaximal.

Protocol 11. *Continued*

4. At the end of the drug incubation period, remove the test solution and replace with fresh drug-free medium; if there are a significant number of floating cells in the medium, centrifuge them, resuspend, and place back in the flask along with fresh medium.

5. Equilibrate the atmosphere in the flask for 1–2 h, then cap tightly and maintain in a 37 °C incubator.

6. Examine 2–3 times weekly for pH change; when obvious pH change has occurred, evaluate as described in Section 8.2. Growth delays of up to several months can be measured.

7. As a reference, use a cell-free flask treated with drug solution, then incubate during the recovery period with fresh drug-free medium.

8. Express recovery as a drug-induced growth delay; quantitate the extent of the pH change if desired by measuring the OD shift at 560 nm or by comparing the colour in test flasks to the colour in a series of reference flasks adjusted to different pH values.

8.3 Colony-forming efficiency on tissue culture plastic

Colony formation on a solid substratum such as tissue culture plastic is the preferred method for determining the CFE of adherent cell populations. Treat cells experimentally, then dissociate and replate them in fresh growth medium. The proper seeding density depends on the growth rate and plating efficiency, but typically ranges from a few tens to a few thousands of cells per T25 flask. The seeding density should be selected so that final end of assay colony counts are in the range of 100–1000. Increased serum sometimes improves the CFE.

The CFE of adherent cells can be determined by counting the number of (1) macroscopic colonies visible by eye; (2) microscopic colonies with more than some threshold number (10, 20, or 50) of cells; or (3) microscopic colonies with two or more cells. These methods are not comparable, and commonly give different results. The last criterion is technically the most rigorous, while the first is the least.

Pure single cell suspensions are impossible to prepare for most cell types. Background levels of multicellular aggregates are typically about 5%, but can be as high as 50% with some cell lines. Aggregates tend to be more viable than single cells, and are more likely to give rise to a progressively growing colony (29). It is critically important to establish the time zero frequency distribution for the number of cells per colony. A colony is any cluster of 1 or more cells. This distribution becomes the reference to which end of assay results are compared. A shift in the distribution towards larger colony sizes is evidence of proliferative recovery. Standard statistical methods can be used to quantitate the shift.

An alternative method is to identify specific colonies at the beginning of an assay and follow them individually throughout the colony-forming process. Make a series of crosses at various points on the bottom of a flask. Align an optical grid successively in each of the four quadrants of each cross. Identify and record the colonies located within the grids, estimating their size at various times throughout the experiment. Colony size can be estimated in several ways: (1) number of cells, (2) colony radius or diameter, (3) colony area from photographs or video images, or (4) frequency histogram of cells per colony (it is extremely difficult to accurately count cells in colonies with more than about 20 cells). This approach works well with non-motile cells. If cells and colonies are motile, the same approach can be used but modified by counting colonies per grid without attempting to follow specific individual colonies.

Protocol 12. Colony-forming efficiency of adherent cells on tissue culture plastic in T25 flasks

Method

1. Prepare a single cell suspension using *Protocol 1* or *2*.
2. Plate cells in a T25 flask containing 5–10 ml of medium.
   (a) Adjust seeding density so that final end of assay colonies are in the range of 100–1000 per flask.
   (b) Elevated serum concentrations (10–15%) sometimes improve cell survivorship.
3. Allow cells to attach; the time required varies with the cell line, and can range from a few minutes to overnight; 0.5–3 h is satisfactory for many cell lines.
4. Change the medium to remove unattached cells.
5. Count the time zero number of colonies containing 1, 2, 3,... cells per colony (Section 8.3).
6. Incubate cultures until colonies are well developed; this period varies with the cell line, and can range from a few days to several months. Change the medium if the colour begins to turn orange.
7. Count colonies; if cultures are not needed for other purposes, they can be TCA-fixed and stained for further quantitation with SRB (*Protocol 3*) or one of the dyes listed in *Table 1*.

8.4 Colony-forming efficiency in soft agar

CFE assays in soft agar are best suited to cells, like those of the haemato-poietic system, that grow naturally in single cell suspensions. While adherent cells will sometimes grow in soft agar, the environment is extremely non-physiological and generates a number of complex artefacts that make quantitative analyses dubious (29).

Soft agar assays contain two separate layers: (1) an underlayer which serves as a barrier to prevent cell adhesion to the bottom of the tissue culture chamber; and (2) an overlayer containing the cells. The underlayer is typically a 0.5% and the overlayer a 0.3% gel. Either agar or agarose can be used. Agar is less pure and contains materials that are growth inhibitory to some cells (29). Methylcellulose, an extremely viscous fluid, can be substituted for agar or agarose in the overlayer.

Protocol 13 describes a survivorship assay in which cells have already been experimentally treated prior to starting the soft agar assay (32). However, the method can also be used as a primary assay in dose–response studies (33). For this purpose, colonies should be grown to observable size before an experimental treatment is begun; their reduction in size or number serves as an index of drug efficacy. The data from soft agar CFE assays are analysed as described in Section 8.3.

Protocol 13. Soft agar colony-forming efficiency in 35 mm dishes

Method

1. Prepare the underlayer.
   - (a) Make solution A: prepare a double strength solution of growth medium (2 × for all components including serum); filter sterilize; divide into two portions; warm to 45°C in a water bath.
   - (b) Make solution B: add 1% (v/v) powdered agar or agarose to distilled water; autoclave for 20 min (slow exhaust) to solubilize and sterilize; place in a 45°C water bath.
   - (c) Make solution C (the underlayer): add equal volumes of solutions A and B; mix well; return to the 45°C water bath to prevent gelling.
   - (d) Add 1 ml of solution C to each 35 mm dish; allow the solution to gel at room temperature, then incubate the dishes at 37°C until further use.

2. Prepare the overlayer.
   - (a) Prepare solution D: add 0.6% (v/v) powdered agar or agarose to distilled water; autoclave for 20 min (slow exhaust) to solubilize and sterilize; place in a 45°C water bath.
   - (b) Prepare solution E: dissociate cells and prepare a 2 × single cell suspension (twice the desired final cell concentration) in double strength growth medium; equilibrate to 37°C.
   - (c) Prepare solution F (the cell overlayer): add equal volumes of

---

solutions D and E; mix well without frothing; transfer immediately to a 37 °C water bath.

(d) Layer 0.5 ml of solution F on top of the underlayer in each 35 mm dish; add immediately before the agar begins to gel.

(e) Leave the dishes at room temperature until the agar overlayer solidifies.

(f) Add 0.5–1.0 ml of 1 × growth medium on top of the overlayer.

(g) Transfer dishes to a 37 °C incubator.

3. Collect time zero plates and construct a time zero frequency histogram of colony size (number of colonies with 1, 2, 3, 4,... cells) as described in Section 8.3.

4. Incubate test cultures at 37 °C; soft agar assays commonly require 1–3 weeks; to prevent medium depletion and to compensate for evaporation, liquid medium should be changed at least twice weekly beginning at the end of the first week, sooner if evaporation is evident.

5. At the end of the assay construct frequency histograms of colony size for test and controls.

6. For counting colonies, contrast can be enhanced by MTT staining (19).

(a) Gently aspirate liquid from the top of the overlayer.

(b) Add 1 ml of 1 mg/ml MTT made by diluting a sterile-filtered 5 mg/ml MTT stock in PBS 1 to 5 in pre-warmed growth medium.

(c) Incubate for 4 h.

(d) Remove MTT.

(e) Add 1.5 ml of 2.5% (w/v) protamine sulfate in normal saline solution; incubate at 4 °C for 16–24 h, then aspirate free solution; repeat. This procedure is optional, and is designed to increase the contrast by reducing background staining of the agar.

(f) Count the colonies.

References

1. Skehan, P., Thomas, J., and Friedman, S.J. (1986). *Cell Biology and Toxicology*, **2**, 357.
2. Monks, A., Scudiero, D., Skehan, P., Shoemaker, R., Paull, K., Vistica, D.T., Hose, C., Langley, J., Cronise, P., Vaigro-Wolff, A., Gray-Goodrich, M., Campbell, H., Mayo, J., and Boyd, M.R. (1991). *Journal of the National Cancer Institute*, **83**, 757.
3. Skehan, P. and Friedman, S.J. (1984). *Cell Tissue Kinetics*, **17**, 335.
4. Skehan, P., Storeng, R., Scudiero, D., Monks, A., McMahon, J., Vistica, D.T., Warren, J.T., Bokesch, H., Kenney, S., and Boyd, M.R. (1990). *Journal of the National Cancer Institute*, **82**, 1107.

5. Skehan, P., Bokesch, H., and Williamson, K. (1993). In *Drug resistance on leukemia and lymphoma. The clinical value of laboratory studies* (ed. G-J. Kaspers), pp. 409–413, Harwood Publishers, Langhorn, PA.
6. Shapiro, H.M. (1988). *Practical flow cytometry*. Alan R. Liss, New York.
7. Rago, R., Mitchen, J., and Wilding, G. (1990). *Analytical Biochemistry*, **191**, 31.
8. Borenfreund, E. and Puerner, J.A. (1984). *Journal of Tissue Culture Methods*, **9**, 7.
9. Vistica, D.T., Skehan, P., Scudiero, D., Monks, A., Pittman, A., and Boyd, M.R. (1991). *Cancer Research*, **51**, 2515.
10. Mossman, T. (1983). *Journal of Immunological Methods*, **65**, 55.
11. Plumb, J.A., Milroy, R., and Kaye, S.B. (1989). *Cancer Research*, **49**, 4435.
12. Jahn, B., Stuben, A, and Bhakdi,S. (1996). *Journal of Clinical Microbiology*, **34**, 2039.
13. AccuMed International, Inc. (1996). AlamarBlue Assay. Instruction brochure.
14. Squatrito, R.C., Connor, J.P., and Buller, R.E. (1995). *Gynecologic Oncology*, **58**, 101.
15. Collins, L. and Franzblau, S.G. (1997). *Antimicrobial Agents and Chemotherapy*, **41**, 1004.
16. Raz, B., Iten, M., Grether-Buhler, Y., Kaminsky, R., and Brun, R. (1997). *Acta Tropica*, **68**, 139.
17. Baker, C.N. and Tenover, F.C. (1996). *Journal of Clinical Microbiology*, **34**, 2654.
18. Larson, E.M., Doughman, D.J., Gregerson, D.S., and Obritsch, W.F. (1997). *Investigative Ophthamology & Visual Science*, **38**, 1929.
19. Ahmed, S.A., Gogal, R.M., Jr, and Walsh, J.E. (1994). *Journal of Immunological Methods*, **170**, 211.
20. Nakayama, G.R., Caton, M.C., Nova, M.P., and Parandoosh, Z. (1997). *Journal of Immunological Methods*, **204**, 205.
21. Lelkes, P.I., Ramos, E., Nikolaychik, V.V., Wankowski, D.M., Unsworth, B.R., and Goodwin, T.J. (1997). *In Vitro Cellular & Developmental Biology*, **33**, 344.
22. Geier, S. (1994). Described in ref. 16.
23. Goegan, P., Johnson, G., and Vincent, R. (1995). *Toxicology In Vitro*, **9**, 257.
24. Kangas, L., Gronroos, M., and Nieminen, A.L. (1984). *Medical Biology*, **62**, 338.
25. Andreotti, P.E., Cree, I.A., Kurbacher, C.M., Hartmann, D.M., Linder, D., Harel, G., Gleiberman, I., Caruso, P.A., Ricks, S.H., Untch, M., Sartori, C., and Bruckner, H.W. (1995). *Cancer Research*, **55**, 5276.
26. Rottman, B. and Papermaster, B.W. (1966). *Proceedings of the National Academy of Sciences USA*, **55**, 134.
27. Vistica, D.T., Scudiero, D., Skehan, P., Monks, A., and Boyd, M.R. (1990). *Journal of the National Cancer Institute*, **82**, 1055.
28. Nygren, P. and Larsson, R. (1991). *International Journal of Cancer*, **48**, 598.
29. Skehan, P. and Friedman, S.J. (1981). In *The transformed cell* (ed. I.L. Cameron and T.B. Pool), pp. 7–65. Academic Press, New York.
30. Skehan, P. (1991). In *Growth regulation and carcinogenesis* (ed. W.R. Paukovits), Vol. 2, pp. 313–325. CRC Press, Boca Raton.
31. Skehan, P., (1995). In *Cell Growth and Apoptosis, A Practical Approach* (ed. G. P. Studzinski), pp. 169–91. IRL Press, Oxford.
32. Freshney, R.I. (1987). Culture of animal cells. A manual of basic technique. Alan R. Liss, New York.
33. Alley, M.C., Pacula-Cox, C.M., Hursey, M.L., Rubinstein, L.R., and Boyd, M.R. (1991). *Cancer Research*, **51**, 1247.

3

Cell growth and kinetics in multicell spheroids

RALPH E. DURAND

1. Introduction

This chapter examines cell proliferation kinetics and mechanisms of cell growth control in a complex tissue culture system: three-dimensional multicell spheroids (1). While spheroids are about to enter their fourth decade of use as an 'in vitro tumour model' (2), they are also very useful as a model for some normal tissues due to the co-existence of heterogeneous microenvironments that collectively influence their growth properties (3). We will, therefore, briefly characterize the system, indicate many of the novel growth perturbations that arise, and then address some of their implications vis-à-vis proliferation and quiescence.

2. Growth of 'spheroids'

2.1 Options

It has long been known that the use of 'semi-solid' culture matrices will lead to the formation of local multicellular colonies derived from each viable cell (4). While such soft agar or methylcellulose cultures have found numerous applications, such as haematopoietic stem cell assays *in vitro*, the inability to easily retrieve the three-dimensional colonies from their supporting structure limits the subsequent analyses that can be undertaken. Consequently, that method of cell propagation will not be considered further here.

At the opposite end of the spectrum, some cells will spontaneously aggregate and grow into multicell clusters when they are placed in a gyratory shaker, or, more commonly, in magnetically stirred liquid suspension cultures (5,6). In this case, the individual clusters, or 'spheroids', are easily observed and can be collected by allowing them to settle to the bottom of the container. Spheroids grown in suspension culture generally show excellent uniformity of structure and growth rates (*Figure 1*).

Intermediate between these two approaches is the so-called 'liquid overlay'

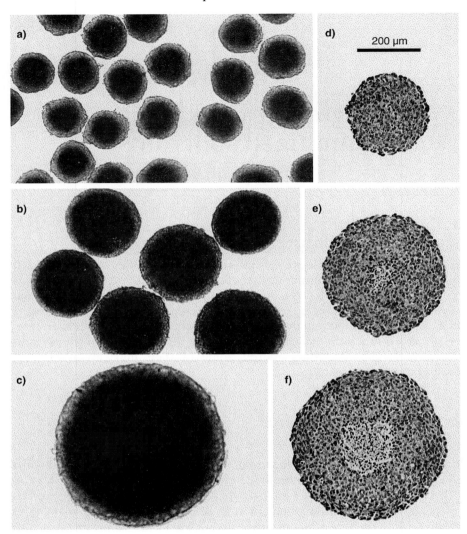

Figure 1. Spheroids of Chinese hamster V79 lung fibroblasts, with intact spheroids of increasing sizes shown under brightfield microscopy at various stages of growth in the left panels, and representative central sections from histological preparations on the right. Note the uniformity of the spheroids as they grow, and the progressive development of central necrosis once a diameter of about 350 μm is exceeded.

method (7,8), where three-dimensional growth is first induced by placing the cells into medium above a non-adherent layer (e.g. agar or agarose). Once all clusters have grown to sufficient sizes, they are typically placed into stirred suspension cultures for subsequent 'bulk' analyses, or into multiwell plates when individual spheroids are to be observed. With combinations of these three techniques, most proliferating cells can be induced to grow as multicell spheroids.

Protocol 1. Recommended procedures for establishing 'accessible' spheroids

Equipment and reagents

- tissue culture flasks, dishes or multiwell plates as desired
- sealable suspension culture vessels with magnetic stir bar (e.g. Bellco 1961-series spinner flasks, or Wheaton Celstir® culture vessels)
- non-heating magnetic stirrer (any supplier)

- agarose (any reputable supplier) prepared at 0.5–1.0% by dissolving in sterile, distilled water and autoclaving
- tissue culture medium and supplements (identical to that normally used for growth of the particular cell line as a monolayer)

Method A: liquid overlay

1. Prepare tissue culture dishes or multiwell plates by adding agarose to a minimum depth of 0.5 mm.

2. Allow agarose to cool and dry; dried dishes can be stored for several weeks before use.

3. Add monodispersed cells directly in normal growth medium (typically $0.5–1 \times 10^4$ cells/ml and sufficient volume to produce a maximum medium depth of 1–2 mm).

4. Observe after 3–5 days depending on growth rate.

5. 'Feed' spheroids as required by careful aspiration/replacement of spent medium; single cells and small clusters can also be removed by this procedure.

6. When 200 μm in diameter or larger, spheroids can be transferred individually to multiwells by Pasteur pipette, or collected and placed in stirred suspension culture (as below) depending on the goals of the experiment.

7. Option: some cell lines will aggregate and grow as spheroids if placed in non-tissue culture dishes; while more convenient, this approach is generally less reliable.

Method B: suspension culture

1. Add monodispersed, exponentially growing cells (final concentration $1–2 \times 10^4$/ml) to the medium in the culture vessel. If CO_2-buffered

---

Protocol 1. *Continued*

medium is used, pre-equilibrate at the optimal pH by flushing the flask with an appropriate gas mixture and then sealing it (typically an atmosphere of 5% CO_2 in air).

2. Maintain a stirring speed of 150–175 r.p.m.

3. 'Feed' the spheroids by removing the flask from the stirrer, allowing the spheroids to sediment to the bottom of the flask, gently removing the spent medium by aspiration, and adding pre-equilibrated fresh medium.

4. Feeding is usually necessary on day 3, day 5 and daily thereafter until the spheroids are used for the experiment.

2.2 Example: the V79 spheroid system

Figure 1 shows the morphological appearance of spheroids grown from Chinese hamster V79 lung fibroblasts, produced by seeding single cells from monolayer cultures into magnetically stirred suspension culture vessels. The left-hand panels show brightfield microscopy images of entire spheroids, whereas the right panels show histological sections from the centre of spheroids of increasing sizes.

The popularity of the spheroid system as an *in vitro* tumour model becomes immediately apparent from its histological appearance. Like many tumour nodules, as the avascular spheroid enlarges, those cells more distant from the nutrient and oxygen supply become visibly smaller, with occasional cells showing a pyknotic or apoptotic appearance. At larger spheroid sizes, a well-defined central area of frank necrosis is always observed. The thickness of the viable rim of cells is a function of the metabolic activity of a particular cell type, the packing density of those cells, and, to some extent, the oxygen and nutrient composition of the supporting growth medium (1). For the V79 spheroids illustrated in *Figure 1*, growth in minimal essential medium supplemented with 5% fetal calf serum uniformly produces a rim thickness of about 170 μm.

Two features of the V79 Chinese hamster cell line are particularly advantageous for many studies. First, the 'time line' illustrated in *Figure 1* is slightly more than one week. Since these cells grow so rapidly in culture (the cycle time in monolayers is about 12 h), large numbers of spheroids can be produced quite economically both in terms of time and media costs. A second feature, which increases the relevance of the model, is shown in *Figure 2* and is the tight 'packing' of the cells in the spheroid structure. The scanning electron micrograph image in *Figure 2a* clearly shows that the cells are not simply growing in apposition to each other, and this is illustrated even better in the transmission section in *Figure 2b*, where very little extracellular space is evident. It should also be noted that *Figure 2b* is an image from the edge of a

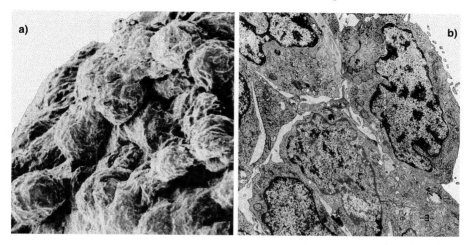

Figure 2. Scanning (a) and transmission (b) electron microscopy of the peripheral cells of large V79 spheroids. Note the tight cell packing, and numerous points of intercellular contact. The edge of the spheroid is at the top right in panel (b).

spheroid; the centre of the spheroid was several cell layers distant from the left side of the image. Even tighter cell packing, including specialized intercellular junctions has been demonstrated in the interior regions of V79 spheroids (5,6).

3. Special features of spheroids

3.1 Metabolite and catabolite gradients

As might be expected, delivery of nutrients to cells in a three-dimensional structure is more difficult than in single cell cultures. Two factors are relevant: *diffusion* and *consumption* of metabolites. In some cases, as for molecular oxygen, diffusion is quite a rapid process. None the less, the cellular consumption of this essential metabolite is so rapid that central hypoxia can be demonstrated in V79 spheroids exceeding about 400 μm in diameter. A practical illustration of the importance of diffusion and consumption processes, albeit with a different 'metabolite', is shown in *Figure 3* where spheroids of various sizes were incubated in the presence of [³H]thymidine for approximately three cell cycles (48 h) prior to sectioning and autoradiography. *Figure 3a* shows that every cell of small spheroids became labelled, that is, the DNA precursor reached and was incorporated into all nuclei present in the section. In the spheroids of increased sizes shown in *Figure 3b* and *3c*, only the peripheral cells were labelled. This could signal either a problem in delivery of the radioactive precursor, or, conversely, that only the external cells remained in the proliferating compartment. While *Figure 3* itself

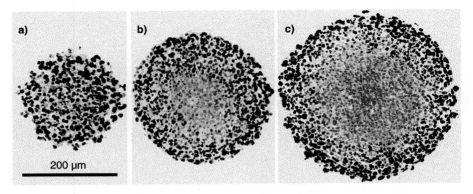

Figure 3. Autoradiographs from central sections of V79 spheroids, prepared after labelling with [³H]thymidine for 48 h (nearly three cell cycles). Whereas all nuclei were labelled in small spheroids (a), proliferating cells were seen only towards the spheroid periphery as the structures enlarged.

does not resolve this issue, other work has clearly demonstrated that the latter explanation is correct (9,10) and can be generalized to other spheroid systems as well (8,11).

3.2 Viability in spheroid cell subpopulations

As spheroids expand in size, only those cells near the periphery remain actively proliferating. However, if the spheroid is reduced to a single cell suspension (usually by enzymatic treatment combined with mechanical agitation), virtually all intact V79 cells retain growth potential (12). Following a brief delay, even cells bordering the necrotic region will re-enter the cell cycle and begin to proliferate (13). Consequently, a large spheroid comprises several subpopulations of cells, including proliferating clonogenic cells, non-proliferating but potentially clonogenic cells, and, ultimately, a non-proliferating non-viable core. Mature spheroids are consequently quite dynamic structures, with these subpopulations maintained under pseudo-steady-state conditions. The feeding schedule employed naturally introduces fluctuations in nutrients that can influence results of some studies; frequent medium changes minimize such problems (14).

4. Unexpected features of spheroids

4.1 Genetic instability

The availability of flow cytometry instrumentation in the mid-1970s led to several unanticipated findings in the V79 spheroids. Perhaps the most important of these was the observation that the V79 line, which had been propagated in tissue culture since the late 1950s as a true diploid cell system under monolayer growth conditions (15), spontaneously became tetraploid in spheroids (10,16). As illustrated in *Figure 4*, a progressively increasing tetraploid cell subpopulation is observed as the spheroids enlarge. Interestingly, this observation is not unlike the 'heterogeneity' in DNA content common in solid tumours. While the normal explanation for this observation is genetic instability, the appearance of the phenomenon in the spheroid system,

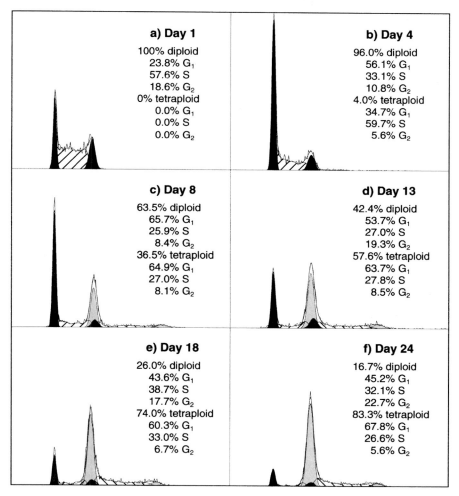

a) Day 1
100% diploid
23.8% G_1
57.6% S
18.6% G_2
0% tetraploid
0.0% G_1
0.0% S
0.0% G_2

b) Day 4
96.0% diploid
56.1% G_1
33.1% S
10.8% G_2
4.0% tetraploid
34.7% G_1
59.7% S
5.6% G_2

c) Day 8
63.5% diploid
65.7% G_1
25.9% S
8.4% G_2
36.5% tetraploid
64.9% G_1
27.0% S
8.1% G_2

d) Day 13
42.4% diploid
53.7% G_1
27.0% S
19.3% G_2
57.6% tetraploid
63.7% G_1
27.8% S
8.5% G_2

e) Day 18
26.0% diploid
43.6% G_1
38.7% S
17.7% G_2
74.0% tetraploid
60.3% G_1
33.0% S
6.7% G_2

f) Day 24
16.7% diploid
45.2% G_1
32.1% S
22.7% G_2
83.3% tetraploid
67.8% G_1
26.6% S
5.6% G_2

Figure 4. Univariate flow cytometry histograms showing the distribution of DNA content in cells from V79 spheroids of increasing sizes. Estimates of cell cycle distribution, using MODFIT software, are shown visually and numerically in each panel. Diploid G1 and G2 cells are shaded black, with the corresponding tetraploid cells a light grey, and the respective S phase population indicated by cross-hatching.

and particularly, in a relatively 'stable' cell line, suggests that an epigenetic or cell microenvironment-related factor is responsible for the development of the higher ploidy cells.

4.2 Aneuploidy—a response to 'architecture'?

This suggestion is based on both the progressive development of tetraploidy illustrated in *Figure 4*, and, additionally, experiments reported by Peggy Olive some years ago in which she found that spheroid size (not age or nutrient status) was the principal determinant of tetraploidy induction (16). Briefly, if spheroids were maintained at a constant size (by successive mild trypsinization to remove only the outer layers of cells), tetraploidy was never observed. Conversely, once the spheroids exceeded about 300 μm in size, external cells tended to become tetraploid, and eventually the proliferation of these tetraploid cells replaced the diploid cells in the spheroids. Interestingly, tetraploid cells, when cloned from spheroids, were completely stable. Spheroids grown from these tetraploid cells did not exhibit further instability. Rather, they remained tetraploid even with continued growth (we have never observed higher levels of ploidy in spheroids initiated with tetraploid cells).

5. Cell kinetics in spheroids

5.1 Problems

As illustrated in *Figure 4*, where each of the DNA distributions was processed by the MODFIT DNA analysis package (Verity Software House, Topsham, Maine), assignment of cells in the $4n$ peak to either diploid G2 or tetraploid G1 populations is quite subjective. The quantitative estimates of cell cycle distribution shown in *Figure 4* are generally reasonable, but closer scrutiny shows that the S:G2 ratio is variable at different spheroid sizes. This observation consequently reduces one's confidence that single parameter analysis (DNA content only) provides the quality of information needed for precise cell kinetic work.

5.2 Approaches

One of the powerful features of flow cytometry analysis is the multiparameter capability of the instrumentation. Following the pioneering work of Darzynkiewicz and his colleagues (see Chapter 7), we have adapted simultaneous staining for DNA precursor uptake and cyclins to permit more detailed cell kinetics analyses. *Figure 5* illustrates this approach in a 'four-colour' experiment, where spheroids were first pulsed with bromodeoxyuridine, 10 h later with iododeoxyuridine, and after an additional 4 h, dispersed into single cells and processed for multicolour flow cytometry, including staining for cyclin B1.

Two-colour analyses of DNA content and cyclin B1 in intact cells are quite routine, and have established the value of cyclin B1 as a probe for late S and

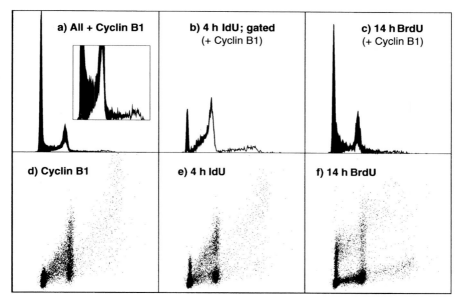

Figure 5. Univariate (top panels) and bivariate (lower panels) flow histograms illustrating four-colour staining in V79 spheroids containing a small tetraploid subpopulation. (a) Shows the DNA distribution of the entire cell population, on which is superimposed the DNA distribution of those cells positive for cyclin B1 (the inset shows the same data with a different vertical scale to more easily visualize the diploid and tetraploid S phase cells). (b) Indicates the DNA distribution of only those cells that incorporated IUdr during a brief exposure 4 h prior to analysis; the superimposed white distribution again shows cyclin B1 labelling. As expected, virtually all late S and G2 cells were positive for cyclin B1. A different picture was seen for the cells that had incorporated BrdU 14 h prior to analysis (c), where only a small fraction of the cells in the second peak (the cells with $4n$ DNA content) also contained cyclin B1 and therefore represented diploid G2 cells. The lower panels show colour-compensated dot plots for the entire cell population, highlighting the binding pattern of each of the antibodies used.

G2 cells (17). However, assessment of the incorporation of halogenated pyrimidines by monoclonal antibodies requires acid hydrolysis, which in turn can reduce cyclin B1 content. None the less, Figure 5 shows that an anti-cyclin B1 antibody is helpful even in relatively small spheroids just beginning to show a tetraploid component. *Figure 5a* shows the DNA distribution from all cells of the spheroid, with the distribution of the cyclin B1 positive cells superimposed in white (and in the inset at an increased magnification to highlight the S phase and tetraploid populations). *Figure 5b* shows the DNA distribution of those cells that had incorporated IUdr 4 h previously, and *Figure 5c* the DNA distribution and cyclin B1 content of cells labelled with BrdU 14 h prior to analysis. For reference, the lower panels show the colour compensated dot plots for each of the probes as a function of DNA content.

As expected, *Figure 5b* shows that all the diploid and tetraploid S phase cells that had progressed to their respective G2 phases stained with cyclin B1. Conversely, for the 14 h time shown in *Figure 5c*, many of the cells in the middle (4n) peak were cyclin B1 negative, therefore indicating that they represented tetraploid cells which had progressed through G2 to their (tetraploid) G1 phase. These data highlight two features: (1) they clearly illustrate the value of multiparameter analyses; and (2) they suggest that kinetics estimates even in a diploid tumour with a small subset of tetraploid cells (i.e. with DNA distributions like *Figure 5a*) are subject to errors that increase with time if all BrdU- or IUdr-labelled cells are ascribed to the G2 population (compare *Figure 5b* and *5c*). Note that these multicolour staining techniques also find utility in tetraploid tumour samples containing significant proportions of normal (diploid) non-malignant cells.

While anti-cyclin antibodies provide the most direct approach for identifying cycling cells in specific phases of the cell cycle, alternative techniques are also available. In some cases, PCNA or Ki67 antibodies can provide similar information. Conversely, hypoxia probes also have some utility, since hypoxic cells are generally not in the active cell cycle (18,19). Ultimately, practical considerations of non-overlapping antibody/stain combinations, availability, and cost will have the major effect on the choice of technique.

Protocol 2. Multiparameter flow cytometry

Equipment and reagents

- flow cytometer capable of multicolour analysis; dual wavelength excitation preferred
- labelling buffer (PBS + 4% serum + 0.1% Triton X-100)
- bromodeoxyuridine (BrdU) and iododeoxyuridine (IUdr)
- anti-BrdU monoclonal antibody (e.g. unconjugated Br3 from Caltag Laboratories)
- anti-IUdr/BrdU monoclonal antibody (e.g. B44–FITC from Becton Dickinson)
- anti-cyclin antibody (e.g. anti-cyclin B1 monoclonal from Upstate Biotechnologies Inc. for three-colour work, or a non-mouse polyclonal, e.g. in rabbit, from Sigma for four-colour studies)

- anti-mouse IgG1–Cy3 (from Caltag Laboratories; used against Br3 for three-colour analyses)
- anti-mouse IgG1–RhoB (from Biosource International; used against Br3 for four-colour analyses)
- anti-rabbit IgG–PE (from Sigma; used against rabbit anti-cyclin B1 for four-colour analyses)
- denaturing solution 2 N HCl + 0.5% Triton X-100
- 4,6-diamidino-2-phenylindole dihydrochloride hydrate (DAPI), prepared as a stock solution of 10 mg/100 ml of 20 mM Tris buffer at pH 7.2

Method

1. Start with $1.5–2 \times 10^6$ BrdU- and/or IUdr-labelled cells, fixed with 70% ethanol for at least 0.5 h.

2. Centrifuge the cells at 300 g for 5 min.

3. remove the supernatant and add 1.0 ml 'denaturing solution' to the cells.

4. Incubate the cells at room temperature for 30 min.

5. Neutralize the cells by centrifuging and resuspending in *cold* tissue culture medium (2–3 times until the indicator in the medium signifies neutrality).

6. Resuspend the cells in the 'labelling buffer' (also used for all subsequent steps).

7. Label the BrdU with the Br3 antibody (typically 1/100 dilution for 2–24 h as convenient).

8. After removing Br3, add the anti-mouse IgG1–Cy3 (typically 1/50 dilution for 2 h).

9. Label the IUdr with the B44–FITC antibody (typically 1/50 dilution for 2 h).

10. If anti-cyclin is also to be used, add the antibody as the last step before DNA staining, followed by the appropriate secondary. Typically, we use a phycoerythrin-conjugated secondary to the anti-cyclin, after substituting a longer wavelength probe (usually rhodamine) to the Br3 primary.

11. Stain the cellular DNA with DAPI at 1.0 μg/ml for 30 min.

5.3 Example: kinetics in V79 spheroids

Figure 6 is a schematic that shows the growth of V79 cells as multicell spheroids, and the underlying reasons for the kinetic changes (9,10). The upper curve reiterates the tumour-growth pattern; the initial phase of rapid exponential doubling is not sustained, as growth progressively decelerates. Underlying those changes are growth characteristics normally ascribed to tumours: during the initial phase of cell aggregation and rapid growth, virtually all cells remain in active cycle, constituting a growth fraction of nearly 100%. This rapidly declines with spheroid age, so that in large spheroids growth fractions of 30% or less are observed. Additionally, development of the central necrotic area leads to cell loss, which further decreases the apparent doubling rate.

Underlying all of this, however, is the observation that those cells that do remain in cell cycle retain a division time that is only marginally changed between the first two or three days of growth and the oldest spheroids we have ever observed (9,10). Thus, the changes in spheroid enlargement rates are not due to a non-specific elongation of the cell cycle as in monolayers (20), but rather, decreases in the number of cells actually dividing, coupled with a progressive increase in cell loss.

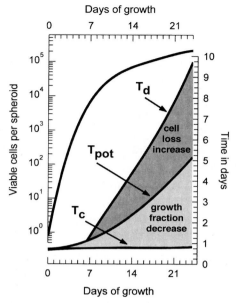

Figure 6. A schematic of the changing cell cycle kinetics seen in V79 spheroids during their initial growth. The cell cycle time, T_c, is essentially invariant with spheroid age or size, whereas the potential doubling time, T_{pot}, progressively increases (due to a decreasing growth fraction), while the net doubling time, T_d, similarly increases as a result of the changing T_{pot} and increasing cell loss.

6. Growth 'compartments' in spheroids

6.1 Proliferation versus quiescence

The model just described for regulation of spheroid growth, that is, a progressively decreasing growth fraction and increasing cell loss, should not be a surprise. Indeed, essentially similar models have been used to describe the growth of solid tumours for many years (21,22). Implicit in those models, however, is an assumption that has not proven to be true in either the spheroid system or the *in vivo* tumour systems we employ in our laboratory. In essence, that assumption is that cells are either in the active growth fraction, or quiescent. Under normal circumstances, quiescent cells will ultimately die without further proliferation, whereas growing cells may continue to proliferate, or may become quiescent. Somewhat surprisingly, our data have seldom suggested that the 'quiescent' cell subpopulation in spheroids behaves with the expected degree of predictability.

Figure 7 shows a basis for the observation just stated, i.e. it tests the hypothesis that most cells in the active growth cycle will continue through an additional cycle. In this experiment, we administered iododeoxyuridine at

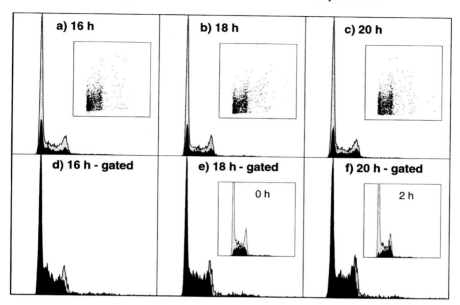

Figure 7. Progression of IUdr pulse-labelled cells in V79 spheroids at relatively long times (about 1 cell cycle time) after labelling. (a)–(c) Show the DNA distribution of all cells, with the superimposed distributions indicating the labelled subpopulation (the insets show the corresponding bivariate histograms with similar shading to indicate labelled and unlabelled cells). (d)–(f) Show the respective labelled subpopulations on an expanded scale, and, additionally, have the non-G2 cells in the 4*n* peaks displayed with light grey shading to more easily visualize the progressing diploid cells. Note that the previously labelled diploid cells were distributed towards early S phase in (d), but later, many had progressed through S phase and into G2 (f). For reference, the insets in (e) and (f) show the distribution of cells in the same spheroids labelled with BrdU 0 or 2 h, respectively, prior to analysis, indicating the expected distributions of S phase cells after one cell cycle time. Note the general similarity of the 0 and 18 h, and 2 and 20 h S phase distributions, but that a large fraction of previously labelled cells did not synchronously enter the subsequent S phase.

various times prior to analysis via flow cytometry. However, rather than preparing cell samples at relatively short times after labelling, as is normally done for T_{pot} determinations (23,24), we harvested the cells approximately one cell cycle time later. With this timing, the cells in S phase at the time of initial labelling should have progressed through a complete cell cycle and again been in S phase one cycle time later. That expectation was found to be both true and false, as illustrated by the data in *Figure 7* (the experiment was performed in V79 spheroids for which the cell cycle time was known to be approximately 18 h).

Figure 7a shows that cells labelled 16 h previously had indeed progressed through G2 and into G1 and/or a second S phase. When double-stained to discriminate between G1 and G2 cells in the 4*n* peak, as in *Figure 7d* (where only the halogenated pyrimidine-labelled cells are shown), it is apparent that the current DNA distribution of the previously labelled cells was heavily weighted towards the early part of S phase. At 18 h, which was the known cell cycle time, the labelled cells that had moved into the second S phase were quite uniformly distributed throughout S (*Figure 7b* and *7e*). By 20 h (*Figure 7c* and *7f*) a picture fairly similar to the 18-h data emerged, although a few more labelled cells were found in G2, causing the DNA distribution of the S phase cells to be somewhat skewed towards late S phase.

The insets in the lower *Figure 7e* and *7f* provide a frame of reference for the expected distribution of the S phase cells, based on the patterns seen for the short-term labelling. If all cells had, in fact, progressed (synchronously) through exactly one cell cycle after 18 h, then a labelling pattern not unlike that seen for a very short duration (0-h) pulse, as in the inset, would be expected. Similarly, after 2 h additional progression, a distribution like that seen in the inset in *Figure 7f* might be expected. In both cases, the DNA distributions *of the S phase cells* were not unlike those anticipated, but surprisingly, a very large G1 population was retained.

6.2 Unusual growth regulation

These data, and others, suggest that there is an additional level of growth control in the spheroids (and, in our experience, in tumour systems) such that some cells do indeed complete consecutive cycles, but the majority show a G1 block (involving the G1 checkpoint?) of variable lengths.

In fact, our data to date suggest that the probability of 'release' from this G1 checkpoint is no greater for a cell that has just gone through a division cycle than for one that has not. Interestingly, this observation is exclusive to three-dimensional systems (spheroids and tumours); the same cell types, when grown as monolayers, continuously progress in a synchronous fashion, as expected. We also find it interesting that the available clinical data describing percentage of labelled mitotic cells as a function of time after injection of radioactive DNA precursors show the same lack of consecutive S phase traversals, although the accepted explanation for those results tends to be 'loss of synchrony' (22).

6.3 Repopulation after cytoreduction

Obviously, the epigenetic and genetic mechanisms that control cell progression in multicell systems are a topic of considerable importance to us. While we do not as yet have definitive evidence as to the nature of the controlling variables, we none the less have gained some insight into the implications of these observations.

In essence, the suggestion that 'quiescence' may be a dynamic rather than static state in three-dimensional systems may explain why tumours often seem

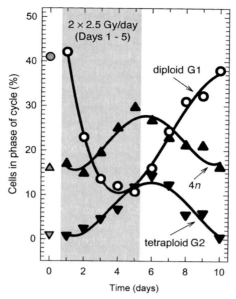

Figure 8. An experiment illustrating that 'quiescent' cells in V79 spheroids can be rapidly recruited into the active cell cycle by multifraction irradiation. Spheroids were irradiated twice daily with 2.5 Gy of X-rays on days 1–5 as indicated (the shorter interfraction interval was 8 h). A radiation-induced G2 block occurred at the expense of the G1 subpopulations, appearing and resolving quite rapidly during and after treatment.

to have a selective growth and survival advantage. By maintaining a quiescent population that retains the capacity to proliferate, both spheroids and tumours would be poised to respond rapidly to any cytotoxic insult, and to therefore repopulate without delay after (or even during?) treatment with anti-proliferative agents. A key feature of this speculation is, however, the need for the systems to rapidly mobilize those quiescent cells given an improvement in growth conditions.

An experiment that illustrates this capacity is shown in *Figure 8*. Here, V79 spheroids were irradiated with 2.5 Gy twice daily for 5 days, and both cell cycle redistribution and cell kinetic parameters were measured throughout the treatment regimen and subsequently. Radiation is a particularly instructive agent to use, since its primary effect on the progression of V79 cells (a p53 mutant) is to induce a G2 block (25,26). That, in turn, means that cells arriving at G2 must have progressed through at least one cell cycle, and, therefore, a differential increase in the G2 fraction indicates that cells must have been recruited from the G1 (largely quiescent) starting phase. As shown in *Figure 8*, the number of cells in the diploid G1 compartment of large spheroids decreases markedly even during radiotherapy, and normalizes with time after the exposure. Concurrently, both tetraploid G2 cells and the 4*n* cells (which comprise both G2 diploid cells being blocked, and G1 tetraploid cells being recruited into cycle) increased with virtually the same kinetics. We interpret these data as indicating that, whereas under normal conditions a large quiescent G1 population (be it diploid or tetraploid) is spontaneously maintained in the spheroids, it can be rapidly reversed by cytoreductive treatment that stimulates proliferation and growth. We therefore conclude that, as previously hypothesized, three-dimensional systems can rapidly respond to cytoreductive therapy to enhance their chance of 'self-preservation'.

7. Clinical extensions: tumour resistance to treatment

7.1 Can tumours 'outgrow' treatment?

In essence, what we have just described is the potential for 'kinetic' resistance (27), a process that may supplement the known and accepted processes of 'genetic' resistance, and which has been suggested even in the clinic (28). In tumours, particularly in clinical data, it is not uncommon for a tumour to initially respond to chemotherapy, but for that tumour to eventually cease to respond and in some cases begin to grow even in the face of continued therapy. Not surprisingly, those observations spawned the operational term 'emergence of drug resistance'. While we have no quarrel with the operational terminology, experiments with the spheroid system have clearly shown that an identical phenomenon can be observed as a result of changing growth kinetics alone, without any development of inherent resistance to the treatment agents.

These observations are illustrated by the data in *Figure 9*, where three typical anti-neoplastic agents, radiation, etoposide, and doxorubicin, were evaluated. In all cases, spheroids were exposed to multiple treatments with each agent. The unique aspect of this experiment, unlike the clinical situation, is that the number of viable cells per spheroid can be measured before and after each and every treatment. This provides unequivocal information on the changing susceptibility of the cells to the drugs or radiation, and importantly, any changes in the number of viable cells per spheroid (regrowth or cell loss) between treatments.

For all regimens shown in *Figure 9*, a clinically relevant response was observed. Radiation led to a progressive decrease in the number of cells per spheroid, whereas the drug treatments initially appeared quite effective, but lost efficacy with time. With doxorubicin, regrowth consistent with development of resistance was observed during the final treatment. However, by measuring the fractional cell kill per treatment, as shown in the middle panels of *Figure 9*, it became evident that resistance (increased survival) was *not* seen for any of the agents. Conversely, little regrowth or repopulation between exposures was observed early in the treatment regimens, whereas later, when spheroid depopulation had begun, repopulation was able to compensate quite

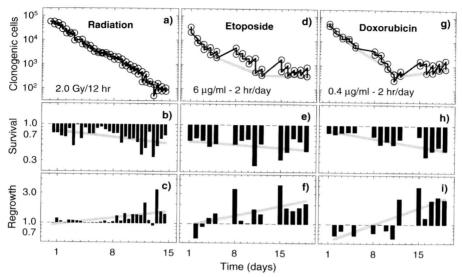

Figure 9. The response of V79 spheroids to multifraction treatments with radiation (a–c), etoposide (d–f), or doxorubicin (g–i). The top panels show the number of clonogenic cells per spheroid immediately before and after each exposure; the relative reduction in surviving fraction for each exposure is indicated in the middle row; and the lower panels show the relative change in clonogenic cell number per spheroid between consecutive treatments. The grey lines in the middle and bottom panels are linear regressions showing the trend of the responses. Despite the apparent development of 'resistance' to the drugs suggested in (d) and (g), a progressive *increase* in cellular sensitivity was actually measured (e and h). Overall, the increasing regrowth (lower panels) over-compensated to produce the apparent resistance to treatment.

effectively. In fact, the regrowth seen at later times was entirely akin to the clinical phenomenon termed 'accelerated repopulation' (29,30).

7.2 Implications and clinical options

Clearly, a reduced efficacy of treatment due to changing growth kinetics as in the spheroid data of *Figure 9* cannot be the only explanation for the clinical resistance of cancers, but is likely to be an additional factor that deserves careful consideration. Interestingly, true (genetic) resistance will be difficult to deal with, as it will be successfully overcome only with better, more specific anti-cancer agents. Conversely, 'kinetic resistance' may be easily handled by solutions as simple as different drug delivery schedules and/or treatment regimens including cytostatic as well as cytotoxic agents. Clearly, to optimally treat a tumour and avoid 'kinetic resistance', one requires a definitive measure of the growth kinetics of that tumour, and a major focus of our laboratory is the development of appropriate predictive assays to provide those data.

8. Conclusions

Our aim in this chapter has been to discuss the interplay between genetic and epigenetic factors that regulate growth in multicell spheroids, and to extrapolate these results to other three-dimensional systems. Clearly, the molecular level growth controls that are now being exhaustively studied in cycling cells can be envisaged as 'downstream' controls, whereas in tumours or other three-dimensional systems factors such as nutrient delivery, cellular packing, interstitial pressure, etc. may be 'upstream' modulators. Taken together, it seems likely that the somewhat unusual growth control mechanisms we have observed in spheroid systems and in transplantable tumours may be generally applicable to spontaneous human neoplasms. Consequently, understanding these observations, and particularly the implications of changing growth fractions in complex systems, may ultimately lead to new designs for clinical anti-cancer therapy.

References

1. Acker, H., Carlsson, J., Durand, R. E., and Sutherland, R. M. (ed.) (1984). *Spheroids in cancer research (recent results in cancer research)*, Vol. 95. Springer-Verlag, Berlin.
2. Inch, W. R., McCredie, J. A., and Sutherland, R. M. (1970). *Growth*, **34**, 271.
3. Sutherland, R. M. (1988). *Science*, **240**, 177.
4. McAllister, R. M., Reed, G., and Huebner, R. J. (1967). *J. Natl. Cancer Inst.*, **39**, 43.
5. Sutherland, R. M., McCredie, J. A., and Inch, W. R. (1971). *J. Natl. Cancer Inst.*, **46**, 113.
6. Sutherland, R. M. and Durand, R. E. (1976). *Curr. Topics Radiat. Res. Q.*, **11**, 87.
7. Yuhas, J. M., Li, A. P., Martinez, A. O., and Ladman, A. J. (1977). *Cancer Res.*, **37**, 3639.
8. Carlsson, J. (1977). *Int. J. Cancer*, **20**, 129.
9. Durand, R. E. (1976). *Cell Tissue Kinet.*, **9**, 403.
10. Durand, R. E. (1990). *Cell Tissue Kinet.*, **23**, 141.
11. Sweigert, S. E. and Alpen, E. L. (1986). *Cell Tissue Kinet.*, **19**, 567.
12. Durand, R. E. (1983). *Radiat. Res.*, **96**, 322.
13. Durand, R. E. (1982). J. Histochem. Cytochem., **30**, 117.
14. Durand, R. E. and Sutherland, R. M. (1984). In *Spheroids in cancer research (recent results in cancer research)*, (ed. H. Acker, J. Carlsson, R. E. Durand, and R. M. Sutherland), Vol. 95, p. 103. Springer-Verlag, Berlin.
15. Ford, D. K. and Yerganian, G. (1958). *J. Natl. Cancer Inst.*, **21**, 393.
16. Olive, P. L., Leonard, J. C., and Durand, R. E. (1982). *In Vitro*, **18**, 708.
17. Darzynkiewicz, Z., Gong, J., and Traganos, F. (1994). *Meth. Cell Biol.*, **41**, 421.
18. Kennedy, A. S., Raleigh, J. A., Perez, G. M., Calkins, D. P., Thrall, D. E., Novotny, D. B., and Varia, M. A. (1997). *Int. J. Radiat. Oncol. Biol. Phys.*, **37**, 897.
19. Durand, R. E. and Raleigh, J. A. (1998). *Cancer Res.*, **58**, 3547.
20. Durand, R. E. and Sutherland, R. M. (1973). *Radiat. Res.*, **56**, 513.

21. Mendelsohn, M. L. (1960). *Science*, **132,** 1496.
22. Steel, G. G. (1977). *Growth kinetics of tumours*. Oxford University Press, Oxford.
23. Begg, A. C., McNally, N. J., Shrieve, D. C., and Kärcher, H. (1985). *Cytometry*, **6,** 620.
24. Durand, R. E. (1993). *Cytometry*, **14,** 527.
25. Durand, R. E. (1993). *Semin. Radiat. Oncol.*, **3,** 105.
26. Sham, E. and Durand, R. E. (1998). *Radiother. Oncol.*, **46,** 201.
27. Durand, R. E. and Vanderbyl, S. L. (1989). *Cancer Comm.*, **1,** 277.
28. Preisler, H. D. and Venugopal, P. (1995). *Cell Prolif.*, **28,** 347.
29. Withers, H. R., Taylor, J. M., and Maciejewski, B. (1988). *Acta Oncol.*, **27,** 131.
30. Fowler, J. F. (1991). *Radiother. Oncol.*, **22,** 156.

4

Assessment of DNA damage cell cycle checkpoints in G1 and G2 phases of mammalian cells

PATRICK M. O'CONNOR and JOANY JACKMAN

1. Introduction

Cell cycle progression is normally tightly regulated such that one cell cycle event does not occur unless the previous cell cycle event has been completed. For example, mammalian cells do not normally enter mitosis while DNA replication is on-going, and cells in mitosis do not normally exit mitosis unless the spindle apparatus is correctly aligned with all the chromosomes. This dependency is controlled by a series of negative-feedback control systems that are called checkpoints (1–4). These checkpoints are being studied in a variety of organisms from yeast to human, and a growing number of studies implicate defective checkpoint control systems in the emergence and evolution of human cancer. One example is the p53 tumour suppressor, the most commonly mutated gene in human cancer (1, 2). p53 is essential for the arrest of cell cycle progression in the G1 phase following DNA damage. Loss of p53 function prevents this arrest mechanism from being activated and also subdues or blocks the induction of apoptosis in DNA damaged cells. In addition, p53 has influences on other cell cycle checkpoints (G2 and M phase) which act together to ensure the fidelity of DNA replication and chromosome segregation. Such actions help to explain why there is a markedly increased likelihood of cancer in mice deficient for p53 function (5).

This chapter focuses on DNA damage checkpoint control systems in mammalian cells (*Figure 1*) and their functional measurement in cultured cell lines. The checkpoints that are considered are the p53-dependent G1 checkpoint and the G2 checkpoint. The procedures described for assessment of G2 arrest apply equally to the assessment of the integrity of a checkpoint that is operational in the S phase which blocks mitosis in the presence of unreplicated DNA. We will describe some background for these checkpoint control systems and techniques available for their assessment. We will also point to literature sources for additional information on the procedures we recommend.

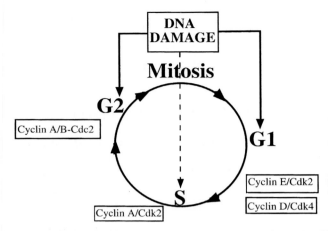

Figure 1. Diagram of points in the cell cycle where arrest occurs following DNA damage. DNA damage-induced G1 arrest requires the function of the p53 tumour suppressor pathway (1, 2, 6) and defects in this pathway prevent p53 from inducing G1 arrest. DNA damage also induces G2 arrest and although this arrest occurs in p53 mutated cells, p53 still appears to influence the integrity of this G2 arrest (9 and refs therein). There is also a delay in the S phase progression following DNA damage (probably due to inhibition of firing of late replication origins) (1, 2).

2. Procedures for assessment of DNA damage checkpoints

The following procedures can be used to assess:

- the integrity of DNA damage cell cycle checkpoints in normal and transformed cells
- the actions of DNA damaging agents and DNA synthesis inhibitors on cell cycle progression
- the actions of chemicals that block DNA damage checkpoint function

2.1 Assessment of the DNA damage checkpoint in the G1 phase of the cell cycle

2.1.1 Background

The p53 tumour suppressor regulates a number of cellular responses to DNA damage, including cell cycle arrest in the G1 phase (1, 2, 6). These responses are lost in the majority of human tumour cell lines and many human primary tumours due to inactivating mutations in the p53 gene or the dominant influence of oncogenes (1, 2). The cellular consequences of p53 deficiency are profound and include genetic instability and reduced apoptosis following DNA damage. These effects can influence cancer cell evolution and sensitivity to current therapies (6). Described below are procedures which can be used to assess the functional integrity of the p53 pathway responding to DNA damage in mammalian cells. These procedures are applicable to assess other activators of p53 function, including signals from ribonuleotide depletion, hypoxia, and the controlled activation of certain oncogene pathways (ras/myc) in normal cells.

2.1.2 Assessment of G1 arrest by flow cytometry

The ability of cells to arrest in the G1 phase of the cell cycle following exposure to 6 or 12 Gray of γ-rays has been successfully used to assay the G1 cell cycle checkpoint in a number of mammalian cell lines (6). A number of procedures can be employed here; however, we will focus on one procedure which we have used most. This procedure is simple and reproducible.

Essentially, exponentially growing cells are irradiated and then incubated in the presence or absence of the mitotic inhibitor, nocodazole, to prevent cells that break through the G2 checkpoint from re-entering the second G1 phase. G1 arrest is quantitated using flow cytometry and can be expressed as the percentage of the original (untreated control) G1 population that remains in G1 for approximately 16 h following irradiation plus incubation with nocodazole (see *Figure 2*).

Protocol 1.

Equipment and reagents
- nocodazole (0.01–0.4 μg/ml) (Aldrich)
- propidium iodide (Sigma)
- RNase A (Sigma)
- PBS
- flow cytometer

Method

1. Twenty-four hours prior to commencement of the G1 arrest assay, plated cells should be trypsinized and replated at 10–30% confluence in a T75cm^2 flask.

   Samples include:
   - control
   - nocodazole, alone
   - radiation, alone (two radiation doses, see below)
   - radiation + nocodazole (two radiation doses, see below).

2. Twenty-four hours later, floating cells should be decanted from the plates and fresh medium added. For suspension cultures, exponentially growing cells should be diluted to 5×10^5/ml.

3. Cells should next be irradiated at room temperature with 6 and 12 Gray of γ-rays using a ^{137}Cs source (1 Gray = 100 Rads) and then cells should be incubated for approximately 16 h at 37 °C in the presence or

---

Protocol 1. *Continued*

absence of the microtubule inhibitor, nocodazole (0.01–0.4 μg/ml). The nocodazole concentration to use should have been determined first using a concentration–response ranging from 0.01–0.4 ug/ml. The goal is to chose the maximum non-toxic concentration of nocodazole that effectively blocks cells in G2/M for 16 h as determined by flow cytometry, leaving few cells in G1 phase (see *Figure 2*).

4. Medium containing floating cells and cells trypsinized from the plates should then be combined and centrifuged to pellet the cells. Cells should then be washed once in ice-cold PBS and finger-vortexed back into suspension in 0.5 ml of PBS before adding 5 ml of 70% ethanol and mixing by hand and storing on ice for a minimum of 2 h. Samples can be stored for up to a week in the fridge.

5. Cells are next rehydrated by resuspending in 5 ml PBS for 5 min. Cells are then pelleted and subsequently resuspended in 0.5 ml of pro-pidium iodide solution (25 μg/ml) (Sigma) that contains RNase A (500 units/ml) (Sigma), before incubation at 37°C for 30 min to destroy RNA.

6. Cell cycle analysis can be performed on a Becton–Dickinson FACSCAN flow cytometer, or similar machine. At least 10000 cells should be used for each analysis.

7. Nocodazole is included in the assay to prevent any cells that might break through the G2 checkpoint from entering the G1 phase of the second cell cycle. Thus, the population of cells in G1 phase approximately 16 h following incubation with nocodazole reflects the cells in the first G1 phase (6). G1 arrest can be quantitated as the percentage of the control G1 population that remained in G1 phase 16 h following irradiation plus nocodazole, compared with the nocodazole alone sample.

8. We have previously ranked G1 arrest responses into three classes: Class 1 cells show strong G1 arrest (>20% of the original G1 population); class 2 cells show intermediate G1 arrest (between 10 and 20% of the original control G1 population); class 3 cells show weak or no G1 arrest (<10% of original G1 population) (6). The bigger the difference between the nocodazole- and the radiation plus nocodazole-treated samples, the more confident one is that a cell has functional p53 activity.

2.1.3 Assessment of p21/WAF1/CIP1 mRNA induction

The p21/WAF1/CIP1 gene is induced in a p53-dependent manner upon exposure of mammalian cells to γ-irradiation and has been used by numerous

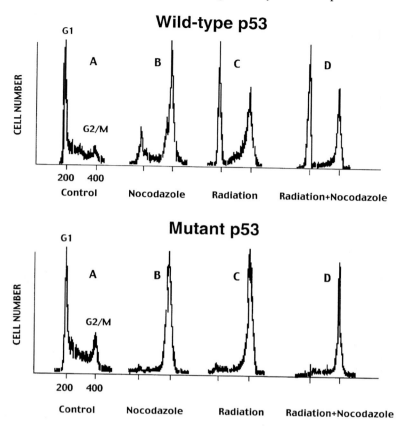

Figure 2. DNA histograms of asynchronously growing cells exposed to nocodazole and radiation. The cell cycle position can be monitored using DNA content per cell. Cells in the G1 phase have 50% of the amount of DNA in G2/M phase cells. The microtubule inhibitor, nocodazole (0.4 μg/ml) arrests cells in G2/M phase. DNA damage induced by radiation arrests wild-type p53 cells in G1 and G2 phases, while p53 mutant cells only arrest in G2 phase. The positions of the G1 and G2/M peaks are highlighted. Sample D has been inserted from an independent experiment for illustrative purposes. The degree of G1 arrest can be determined by first calculating the percentage of cells in the G1 phase following radiation and incubation with nocodazole (sample D) and then dividing this value by the percentage of cells G1 phase in the untreated control sample (sample A). Multiplying the outcome of this calculation by 100% provides the percentage of the original G1 population (untreated control sample A) that remained in the G_1 phase following radiation and incubation with nocodazole (sample D).

workers to assess the integrity of the p53 tumour suppressor pathway as it relates to DNA damage signalling (1, 2, 6, 7). A number of workers have also assessed p21 induction by Western blotting for the protein. This latter analysis works reasonably well and is straightforward. However, more quantitative results are obtainable from mRNA induction profiles (6, 7).

Protocol 2.

Reagents

- PBS
- 4 M guanidine isothiocyanate
- Nytran filters (Schleicher and Schuell)
- Hybri-dot manifold (BRL, MD).
- ³⁵S-labelled polythymidylate probe

- p21/WAF1/CIP1 cDNA ³²P-labelled probes (probe kindly provided by Dr Bert Vogelstein, Johns Hopkins University, MD)
- ³²P-labelled probe to glyceraldehyde-3-phosphate dehydrogenase

Method

1. The induction of p21/WAF1/CIP1 mRNA should be assessed within 4 h following exposure to 10–20 Gray of γ-rays.

2. Exponentially growing and irradiated cells are harvested by trypsinization and centrifugation and then washed in ice-cold PBS.

3. Polyadenylated mRNA can be prepared from total RNA extracted from cells using 4 M guanidine isothiocyanate and then using oligo(dT)–cellulose chromatography according to standard procedures supplied with commercially available kits.

4. Eight incremental 1:2 dilutions of the polyadenylated mRNA samples should be blotted on to Nytran filters using a Hybri-dot manifold.

5. Membranes should be irradiated with UV light (700 J/m² at 254 nm).

6. The membranes should be cut so that the first four dilutions can be probed with p21/WAF1/CIP1 cDNA ³²P-labelled probes.

7. The last four dilutions should be hybridized with a ³⁵S-labelled poly-thymidylate probe and also a ³²P-labelled probe to glyceraldehyde-3-phosphate dehydrogenase which enables one to confirm equivalent loadings between individual samples.

8. The level of mRNA induction, relative to unirradiated control samples, can be grouped into three classes: Class 1: strong mRNA induction (> fourfold above basal levels); class 2: intermediate mRNA induction of (between two- and four-fold); class 3: weak or no mRNA induction (< twofold). The greater the degree of p21/WAF1/CIP1 mRNA induction following irradiation the more confident one will be as to the integrity of the p53 checkpoint in the G1 phase.

2.1.4 Assessment of the inhibition of cyclin E/Cdk2

Transcriptional induction of the p21 gene product has been linked to G1 arrest through inhibition of cyclin E/Cdk2 kinase following DNA damage in wild-type p53-containing cells (7, 8). Cyclin E/Cdk2 is a kinase that promotes

S phase entry, at least in part, through phosphorylation and inactivation of the retinoblastoma family of proteins. Inactivation of cyclin E/Cdk2 is required for p53 to induce G1 arrest and the p21 gene product is almost exclusively responsible for inhibition of cyclin E/Cdk2. A procedure to isolate and measure cyclin E/Cdk activity is provided below (see *Figure 3*).

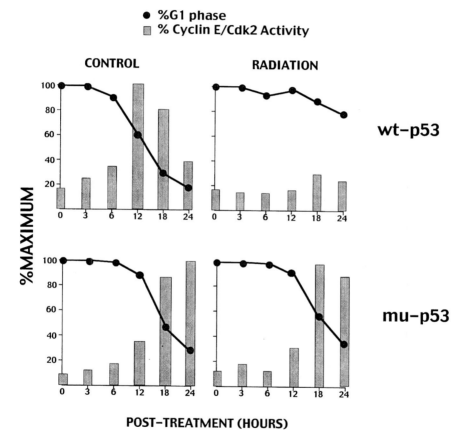

Figure 3. Effect of radiation on the progression of cells from G0/G1 into S phase. Exponentially growing cells were arrested in G0/G1 phase by isoleucine deprivation (12) and then released back into the cycle in the presence or absence of DNA damage. Cell cycle progression was monitored by flow cytometry and cyclin E/Cdk2 activity was assayed as described in *Protocol 3*. The results show that radiation-induced G1 arrest occurs in cells with wild-type but not mutant p53 cells, and that associated with the G1 arrest was an inhibition of cyclin E/Cdk2 kinase activity.

Protocol 3.

Equipment and reagents

- lysis buffer: 1 ml of 50 mM Tris–HCl, pH 8.0, 150 mM NaCl, 50 mM NaF, 10 μg/ml aprotinin, 0.1 mg/ml phenylmethylsulfonyl fluoride, 1% bovine serum albumin (BSA), 1% NP-40
- rabbit polyclonal cyclin E antibody (UBI, New York)
- protein A–agarose (Oncogene Science, New York)

- [γ-^{32}P]ATP (4500 Ci/mmole, NEN)
- kinase reaction mixture: 10 mM Tris–HCl, pH 7.5, 10 mM MgCl$_2$, 3 μg histone H1, 5 μM ATP
- 12% SDS–PAGE gel
- histone 1 (Calbiochem)
- Betascope Analyser (Betagen) or phospho-imager (Molecular Dynamics)

Method

1. For each sample a minimum of two million cells should be lysed on ice for 30 min in lysis buffer. Lysis is initiated by pipetting up and down the 1 ml solution.

2. Samples should next be clarified in a microcentrifuge for 15 min at 4 °C and supernatants incubated for approximately 3 h at 4 °C with a rabbit polyclonal cyclin E antibody and protein A–agarose according to the manufacturers' recommendations.

3. Immune complexes are isolated by briefly microfuging (20 sec) and decanting the supernatant using a 1ml pipette tip. Afterwards, samples are washed three times with 1 ml of lysis buffer and then three times with 1 ml of lysis buffer minus BSA.

4. Samples should then be resuspended in 25 μl of the kinase reaction mixture and approximately 8 μCi of [γ-^{32}P]ATP. Samples are next incubated for 20 min at 37 °C.

5. Following reaction, samples are boiled in SDS gel loading buffer (25 μl) for 5 min and then subjected to electrophoresis on 12% SDS–polyacrylamide gels. Quantitation of ^{32}P incorporated into histone H1 (Calbiochem) is measured by autoradiography and using a Betascope Analyser or phosphoimager.

2.2 Assessment of the DNA damage checkpoint in the G2 phase of the cell cycle

2.2.1 Background

The G2 to M phase transition is regulated by the cyclin B/Cdc2 kinase which is kept inactive until the completion of DNA replication and repair of DNA damage by the imposition of inhibitory phosphorylations on the Cdc2 kinase (threonine-14 and tyrosine-15) (2, 9, 10). At the G2/M transition these Cdc2-inhibitory phosphorylations are removed by a dual specificity phosphatase called Cdc25C which promotes entry into mitosis. DNA damage prevents the removal of these inhibitory phosphorylations, at least in part, by suppression of Cdc25C function through a complex pathway that involves a checkpoint kinase called Chk1 (2, 3, 10).

2.2.2 Assessment of G2 arrest by flow cytometry

The ability of cells to arrest in the G2 phase of the cell cycle following exposure to 6 or 12 Gray of γ-rays has also been successfully used to assay the G2 cell cycle checkpoint in a number of mammalian cell lines (6, 9, 10). Essentially, exponentially growing cells are irradiated and then incubated in the presence or absence of the mitotic inhibitor, nocodazole (see *Figure 2*). One can distinguish between cells arrested in the G2 phase versus those that enter mitosis using a combination of flow cytometry and mitotic index measurements (see below).

Protocol 4.

Samples include:

- control
- nocodazole, alone
- radiation, alone (two radiation doses, see below)
- radiation + nocodazole (two radiation doses, see below)

Equipment and reagents

- nocodazole (0.01–0.4 μg/ml) (Aldrich)
- propidium iodide (Sigma)
- RNase A (Sigma)
- PBS
- flow cytometer

Method

1. Twenty-four hours prior to commencement of the G2 arrest assay, plated cells should be trypsinized and replated at 10–30% confluence in a T75cm^2 flask.

2. Twenty-four hours later, floating cells should be decanted from the plates and fresh medium added. For suspension cultures, exponentially growing cells should be diluted to 5×10^5/ml.

3. Cells should next be irradiated at room temperature with 6 and 12 Gray of γ-rays using a ^{137}Cs source (1 Gray = 100 Rads) and then cells should be incubated for approximately 16–24 h at 37 °C in the presence or absence of the microtubule inhibitor, nocodazole (0.01–0.4 μg/ml) (Aldrich). As described earlier, the nocodazole concentration to use should have been determined first using a concentration–response ranging from 0.01–0.4 μg/ml. The goal is to chose the maximum non-toxic concentration of nocodazole that effectively blocks cells in G2/M

Protocol 4. *Continued*

by flow cytometry, leaving few cells in the G1 phase. The mitotic index measurement procedure is described below.

3. Medium containing floating cells and cells trypsinized from the plates should then be combined and centrifuged to pellet the cells. Cells should then be washed once in ice-cold PBS and finger-vortexed back into suspension in 0.5 ml of PBS before adding 5 ml of 70% ethanol and mixing by hand and storing on ice for a minimum of 2 h. Samples can be stored for up to a week in the fridge.

4. Cells are next rehydrated by resuspending in 5 ml PBS for 5 min. Cells are then pelleted and subsequently resuspended in 0.5 ml of propidium iodide solution (25 μg/ml) (Sigma) that contains RNase A (500 units/ml) (Sigma) before incubation at 37°C for 30 min to destroy RNA.

5. Cell cycle analysis can be performed on a Becton–Dickinson FACSCAN flow cytometer or similar machine. At least 10 000 cells should be used for each analysis.

6. The population of cells in G2/M phase by flow cytometry is calculated from the manufacturer's software and compared with the untreated control and nocodazole alone-treated sample (see *Figure 2*).

2.2.3 Assessment of G2 arrest by mitotic index measurements

The mitotic index can be used to monitor the progression of cells into mitosis where chromosome condensation is clearly a visible determinant (10). DNA damage-induced G2 arrest should block mitotic entry. G2 arrest will be easily distinguishable, over the short course (approximately 24 h) in most cells in comparison with nocodazole-treated cultures that did not receive DNA damage and thus go on to arrest in mitosis.

Protocol 5.

Reagents

- PBS
- Ethanol
- Giemsa (Sigma)
- Acetic acid

Method

1. At least 1×10^3 cells are needed (and preferably work with 10 times that amount for good preparations) for mitotic assessments.

2. Cells are washed once with 2 vols of ice-cold PBS and then resuspended in 0.5 ml of 50% strength ice-cold PBS for 10 min.

3. Cells are then centrifuged and resuspended in a 2% solution of 3:1 ethanol/acetic acid and stored at 4°C overnight.

4. Cells are then dropped (or smeared using a coverslip) on to glass slides, air-dried for 0.5–2 h and subsequently stained with Giemsa for 10 min. Giemsa is gently rinsed from the slide by dipping the slide in PBS. Slides are then permitted to air-dry.

5. Cells are next evaluated for the presence of mitotic figures [condensed nuclear material (chromosomes) and the lack of nuclear membrane]. Generally, 500 cells are scored for each sample and are represented as a percentage of the total.

2.2.4 Procedure for assessment of cyclin B/Cdc2 inhibition

Inhibition of the cyclin B1/Cdc2 kinase has been used by a number of workers to assess the integrity of the G2 checkpoint. This works particularly well if the DNA-damaged cell samples are compared with an equivalent number of cells trapped in mitosis with nocodazole, where Cdc2 activity is at a maximum. A procedure to isolate and measure cyclin B1/Cdc2 activity is provided below. This is a modification of earlier procedures (9, 10).

Protocol 6.

Reagents

- lysis buffer: 50 mM Tris–HCl, pH 8.0, 150 mM NaCl, 50 mM NaF, 10 μg/ml aprotinin, 0.1 mg/ml phenylmethylsulfonyl fluoride, 1% bovine serum albumin (BSA), 1% NP-40 (Calbiochem)
- rabbit polyclonal cyclin B1 antibody (UBI, New York)
- Cdc2 antibody (UBI, New York)
- protein A–agarose (Oncogene Science, New York)

- kinase reaction mixture: 10 mM Tris–HCl, pH 7.5, 10 mM $MgCl_2$, 3 μg histone H1, 5 μM ATP
- [γ-^{32}P]ATP (4500 Ci/mmole, NEN)
- 12% SDS–PAGE gel
- histone 1 (Calbiochem)
- Betascope Analyser (Betagen) or phospho-imager (Molecular Dynamics)

Method

1. For each sample a minimum of two million cells should be lysed on ice for 30 min in 1 ml of lysis buffer. Lysis is initiated by pipetting up and down the 1 ml solution.

2. Samples should next be clarified in a microcentrifuge for 15 min at 4 °C and supernatants incubated for approximately 3 h at 4 °C with a rabbit polyclonal cyclin B1 antibody and/or Cdc2 antibody and protein A–agarose according to the manufacturer's recommendations.

3. Immune complexes are isolated by briefly microfuging (20 sec) and decanting the supernatant using a 1ml pipette tip. Afterwards, samples are washed three times with 1 ml of lysis buffer and then three times with 1 ml of lysis buffer minus BSA.

4. Samples should then be resuspended in 25 μl of a kinase reaction

---

Protocol 6. *Continued*

mixture and approximately 8 μCi of [γ-^{32}P]ATP. Samples are next incubated for 20 min at 37 °C.

5. Following reaction, samples are boiled in SDS gel loading buffer (25 μl) for 5 min and then subjected to electrophoresis on 12% SDS–polyacrylamide gels. Quantitation of ^{32}P incorporated into histone H1 (Calbiochem) can be measured by autoradiography and using a Beta-scope Analyser (Betagen) or phosphoimager (Molecular Dynamics).

3. Summary

The above procedures enable one to assess the function of the G1 and G2 checkpoints in asynchronously growing mammalian cells and can be applied with procedures for cell synchronization described elsewhere (e.g. ref. 11) to provide a picture of the functionality of these checkpoint pathways. Careful assessments of G1 and G2 checkpoints are uncovering differences between normal and neoplastic cells and potential routes through which tumour cells develop unstable genomes and evade cell killing with current cancer chemotherapy and radiotherapy regimens. These same checkpoints are under intense investigation to better understand their operation in mammalian cells and for their exploitation using novel approaches for cancer treatment.

References

1. Hartwell, L.H. and Kastan M.B. (1994). *Science*, **266**, 1821.
2. O'Connor, P.M. (1997). *Cancer Surveys*, **29**, 151.
3. Nurse, P. (1997). *Cell*, **91**, 865.
4. Hartwell, L.H. and Weinert, T.A. (1989). *Science*, **246**, 629.
5. Donehower, L.A., Harvey, M., Slagle, B.L., McArthur, M.J., Montgomery, C.A., Jr, Butel, J.S. and Bradley, A. (1992). *Nature*, **356**, 215.
6. O'Connor, P.M., Jackman, J., Bae, I., Myers, T., Fan, S., Scudiero, D., Monks, A., Sausville, E., Weinstein, J., Friend, S., Fornace, A. J., Jr and Kohn, K.W. (1997). *Cancer Res.*, **5**, 4285.
7. El-Deiry, W.S., Harper, J.W., O'Connor, P.M., Velculescu, V.E., Canman, C.E., Jackman, J., Pietenpol, J.A., Burrell, M., Hill, D.E., Wang, Y., Wiman, K.G., Mercer, W.E., Kastan, M.B., Kohn, K.W., Elledge, S.J., Kinzler, K.W. and Vogelstein, B. (1994). *Cancer Res.*, **54**, 1169.
8. Dulic, V., Kaufmann, W.K., Wilson, S.J., Tlsty, T.D., Lees, E., Harper, J.W., Elledge, S.J. and Reed, S.I. (1994). *Cell*, **76**, 1013.
9. Wang, Q., Fan, S., Eastman, A., Worland, P.J., Sausville, E.A. and O'Connor, P.M. (1996). *J. Natl. Cancer Inst.*, **88**, 956.
10. Yu, L., Orlandi, L., Wang, P., Orr, M.S., Senderowicz, A.M., Sausville, E.A., Silvestrini, R. Watanabe, N., Piwnica-Worms, H. and O'Connor, P.M. (1998). *J. Biol. Chem.*, **273**, 33455.

11. Stein, G.S., Stein, J.L., Liam, J.B., Last, T.J., Owen, T. and McCabe, L. (1995). In *Cell Growth and Apoptosis*: *A Practical Approach* (ed. G.P. Studzinski), pp. 193–203. IRL Press, Oxford.
12. Jackman, J. and O'Connor, P.M. Cell Cycle Analysis. In *Current protocols in cell biology*. Chapter 8, pages 8.0.1 to 8.3.20. Editors: Bonifacino, J. S., Dasso, M., Lippincott-Schwartz, J., Harford, J. B. and Yamada, K. M. John Wiley and Sons, New York, 1999 (January).

Assessment of the role of growth factors and their receptors in cell proliferation

SUBAL BISHAYEE

1. Introduction

The growth of cells in culture and, most likely , in the animal, is subject to control by various agents, such as extracellular signalling proteins (growth factors). These growth factors mediate their pleiotropic actions by binding to and activating high affinity cell surface receptors. At least five important classes of polypeptide growth factors have been identified and their inter-actions with cell surface receptors well characterized. These are: (a) epidermal growth factor (EGF) and EGF-like factors, a family of 6–7 kDa polypeptides that share a common receptor (EGF receptor); (b) platelet-derived growth factor (PDGF), a disulfide-bonded dimeric protein (25–30 kDa) composed of two non-identical but highly homologous polypeptide chains (in different combinations); (c) insulin and insulin-like growth factors, a class of structurally related 6–7 kDa polypeptides which interact with multiple receptors; (d) transforming growth factor-β (TGF-β; also acts as a differentiation factor), a family of 25 kDa proteins that act through multiple receptors; and (e) the interleukins and their respective receptors (reviewed in ref. 1). It is recognized that ectopic synthesis or overproduction of these growth factors can lead to an autocrine stimulation of growth and expression of the transformed phenotype (reviewed in ref. 2). In addition, aberrant expression of a growth factor receptor due to gene deletion or mutation and overexpression of a normal growth factor receptor gene can also produce transformation (3). This chapter will focus only on growth factors whose receptors display intrinsic tyrosine kinase activity.

2. Most growth factors are synthesized as transmembrane proteins

A number of growth factors are synthesized as membrane-anchored proteins (reviewed in ref. 4). EGF and at least nine other members of the EGF gene family are synthesized as large transmembrane proteins. Mature EGF is a 53-amino acid polypeptide. However, human and mouse EGF precursors (proEGF) have more than 1200 amino acids with a total of nine EGF units in the extracellular region. The EGF unit nearest to the transmembrane domain corresponds to mature EGF. The function of the other EGF units is not known. ProEGF is synthesized as a glycoprotein of 140–170 kDa. Following its synthesis, proEGF translocates to the cell surface and within a few hours is proteolytically cleaved to generate the soluble EGF. Like the soluble EGF, the intact membrane-anchored proEGF is biologically active.

Another member of the EGF family is TGF-α. This mitogenic 50-amino acid polypeptide which also acts through the EGF receptor (EGFR) is synthesized as a 160-amino acid proTGF-α. The involvement of TGF-α in cellular transformation and tumorigenesis is suggested by its expression in transformed cells, tumour-derived cell lines, and different tumours. The vaccinia virus growth factor (VVGF), a 77-amino acid soluble glycoprotein with an EGF unit, is synthesized in virally infected cells as a 140-amino acid membrane precursor. It also acts through the EGFR.

The macrophage colony-stimulating factor-1 (CSF-1/ M-CSF) is a factor produced by bone marrow stromal cells and other sources; it regulates the survival, proliferation, and differentiation of mononuclear phagocytic cells. CSF-1 is specific for cells of the macrophage lineage, although it can also stimulate granulocytic cells at higher concentrations, and may also be involved in placental development. It is synthesized as a 550-amino acid protein that dimerizes via disulfide bonds early during its biosynthesis. The extracellular domain of the precursor is proteolytically cleaved, giving rise to soluble, secreted homodimers of 86 kDa. The lower molecular weight product (44 kDa homodimer) is generated as a result of alternative splicing.

Stem cell factor, also known as *steel* factor or Kit ligand (KL) interacts with its receptor, Kit, and mediates diverse responses, including proliferation, survival, chemotaxis, migration, differentiation, and adhesion to the extra-cellular matrix. It is encoded at the *Sl* locus on mouse chromosome 10. Although human and mouse KL have more than 80% sequence identity at the amino acid level, the human protein has limited activity on mouse cells; how-ever, murine KL is active on human cells. The murine and human KL proteins are synthesized as transmembrane glycoproteins containing 273 amino acids including the signal peptide. Following cleavage of the signal peptide, the mature KL protein is further proteolytically processed to generate soluble, biologically active 165-amino acid glycoprotein. KL forms non-covalently linked homodimers in solution. A membrane-bound form of KL, detected in both mouse and human, originates from alternative splicing of exon 6. This cell-bound form of KL appears to be needed for normal development in mice.

PDGF is a powerful mitogen for cells of mesenchymal and glial origin and has been implicated in the regulation of cell proliferation during normal as well as pathological conditions, including malignancies (reviewed in ref. 5). It is synthesized in the megakaryocytes of bone marrow and stored in alpha granules of platelets. It is a ~30 kDa protein composed of two polypeptide chains, A and B, that give rise to three disulfide-linked dimers, AA, AB, and BB. The A chain and the B chain are encoded by separate but related genes. The B chain is almost identical to a part of p28^{sis}, the transforming protein of simian sarcoma virus (SSV). This provides a link between growth factors and oncogene products and suggests a mechanism whereby oncogene products transform cells, i.e. by subversion of the mitogenic pathway of growth factors. In SSV-transformed cells, the v-sis product is cell associated and undergoes further proteolytic processing to a 24 kDa form. In fact PDGF BB is the first growth factor that is known to have sequence identity with an oncogene product.

3. Growth factor receptors

3.1. Structural features

So far, more than 50 receptor tyrosine kinases have been identified. These multi-sited and multifunctional receptors have similar structural features—a single hydrophobic transmembrane region of 20–25 amino acids separates the large extracellular domain from the cytoplasmic region. The exoplasmic domain contains the ligand-binding site, whereas the intracellular domain contains a tyrosine kinase catalytic site and also other sites that are important for signal transduction. These receptors can be broadly classified into subfamilies on the basis of the differences in the structural motifs of their extracellular domains: some receptor subfamilies contain cysteine-rich repeat units (EGFR), some contain immunoglobulin-like domains (PDGFR), whereas others express both the cysteine-rich and immunoglobulin-like domains (reviewed in refs 6 and 7). The importance of these repeat units in the biological functioning of the receptors is not clear. In addition to the exoplasmic domains, the structural motifs of the kinase domain of different receptors also differ considerably. Unlike the receptors for EGF and insulin, the kinase domain of the PDGFR (both α- and β-type), the fibroblast growth factor receptor (FGFR), c-Kit, and the CSF-1R is interrupted by a span of 14–100 amino acids known as the kinase insert domain. In some of the receptors this domain is required for mitogenic activity. The structural characteristics of these receptor kinases are described below.

3.1.1 The epidermal growth factor receptor family

The EGFR (ErbB1) and the other three members (ErbB2/neu, ErbB3 and ErbB4) of this family have similar structural characteristics—two cysteine-rich domains in the extracellular region, an uninterrupted kinase domain, and multiple autophosphorylation sites clustered at the C-terminal tail. Among these receptors, ErbB3 has virtually no kinase activity, although it is capable of binding ATP. The ligands for ErbB1 include EGF, transforming growth factor-α, heparin-binding EGF, betacellulin, amphiregulin, and epiregulin. Neuroregulins (NRG) which are predominantly expressed in parenchymal organs and in embryonic central and nervous systems bind to both ErbB3 and ErbB4. No ligand has yet been identified for ErbB2. All the receptors of the ErbB family are capable of forming ligand-induced homo- or heterodimers resulting in recruitment of a distinct set of signalling molecules to the receptor complexes. For example, Cbl (<u>c</u>asitas <u>B</u>-lineage <u>l</u>ymphoma) is recruited by the EGFR-containing complexes, but not by other dimers. In addition, PI-3 kinase (PI3K) has a much higher affinity for phosphorylated ErbB3 compared with the EGFR. Such differential association of the signal transducers with receptor complexes may explain the diverse signalling by homo- or hetero-dimeric complexes (reviewed in ref. 8).

3.1.2 Insulin receptor family

Three members of this family have so far been identified: the insulin receptor (IR), the insulin-like growth factor-1 receptor (IGF-1R), and the insulin-related receptor. All these receptors are heterotetrameric proteins composed of two α chains and two β chains linked by disulfide bonds. The α chains represent most of the extracellular domain of the receptor and contain a cysteine-rich domain and two fibronectin type III domains, whereas the β chains contain the transmembrane domain, the cytoplasmic tyrosine kinase domain, and the autophosphorylation sites.

3.1.3 The platelet-derived growth factor receptor family

Because of the similar structural organization in the extracellular domain and the kinase site, PDGFR (both α- and β-type), CSF-1R, the stem cell factor (SCF) receptor (Kit), and Flt3/Flk2 can be grouped together in this family (reviewed in ref. 6). Each member of this family has five immunoglobulin-like domains in the exoplasmic region and a kinase domain which is split into two parts by a span of 14–100 amino acids. The kinase insert domain is required for receptor functioning. Unlike the EGFR family, the autophosphorylation sites in this receptor family are distributed throughout the intracellular domain. The ligands for all the receptors are dimeric proteins with two receptor-binding sites per molecule. The PDGFRα binds to all three isoforms of PDGF (AA, BB, and AB) and PDGFRβ binds to PDGF BB and not to PDGF AA or PDGF AB; such binding results in receptor dimerization and kinase activation. However, in the presence of PDGFRα, the β receptor can be activated by PDGF AB (9, 10). PDGFR is expressed in mesenchymal cells, whereas its ligands are synthesized in epithelial and endothelial cells.

3.1.4 Eph (<u>e</u>rythropoietin-<u>p</u>roducing <u>h</u>epatocellular) receptor family

This is by far the largest tyrosine kinase receptor family with at least 14 distinct members that have been identified in different species, including humans, rodents, chickens, and *C. elegans*, and these receptors are predominantly expressed in the central nervous system (11). Eph receptors and their ligands are the newest members of the receptor tyrosine kinase family. All the members of this receptor family have highly similar structural features—an immunoglobulin-like motif, a single cysteine-rich module, two fibronectin type III repeats in the extracellular domain, and a highly conserved tyrosine kinase site in the cytoplasmic domain. The first member of this receptor family was pulled out in a screen from an erythropoietin-producing hepatocellular carcinoma cell line and hence was termed 'Eph'. So far, eight ligands for this receptor family, known as ephrins, have been identified.

Ephrins are membrane-anchored proteins; five of the ligands are bound to the membrane via glycosyl-phosphotidyl inositol (known as ephrin-A subclass), whereas the other three are transmembrane proteins (ephrin-B subclass). All the ligands of the ephrin-A subclass bind to and activate one group of receptors (EphA subclass), whereas ephrin-B ligands bind to and activate the other major subclass of receptors (EphB subclass). These receptors are mediators of diverse biological processes, such as axonal guidance, the formation of topographic neural projections, axon fasciculation, and the regionalization of the central nervous system. In addition, because of the expression of the Eph receptors and their ligands in complementary domains of tissues, both the receptors and the ligands have also been implicated in mediating contact repulsive interactions that restrict cell and axon migration. Unlike the other receptor tyrosine kinases, ephrins and their receptors have a number of special features: (a) nearly every ligand of a certain subclass can interact with almost all receptors of the corresponding subclass; (b) although a soluble ligand can bind to its receptor, the membrane-anchored ligand can only activate the cognate receptor; (c) receptor activation does not lead to mitogenic responses; (d) in addition to the receptors, ligands of ephrin-B subclass which are transmembrane proteins, also directly participate in signal transduction by receptor-mediated tyrosine phosphorylation in their cytoplasmic domain; and (e) unlike other receptor tyrosine kinases, there is no evidence that Eph receptors play a role in tumorigenesis, although these receptors are overexpressed in certain tumours.

The structural characteristics and the autophosphorylation sites of the receptors described above, and of other receptor tyrosine kinases, are summarized in *Table 1*.

3.2 Receptor assays

The quantification of receptors can be carried out by a number of methods. The most widely used and simple method is by using ^{125}I-labelled ligand. Ligand binding can be conducted with (a) living cells, (b) isolated membranes, and (c) detergent-solubilized receptor preparations; the other method for

Table 1. Structural features and autophosphorylation sites in different receptor tyrosine kinases

Receptor type	Domain structure	Phosphate acceptor sites	Reference
EGFR (h)	2 cysteine-rich; uninterrupted kinase	992, 1068, 1086, 1148, 1173	Cited in 12
IR (h)	2 cysteine-rich; 8 fibronectin-type; uninterrupted kinase	953, 960, 972, 1146, 1150, 1151, 1316, 1322	Cited in 13
PDGFRβ (h)	5 Ig-like; split kinase	579, 581, 716, 740, 751, 763, 771, 775, 778, 857, 1009, 1021	Cited in 10
PDGFRα (h)	5 Ig-like; split kinase	572, 574, 720, 731, 742, 754, 762, 768, 849, 988, 1018	Cited in 10
CSF-1R (h)	5 Ig-like; split kinase	561, 699, 708, 723, 809	Cited in 14
C-Kit (h)	5 Ig-like; split kinase		Cited in 15
FGFR1 (h)	3 Ig-like; acid box; split kinase	463, 583, 585, 653, 654, 730, 766	16
Trk A (NGFR) (h)	2 Ig-like; leucine-rich; uninterrupted kinase	490, 670, 674, 675, 785	Cited in 17
EphR (c)	Cysteine-rich; Ig-like; 2 fibronectin type III, uninterrupted kinase	605 (EphB2), 611 (EphB2), 929 (EphB1)	11, 18

All the receptors are monomeric except IR which is a heterotetrameric protein composed of two α chains and two β chains. Abbreviations: c, chicken; h, human; Ig, immunoglobulin.

receptor quantification is by kinase assay. Both ligand-binding and kinase assays are possible only with the functionally active receptor.

3.2.1. By ^{125}I-labelled ligand binding (using intact cells, isolated membranes, or detergent-solubilized cell lysates)

Although *Protocols 1* and *2* describe the EGF receptor assay in intact cells and isolated membranes using radiolabelled ligand, similar methods can also be used for other receptors. As highly basic proteins, such as PDGF, tend to bind to glass surfaces very tightly, use of glass containers either for storage of the proteins or for receptor assays, should be avoided. ^{125}I-labelled growth factors are commercially available. However, in most cases, radiolabelled growth factors can also be prepared in the laboratory (19). The most commonly used method for iodination employs chloramine-T or modifications of this reagent, such as iodogen (1, 3, 4, 6-tetrachloro-3α, 6α-diphenylglycouril) and iodobeads (chloramine-T linked to polystyrene beads). The biological activities of a number of growth factors are highly susceptible to iodination by chloramine-T. A milder method, such as the use of lactoperoxidase, should be considered in those cases. Also, as the tyrosine residues in a protein are iodinated by chloramine-T, this reagent cannot be used to label PDGF BB which lacks this amino acid. In this case, the method of choice is Bolton–Hunter reagent [*N*-succinimidyl-3-(4-hydroxyphenyl)propionate] (19). *Protocols 1* and *2* described below are routinely used in my laboratory.

Protocol 1.   Binding of [^{125}I]EGF to living cells

Equipment and reagents

- binding buffer: Earle's balanced salt solution (without bicarbonate) containing 20 mM Hepes, pH 7.5, and 2.5 mg/ml bovine serum albumin (BSA)
- [^{125}I]EGF (~3 × 10^5 cpm)[a]

- unlabelled EGF[a]
- rocker platform (Bellco Biotechnology, Vineland, NJ, USA)
- 1 mg/ml BSA containing 1% Triton X-100

Method

1. Grow cells for 24–48 h to a confluency of 80–90% in 2 cm^2 24-well plates.

2. Remove the medium by aspirating gently with a pasteur pipette connected to an in-house vacuum (care should be taken to avoid dislodging cells during this and subsequent steps).

3. Wash cell monolayers carefully three times (0.5 ml/wash) with binding buffer at room temperature (20°C).

4. Add 200 μl of binding buffer containing 2 ng of [^{125}I]EGF (~3 × 10^5 cpm) and unlabelled EGF (25 ng/ml for cells expressing less than 1 × 10^5 receptor sites/cell and 100 ng/ml for cells expressing more than 1 × 10^5 receptor sites/cell) to some wells to determine total binding.[b,c]

5. Incubate cells at room temperature for 90 min on a rocker platform with a speed adjusted between 2 to 3.

6. Determine non-specific binding, i.e. the binding which is not displaced by excess of unlabelled EGF, by incubating cells with labelled EGF in the presence of 100–200 ng of unlabelled EGF.

7. Wash cell monolayers three times with ice-cold binding buffer to remove the unbound radioactivity.

8. Add 0.5 ml of 1 mg/ml BSA containing 1% Triton X-100 to each well and incubate at 37°C for 15 min to solubilize the cells.

9. Transfer the solubilized cell-associated radioactivity to tubes and count in a gamma counter.

10. Subtract the non-specific binding from the total binding to determine the specific binding.[d]

[a] Both [^{125}I]EGF and receptor-grade EGF are commercially available.
[b] An aliquot of the binding buffer should be counted to determine the specific radioactivity (cpm/ng of EGF).
[c] Since the saturating concentration of EGF has been used in this protocol, B_{max}, the maximum number of EGF molecules bound per cell can be determined by this assay without Scatchard analysis.
[d] To express the results as receptor sites/cell, cell numbers in wells that have been washed and incubated with the binding buffer (in the absence of labelled EGF) should be determined.
[e] To avoid endocytosis and ligand degradation, it is always advisable to carry out the binding assay at 4 or 20°C instead of at 37°C.
[f] This basic binding assay can be used to determine the K_D for the ligand–receptor interaction. For this purpose, cells are incubated with a fixed amount of labelled EGF in the presence of increasing concentrations of unlabelled EGF. Both K_D and B_{max} can be determined by Scatchard analysis (20).

Protocol 2. Binding of [^{125}I]EGF to isolated membranes

Reagents
- Hepes, pH 7.4
- BSA
- [^{125}I]EGF
- unlabelled EGF

Method

1. Add 2–10 μg (depending on the concentration of the receptors in original cells) of isolated membrane proteins to 100 μl of a solution containing Hepes, pH 7.4, and BSA and adjust the concentrations of these two components to 20 mM and 1 mg/ml, respectively.

2. Initiate the binding reaction by adding [^{125}I]EGF (for total binding) or [^{125}I]EGF plus a 50-fold excess of unlabelled EGF (for non-specific binding) and incubate at 4 or 20°C for 60–90 min with continuous vortexing.

3. Place cellulose acetate filters (25 mm, 0.45 μm pore size) in a Petri dish and soak the filters in 5 mg/ml BSA solution while the incubation is in progress.

4. Terminate the reaction by adding 900 μl of ice-cold 20 mM Hepes, pH 7.4, containing 1 mg/ml BSA to the tubes.

5. Separate the membrane-bound [^{125}I]EGF from free [^{125}I]EGF by rapidly filtering under vacuum using pre-soaked filters placed on a multisample filtration manifold (filters are pre-soaked to reduce the background binding).

6. Wash each filter 3–4 times with 20 mM Hepes, pH 7.4, containing 1 mg/ml BSA (1 ml/wash), remove from the manifold, and count in a gamma counter.

7. Conduct the binding reaction in duplicate or triplicate to get reproducible results.

Protocol 3. Binding of [^{125}I]EGF with soluble receptor (21)

Reagents
- buffer: 20 mM Hepes, pH 7.4, 10% glycerol, 1% Nonidet P-40 (NP-40), 1 mM PMSF, 10 μg/ml leupeptin, and 0.15 unit/ml aprotinin
- BSA
- 0.1 M phosphate buffer, pH 7.4
- γ-globulin
- [^{125}I]EGF
- unlabelled EGF
- 20.4% polyethylene glycol
- 8.5% polyethylene glycol in 0.1 M phosphate buffer, pH 7.4

Method

1. Add 1 ml of the buffer to 2 mg of plasma membrane proteins.

2. Incubate at 4°C for 20 min to solubilize the membrane proteins and centrifuge at 100 000 *g* for 30 min.

3. Collect the supernatant and use it immediately for receptor assay or store at −70°C in small aliquots for future use.

4. Add 0–40 μl of the supernatant to glass tubes (12 cm × 75 cm) containing Hepes, pH 7.4, NP-40, and bovine serum albumin (BSA), and adjust the volume to 200 μl and the concentrations to 20 mM, 0.2%, and 1 mg/ml, respectively.[a]

5. Initiate the binding reaction by adding [^{125}I]EGF alone (for total binding) or [^{125}I]EGF plus a 50-fold excess of unlabelled EGF (for non-specific binding).

6. Incubate at 4°C or at room temperature for 60–90 min.

7. Add, at room temperature, 500 μl of 0.1 M phosphate buffer, pH 7.4, containing 1 mg/ml γ-globulin to all tubes, followed immediately by 500 μl of 20.4% (w/v) of polyethylene glycol 6000 (or 8000) in water.[b]

8. Vortex and filter quickly under vacuum using cellulose acetate filters (25 mm, 0.45 μm pore size) pre-soaked in 5 mg/ml BSA.

9. Wash the filters three to four times with a solution of 8.5% polyethylene glycol in 0.1 M phosphate buffer, pH 7.4 (1.5 ml/wash), and count the filters in a gamma counter.

[a] control tube without membrane proteins should always be included.
[b] The precipitation of the ligand–receptor complexes by polyethylene glycol in the presence of phosphate-γ-globulin solution should be conducted at room temperature.

3.2.2 Kinase assay by receptor autophosphorylation

The basic kinase assay as described for the ligand-dependent autophosphorylation of the PDGF receptor can be conducted with solubilized cell lysates, isolated membranes or purified receptor preparations (9). Since high concentrations of the non-ionic detergents are known to interfere with the kinase activity, the detergent concentrations during the assay should not exceed 0.2%. When crude receptor preparations are used, great care should be exercised to protect the receptor from cleavage by proteases by carrying out the assay at 4°C in the presence of protease inhibitors, such as aprotinin, leupeptin, and PMSF. In addition, because of the presence of tyrosine phosphate phosphatases in the receptor preparations, sodium vanadate, an inhibitor of the phosphatases, should be added to the reaction mixture. To discount the likelihood that part of the receptor phosphorylation is mediated by other protein kinases, the contaminating kinases can be removed by immunoprecipitating the receptor by using a receptor-specific antibody and conducting the kinase assay in the immunoprecipitates. The antibody used for this purpose should not inhibit the kinase function or interfere with the ligand binding. Immobilized ligand can also be used as an affinity matrix to purify a receptor. Although *Protocol 5* (22) describes the purification of the EGF receptor using solid-supported EGF, a similar method can also be used for the purification of other receptors, provided the appropriate ligand is readily available.

103

Protocol 4. Autokinase assay of the β PDGF receptor

Reagents

- 0.25% NP-40 (in 20 mM Hepes, pH 7.4, 10% glycerol, 0.15 M NaCl, 1 mM sodium vanadate, 2.5 mM MgCl$_2$, 0.5 mM MnCl$_2$, 1 mM PMSF, 10 μg/ml leupeptin, and 0.15 unit/ml aprotinin)
- 10 nM PDGF BB

- 100 μM [γ-^{32}P]ATP (specific radioactivity: 40–60 Ci/mmol)
- 10 mM EDTA, pH 7.4
- 0.5 mM unlabelled ATP
- solid phase anti-phosphotyrosine antibody

Method

1. Isolate the plasma membranes from cells expressing β-type PDGF receptors, such as human glioblastoma cell line, A172.
2. Solubilize the membranes with 0.25% NP-40 in 20 mM Hepes, pH 7.4, 10% glycerol, 0.15 M NaCl, 1 mM sodium vanadate, 2.5 mM MgCl$_2$, 0.5 mM MnCl$_2$, 1 mM PMSF, 10 μg/ml leupeptin, and 0.15 unit/ml aprotinin.
3. Collect the clarified supernatant by centrifugation in a microfuge for 15 min in the cold.
4. Incubate 20 μl of the supernatant in the absence or presence of 10 nM PDGF BB for 30 min at 4°C.
5. Initiate the phosphorylation reaction by adding 5 μl 100 μM [γ-^{32}P]ATP (specific radioactivity: 40–60 Ci/mmol).
6. Terminate the reaction after 30 min at 4°C with 10 mM EDTA, pH 7.4, containing 0.5 mM unlabelled ATP and incubate the reaction mixture with 20 μl of solid phase anti-phosphotyrosine antibody (5–10 mg/ml protein) to isolate the tyrosine-phosphorylated receptors.
7. Wash the gel beads three times (200 μl per wash) with 20 mM Hepes, pH 7.4, containing 0.2% NP-40, 0.15 M NaCl, and 1 mM vanadate and elute the bound protein by incubating at 4°C for 30 min with 50 μl of 40 mM phenyl phosphate in the above buffer.
8. Repeat the elution process one more time and pool both supernatants.
9. Analyse the extent of autophosphorylation by SDS–PAGE/ autoradiography and scan the autoradiogram.

Protocol 5. Purification of the EGF receptor by EGF–Affi–Gel[a] chromatography

Reagents

- solubilizing buffer: 20 mM Hepes, pH 7.4, 10% glycerol, 1% NP-40, protease inhibitors (aprotinin, leupeptin, and phenylmethyl-sulfonyl fluoride)
- EGF-Affi-Gel 15[a,b,c] (EGF–agarose containing 0.2 mg of covalently bound EGF)

- 5 mM ethanolamine solution, pH 10
- 10% glycerol
- 0.2% NP-40
- 0.05 M HCl
- PM-30 filters

Method

1. Solubilize isolated membranes (1 mg of protein) from human carcinoma cell line, A431, by stirring at 20°C for 30 min with 0.5 ml of a solution of solubilizing buffer.

2. Centrifuge at 100,000 g (45000 r.p.m.) for 60 min in a Ti50 rotor and collect the supernatant.

3. Add the supernatant to 0.25 ml of EGF-Affi-Gel 15 (EGF–agarose containing 0.2 mg of covalently bound EGF) and incubate at 4°C for 60 min with gentle stirring.

4. Centrifuge the suspension at a maximum speed in a microfuge for 5 min at 4°C.

5. Wash the gel pellet four times at 4°C with the solubilizing buffer (1 ml per wash).

6. Elute the EGF receptor by stirring the washed gels three times (10 min at 4°C each time) with 0.5 ml of 5 mM ethanolamine solution, pH 10, containing 10% glycerol, and 0.2% NP-40.

7. Adjust immediately the pH of the eluted receptor to 7.0 by adding 0.05 M HCl.

8. Concentrate the solution by ultrafiltration using PM-30 filters.

9. Adjust the buffer to 20 mM Hepes, pH 7.4, 10% glycerol, and 0.2% NP-40 during the concentration.

10. Determine the receptor concentration by Scatchard analysis of [^{125}I]EGF (see *Protocol 3*).

11. Receptor recoveries vary from 40 to 60% of the input solubilized activity.

---

[a] Affi–Gel is an *N*-hydroxysuccinimide ester of a derivatized, cross-linked agarose gel bead support (Bio-Rad Laboratories).
[b] These gels are available as Affi–Gel 10, which contains a neutral 10-atom spacer arm, and Affi–Gel 15, which contains a cationic charge in its 15-atom spacer arm.
[c] Affi–Gel 10 and 15 are suitable for coupling basic (with isoelectric point between 6.5 to 11.0) and acidic (isoelectric point below 6.5) proteins, respectively.
[d] EGF–Affi–Gel 15 is prepared by reacting Affi–Gel 15 with EGF in 0.1 M NaHCO$_3$ buffer, pH 8 according to manufacturer's directions. After coupling, the gel beads should be loaded on to a column and extensively washed successively with water, 1 M NaCl, 0.1 M acetic acid, 1 M urea, water, and finally with 20 mM Hepes to remove the unbound ligand. It is stored at 4°C in 20 mM Hepes, pH 7.4, containing 0.05% sodium azide.

4. Consequences of growth factor–receptor interaction

4.1 Dimerization/oligomerization of the receptor

The earliest consequence of the interaction of a growth factor with the extracellular ligand-binding domain of its receptor is the ligand-induced dimerization/oligomerization of the receptor, and the dimeric receptor becomes an active kinase. In the case of the dimeric ligands, such as PDGF, CSF-1, and SCF (stem cell factor; also known as Kit ligand), each subunit of the ligand binds with one molecule of the receptor, leading to receptor dimerization. Such dimerization is further stabilized by receptor–receptor interaction, as has been shown recently with c-Kit and the PDGFR (23, 24). The extracellular domain of both receptors has five immunoglobulin-like sequences. Ig-like domains 1–3 are involved in ligand binding, whereas Ig-like domain 4 is the dimerization site involved in receptor–receptor interaction. These results suggest that blocking receptor–receptor inter-actions by targeting Ig-like domain 4 may serve as a strategy to interfere with signalling through PDGFR and c-Kit (25). However, the mechanism by which binding of a monomeric ligand, such as EGF, results in the dimeric receptor is not clearly understood. It is possible that two EGF molecules interact with two receptors in a symmetrical fashion, as suggested by spectrophotometric studies. Human growth hormone (hGH), which is a single chain protein, binds with very high affinity to one receptor and, through a different epitope, with lower affinity to another receptor. The binding of the second receptor is stabilized by direct interactions between the receptors (26).

4.1.1 Methodologies to study receptor dimerization/oligomerization

A number of methods have been developed to study the ligand-induced dimerization of the growth factor receptors in living cells and in isolated membranes, as well as in solubilized preparations. When the receptor forms a stable dimer, such dimeric receptor can be isolated by sucrose density gradient centrifugation (27) or by HPLC (28) in a biologically active form, and the kinase activity of the receptor can be assayed by auto-phosphorylation or by phosphorylation of exogenous substrates. Ligand-dependent receptor association can also be followed by cross-linking analysis (27). 3,3′-Bis(sulfosuccinimido)suberate (BS^3), a non-cleavable reagent with a linker length of 11.4 Å, which is capable of creating covalent linkage between tightly associated polypeptides, is routinely used in my laboratory. Cross-linking experiments can be conducted with radiolabelled ligand bound to intact cells or to isolated membranes or with the ^{32}P-labelled receptors phosphorylated in the presence of the ligand, and the cross-linked complexes (monomeric and the dimeric) can be identified by SDS–PAGE (3.5–10% gel). If crude receptor preparation is used for ^{32}P-labelling, tyrosine-phosphorylated receptor should be purified by anti-phosphotyrosine anti-body before subjecting to chemical cross-linking. Such purification is needed to avoid non-specific cross-linking.

Protocol 6. Ligand-induced dimerization of the PDGF receptor in living cells

Reagents

- binding buffer [e.g. Earle's balanced salt solution (without bicarbonate) containing 20 mM Hepes, pH 7.5, and 2.5 mg/ml bovine serum albumin (BSA)]
- 0.3 nM [^{125}I]PDGF BB (~100000 cpm/ng)
- BSA
- 10 mM Tris, pH 7.5
- 0.5% NP-40
- protease inhibitors

Method

1. Grow cells for 24–48 h to a confluency of ~80% in 6 cm dishes.
2. Wash cell monolayers carefully three times (3 ml/wash) with a binding buffer at 4°C.
3. Add 1 ml of the binding buffer containing 0.3 nM [^{125}I]PDGF BB to each of the dishes and incubate at 4°C for 90 min on a rocker platform at a low speed.
4. Wash the cell monolayers twice with the binding buffer containing BSA and twice with the binding buffer in the absence of BSA to remove unbound radioactivity.
5. Incubate the cells with 1 ml of Earle's balanced salt solution (without bicarbonate) containing 20 mM Hepes, pH 7.5, and 0–1 mM BS^3, a water-soluble homobifunctional cross-linker, for 10 min at 4°C.[a,b,c]
6. Quench the reaction with 10 mM Tris, pH 7.5, wash the monolayers twice and solubilize the cells with a buffer containing 0.5% NP-40 and protease inhibitors.
7. Centrifuge at high speed and immunoprecipitate the clarified supernatant with non-immune serum or PDGF receptor-specific antibody and subject the immnoprecipitate to non-reducing SDS–PAGE/ autoradiography using a 3.5–10% gradient gel.

[a] The presence of a 370–390 kDa band in the immunoprecipitates from samples treated with the cross-linking reagent and its absence in samples not incubated with BS^3 suggest that ligand binding results in receptor dimerization.
[b] To demonstrate the specificity of the cross-linking reaction, cells should also be incubated with [^{125}I]PDGF BB in the presence of excess unlabelled PDGF BB. In this case, both the dimeric 370–390 kDa and the monomeric 180–190 kDa bands will be absent.
[c] BS^3 is available from Pierce Chemical Co.

4.2 Interaction between activated (phosphorylated) receptor and signalling molecules

4.2.1 Signal transducers with the SH2 domain

Under normal circumstances the tyrosine kinase activity of a receptor is dormant (due to negative regulation exerted by the exoplasmic domain). The residual kinase activity in the dimeric receptor is sufficient for cross-phosphorylation. This initial phosphorylation somehow activates the kinase. The consequences of the activation of the receptor kinase are the phosphorylation of its own tyrosine residues (known as autophosphorylation) followed by phosphorylation of the target molecules. In the EGFR, there are five phosphate acceptor sites that are clustered in the C-terminal tail of the receptor. However, the autophosphorylation sites are distributed throughout the cytoplasmic domain of the PDGFR. The basic strategy to identify the acceptor sites involves the phosphorylation of the highly purified receptor, tryptic digestion of the phosphorylated receptor, separation of the phosphopeptides by reverse phase HPLC, and sequencing the individual peptides. A basic method for the phosphopeptide analysis of the EGF receptor, which is routinely used in my laboratory, is described in *Protocol 7*.

Protocol 7. Phosphopeptide analysis of the EGF receptor by HPLC

Equipment and reagents
- solubilizing buffer (see *Protocol 3*)
- EGF–Affi–Gel 15
- 20 mM Hepes, pH 7.4 (containing 10% glycerol, 0.15 M NaCl, 0.2% NP-40, 1 mM sodium orthovanadate, and protease inhibitors)
- [γ-^{32}P]ATP (20 μCi/μl)
- 50 μM unlabelled ATP
- ^{32}P-labelled EGF receptor

- 5% acetic acid
- 25% methanol
- 0.05 M ammonium bicarbonate
- trypsin
- 0.1% TFA (trifluoroacetic acid)
- reverse phase HPLC C$_{18}$ column (column size : 3.9 mm 150 mm)
- X-ray film
- scintillation counter

Method

1. Solubilize membranes (50 μg of protein) with 100 μl of solubilizing buffer (see Protocol 3) and prepare the high speed supernatant as described.

2. Incubate the supernatant with 20 μl of EGF–Affi–Gel 15 at 4°C for 60 min with gentle stirring.

3. Wash the gel beads three times with the solubilizing buffer (100 μl per wash) to remove unbound proteins.

4. Wash the gel beads one more time with 100 μl of 20 mM Hepes, pH 7.4, containing 10% glycerol, 0.15 M NaCl, 0.2% NP-40, 1 mM sodium orthovanadate, and protease inhibitors.

5. Suspend the beads in 15 μl of the above buffer. Add 5 μl of [γ-^{32}P]ATP and 1 μl of 50 μM unlabelled ATP.

6. Incubate at 4°C for 60 min with gentle stirring.

7. Centrifuge and wash the beads four times (100 μl per wash) with 20 mM Hepes, pH 7.4, containing 0.15 M NaCl, 0.1% NP-40, and 1 mM vanadate.

8. Elute the ^{32}P-labelled EGF receptor by suspending the beads in 100 μl of SDS–sample buffer and heating at 100 °C for 3 min.

9. Centrifuge and analyse the supernatant by SDS–PAGE (7% gel). Run pre-stained standard proteins on each side of the labelled receptor.

10. Fix the gel overnight at the end of the run in a large volume of 5% acetic acid containing 25% methanol.

11. Discard the fixing solution and rinse the gel once with fresh fixing solution.

12. Remove the gel from the glass tray and wrap it carefully with Saran wrap (cling film).

13. Expose the gel for a very short time (15–45 min) at room temperature to an X-ray film to identify the labelled receptor.

14. Excise the receptor band from the wet gel and transfer it on to a Petri dish containing ~10 ml of 25% methanol.

15. Transfer the gel after incubation with methanol for 60 min to an Eppendorf tube (1.5 ml), lyophilize in a Speed Vac, and determine the Cerenkov counts (counts in the absence of scintillation fluid) in a scintillation counter.

16. Add 0.5 ml of .05 M ammonium bicarbonate, pH 7.8, and mash the gel with a disposable pestle that fits snugly in the Eppendorf tube.[a]

17. Add sequencing grade trypsin to the suspension to a final concentration of 10–20 μg/ml and incubate at 37 °C for 20 h.

18. Add fresh trypsin to the reaction mixture and incubate for another 20 h.

19. Centrifuge in a microfuge for 10 min and save the supernatant.

20. Extract the pellet three times with water (0.3 ml/extraction) to recover the radioactivity that is trapped in the pellet and pool the supernatants with the first supernatant.

21. Centrifuge the pooled supernatant to remove trace amount of gel beads that might be present in the gel extract.

22. Lyophilize the supernatant and determine the Cerenkov count to estimate the % of recovery (~65–80% in our hands).

23. Dissolve the lyophilized material in 0.1% trifluoroacetic acid in water and inject an aliquot on to a reverse phase HPLC column previously equilibrated in the same solvent system.

24. Wash the column with 10 ml of 0.1% trifluoroacetic acid in water and elute the phosphopeptides with a 0–60% acetonitrile gradient containing 0.1% trifluoroacetic acid with a flow rate of 1 ml/min.

Protocol 7. *Continued*

25. Collect 0.5 ml fractions, and determine the Cerenkov counts after drying the fractions by lyophilization.

[a]Ammonium bicarbonate solution and all other solutions used for HPLC analysis should be filtered using 0.2–0.5 μm pore size to remove dust particles.
Trapped air from all the solutions and solvents should be removed under vacuum.
Another method to purify the receptor is by immunoaffinity chromatography using antibodies that do not block either EGF binding or kinase activity.

The phosphotyrosine residues of the activated receptor kinases act as docking sites for the target molecules. If cells in culture are treated with a growth factor (GF), and then its receptor is immunoprecipitated with a receptor-specific antibody, the immunoprecipitate contains not only the receptor but also a host of other proteins. However, these proteins are not present when immunoprecipitation is carried out from control cell lysate that has not been treated with the GF. This suggests that these proteins are specifically associated with the activated and hence phosphorylated receptor and not with the dormant receptor. Most of these cellular proteins are enzymatic in nature, such as phospholipase C_γ (PLC$_\gamma$), the regulatory subunit (p85) of PI3K, GTPase-activating protein of *ras* (GAP), as well as c-*raf*, and members of the Src family of protein tyrosine kinases, such as pp60^{c-src}, p62^{c-yes}, and p59^{fyn} (7). Association of these proteins, also known as signal transduction molecules, with an activated kinase results either in the phosphorylation or conformational alteration of these molecules. Such modifications result in activation of these enzymes, leading to changes in gene expression and a change in the phenotypic state of the cell. Although these proteins are not structurally or functionally related to each other, they all contain a common domain known as the Src homology 2 (SH2) domain. This domain is ~100 amino acids long and is responsible for binding with phosphotyrosine. In addition to phosphotyrosine, the specifity of the interaction is also determined by the amino acid sequence C-terminal to phosphotyrosine. The target molecules inside the cell are distributed throughout the cytoplasm, and hence their local concentration is very low. By binding to the activated transmembrane receptor, the cytosolic signalling molecules are translocated to the plasma membrane. Under normal circumstances, the receptor proteins are distributed throughout the plasma membrane. However, in the presence of the ligand, these receptors are clustered in the coated pits. In other words, the local concentration of the receptors is several fold higher in the GF-treated cells. Hence, the local concentration of the receptor-associated target molecules is also very high. In addition, the physical association of the signalling molecules with a receptor tyrosine kinase can markedly lower the K_m for phosphorylation, making the SH2-containing proteins preferred substrates. Furthermore, such association may alter the conformation of the signalling proteins, directly stimulating their activity, as has been shown with the Src family kinases and PI3K.

4.2.2 Signal transducers with the SH3 domain

In addition to SH2 domains, another conserved motif of 50–75 residues, the SH3 domain, has also been identified in some proteins. The SH2 and SH3 domains are frequently found within a single protein, although certain proteins contain only one of them. Proteins with enzymatic activities such as PLC_γ and GAP contain only a single SH3 domain, but the adaptor proteins Crk and Grb2, which lack catalytic sequences, contain two SH3 domains. The SH3 domain binds to a proline-rich region of a protein. This domain of cytoplasmic tyrosine kinases acts as a negative regulator of kinase activity (29). In all Src family kinases, the N-terminal myristoylation site, which is needed for membrane anchoring, is followed by one SH3 domain, one SH2 domain, and a C-terminal kinase site and a tyrosine residue (527 in $pp60^{Src}$ and 505 in $pp56^{lck}$) at the extreme C-terminus tail. Phosphorylation of this tyrosine residue, thought to be mediated by a distinct tyrosine kinase, Csk of M_r 50 k, leads to intramolecular interaction between the phosphotyrosine residue and the SH2 domain. As a result of this interaction and subsequent conformational alteration, the SH3 domain comes in contact with a small lobe of the kinase site and thereby causes its inhibition. Dephosphorylation of this tyrosine, such as by phosphatase CD45 in activated T cells, reverses these events and frees the SH2 domain to bind to activated receptor kinase. The association between the activated receptor and Src results in the phosphorylation of tyrosine-414 in $pp60^{Src}$ (and 394 in $pp56^{lck}$) located in the kinase domain and subsequent activation of Src kinase. In v-Src, the C-terminal tail along with its negative regulatory tyrosine-527 is deleted, and as a result it is a constitutively active kinase.

4.2.3 Other molecules in signal transduction

It has been known for quite some time that a series of Ser/Thr kinases, including mitogen-activated protein (MAP) kinases, are stimulated by growth factors and such activation results in the activation of a number of transcription factors such as *fos*, *jun*, *myc*. A small GTPase, known as Ras GTPase, plays a major role in the MAP kinase activation. Ras is a 21 kDa membrane-anchored protein originally identified as an oncogene in human bladder, lung, and colon carcinomas. Ras exits in two forms—an inactive GDP-bound form and active GTP-bound form. The molecular basis for this activation process became clear with the identification of the second generation signal transducers, such as Grb2, Sos, and Shc.

Grb2 (growth factor receptor-binding protein 2) is a mammalian homologue of a protein encoded in *C. elegans* by *Sem-5*. It is a small protein (23–26 kDa) consisting entirely of two SH3 domains flanking an SH2 domain and it functions as an adaptor that assembles other proteins into a multiprotein complex. *Sos* (son of sevenless) gene has been shown to function downstream of *Sevenless*, a receptor tyrosine kinase gene critical for induction of R7 photoreceptor neurons in *Drosophila*. A mammalian homologue of the *Drosophila Sos* gene has been identified. It encodes a protein of 150 kDa with a proline-rich sequence in the C-terminus and a catalytic domain in the middle of the molecule. The N-terminus of Sos contains two potentially important motifs, a Dbl homology domain and a pleckstrin homology (PH) domain (30). The catalytic domain acts as a guanine nucleotide exchange factor (GNEF) for small GTPases, such as Ras. In the cytoplasm, two SH3 domains of Grb2 constitutively form a stable complex with the proline-rich C-terminal region of mammalian Sos. The SH2 domain of the Grb2–Sos complex binds to the phosphorylated receptor, thus recruiting soluble Sos to the plasma membrane. Therefore, translocation of Sos to the plasma membrane draws Sos into proximity with the inactive Ras protein, which is membrane bound. Sos then activates Ras by catalysing GDP/GTP exchange.

The alternative pathway to Ras activation is through Shc (Src homology 2 with collagen-like motif), recently renamed as ShcA. Three isoforms of Shc have been specifically identified; p52 and p46 result from differential translation initiation at two proximal ATG sites, while p66 is a splicing isoform, containing the p52 sequences and a unique glycine/proline-rich amino terminal region. These proteins are modular molecules characterized by an N-terminal phosphotyrosine-binding domain (PTB), a central proline/glycine-rich region (CH1) containing the unique Grb2-binding site, and a C-terminal SH2 domain (31). Although PTB and SH2 domains bind to phosphotyrosine residues of a peptide (or protein), their binding specificities are determined by the amino acid sequence surrounding the phosphotyrosine residue. In the case of the SH2 domain, the binding to phosphotyrosine is influenced by the sequence downstream of the tyrosine residue. However, the PTB domain has high specificity for sequences that are N-terminal to the tyrosine residue (Asn-Pro-X-Tyr motif). Thus, Shc binds to the phosphotyrosine of the activated receptor and becomes phosphorylated at Tyr-317 in the CH1 region. The phosphotyrosine of Shc then function as an alternative binding site for the SH2 domain of the Grb2 molecule. Formation of the Shc–Grb2–Sos complex results in Ras activation. Several lines of evidence suggest that Shc–Grb2 association after GF stimulation is involved in Ras activation. (i) Overproduction of Shc in PC12 cells induces neurite extension, a response that is also produced by NGF stimulation and is normally dependent on the Ras pathway; (ii) the Shc effect on PC12 cells is blocked by synthesis of a dominant negative Ras mutant; (iii) Shc overproduction induces a transformed phenotype in mouse fibroblasts; and (iv) mutant Shc lacking Y317 (major phosphorylation site) cannot bind to the SH2 domain of Grb2 *in vivo* and *in vitro*, and does not induce neoplastic transformation. The domain structures of Grb2, Sos, and ShcA are shown in *Figure 1*.

Ras activation by two different adaptor molecules suggests that two pro-

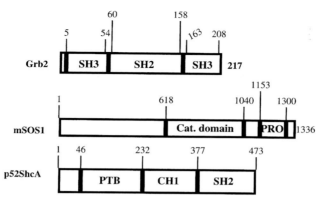

Figure 1. Schematic structures of Grb2, Sos, and ShcA. Cat. Domain: catalytic domain (GDP→GTP exchange); CH1: collagen homology 1; PTB: phosphotyrosine-binding domain; PRO: proline-rich sequence; SH 2/3: Src homology domain 2/3. m= mammalian

teins induce the elevation of Ras–GTP formation to different extents, which results in either persistent or transient MAP kinase activation. EGF stimulates the direct binding of the Grb2–Sos complex to the receptor and NGF activation of the corresponding receptor (Trk) leads to Shc phosphorylation and formation of a Shc–Grb2–Sos complex without direct binding of Grb2 to the receptor. An alternative explanation for the existence of two adaptor molecules might be that Grb2 and Shc have other roles in addition to Ras activation. This hypothesis is strengthened by recent experiments (31).

In addition to Grb2, three other SH2 domain-containing proteins that are highly related to each other have been identified in the last few years. Grb7, Grb10, and Grb14 have highly homologous amino acid sequences and similar modular structures, including a C-terminal SH2 domain, a conserved proline-rich region within the N-terminus that fits the consensus sequence for the SH3-binding domain, a central PH domain involved in protein–protein and protein–lipid interactions, and a ~50-amino acid BPS domain (between the PH domain and the SH2 domain) (13; *Figure 2*). The amino acid sequence of the BPS domain is unique to these three proteins. All three adaptor proteins were originally cloned by screening cDNA expression libraries with a labelled tyrosine-phosphorylated C-terminal peptide of the EGF receptor. Grb7 is known to interact with the EGFR, ErbB2/neu, PDGFR, the Syp phosphatase, and the Shc adaptor protein; Grb14 binds *in vitro* with the PDGFR (cited in ref. 32). Grb10 binds very strongly with the IR and weakly with the EGFR and the IGF-1R (13). Such differential binding of Grb10 is due to the interaction of both the BPS and the SH2 domains with the IR, the SH2 domain alone with the EGFR, and the BPS domain alone with the IGF-1R (13). The roles played by these adaptor proteins in receptor tyrosine kinase-mediated signal transduction are not clear at present. However, studies conducted with

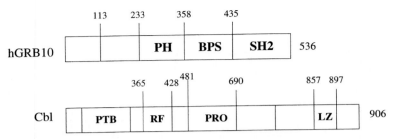

Figure 2. Domain structures of Grb10 and Cbl. BPS: between PH and SH2 domain; LZ: leucine zipper; PH: pleckstrin homology domain; RF: RING Finger.

a CHO cell line overexpressing one of the spliced variants of Grb10 (Grb10$_\alpha$), and rat1 fibroblasts overexpressing IR and microinjected with a fusion protein containing the BPS and the SH2 domain, appear to suggest that Grb10 acts as a negative regulator of insulin action (33, 34). Since Grb10 binds to two phosphotyrosine residues at 1150 and 1151, located in the activation loop of the insulin receptor kinase domain, it is likely that access of the substrates or ATP to the catalytic site is blocked by Grb10 binding (13). It remains to be seen how these results fit in with the recent finding that Grb10 interacts constitutively with Raf1 and, in an insulin-dependent fashion, with MEK1 (32).

4.2.4 PI3K in signal transduction

PI3K plays a major role in mitogenesis, actin reorganization, vesicular trafficking, and apoptosis. It is a heterodimeric protein composed of one catalytic subunit ($\sim$110–130 kDa) and one regulatory/adaptor subunit ($\sim$85 kDa). The p85 subunit contains a proline-rich sequence and two SH2 domains—one in the N-terminus and the other one in the C-terminus. Its activity is regulated by a number of receptor tyrosine kinases and cytokine receptors. The two SH2 domains of the adaptor subunit bind to the phosphotyrosine residues of the activated receptor and, as a result, the catalytic subunit is recruited to the plasma membranes. There it catalyses the phosphorylation of its lipid sustrates, producing phosphatidylinositol-3,4,5-trisphosphate [(PI)P3] which is then converted to phosphatidylinositol-3,4-biphosphate [(PI)P$_2$] by 5′-phosphatase. These two products then act as second messengers. Recent studies suggest that PI3K acts through a Ser/Thr kinase known as PKB/Akt (reviewed in refs 10 and 35). Because of the high homology of this protein of $\sim$60 kDa with PKC and PKA, it is known as PKB. This kinase was also identified as the product of the oncogene v-*akt* of the acutely transforming retrovirus AKT8, found in a rodent T cell lymphoma, and hence the name Akt. The retroviral oncogene encoded a fusion of the cellular Akt protein to the viral Gag structural protein. PKB/Akt is a multidomain protein of 480 amino acids and contains an Akt homology (AH) domain (amino acids 1–147) and a pleckstrin homology (PH) domain (1–106) in its N-terminus, the kinase domain (148–411), and a carboxy-terminal tail (412–480). The PH domains are sequence motifs that share homology with an internal repeat in pleckstrin, the major substrate of PKC in platelets, and are present in a large variety of proteins involved in intracellular signalling or cytoskeletal organization. The PH domains have been implicated in interactions with other proteins, including G-protein βγ subunits, and with lipids such as phosphoinositides. The membrane-bound products of PI3K, namely [(PI)P$_2$] and [(PI)P$_3$] bind to the PH domain of PKB/Akt, thereby partially activating Akt and recruiting it to the plasma membranes. The kinase activity of Akt is further increased by phosphorylation of Thr-308 in the kinase domain and Ser-473 in the C-terminal tail. These phosphorylations are mediated by PDK1 (phosphatidylinositide-dependent protein kinase 1) and PDK2, respectively (36). Activated PKB/Akt phosphorylates substrates, such as BAD, a protein involved in apoptosis, and glycogen synthase kinase (GSK3), resulting in a variety of biological effects, including suppression of apoptosis (cell survival) and tumorigenesis (10, 35, 36). The ability of PKB/Akt to promote cell survival may also contribute to its function as an oncogene. It should be mentioned in this context that PKB/Akt undergoes activation in some human tumours; 12% of ovarian cancer, 3% of breast carcinomas, and 10–20% of pancreatic carcinomas undergo amplification in PKB/Akt gene (37).

4.2.5 Role of Cbl in receptor tyrosine kinase regulation

Uncontrolled activation of the receptor tyrosine kinases has been implicated in oncogenesis; conversely, lack of kinase activation results in developmental abnormalities and immunodeficiencies. Therefore, the tight regulation of tyrosine kinase is of paramount importance for appropriate signalling. Two biochemical mechanisms have previously been identified for negative regulation of receptor kinases. One way to control tyrosine kinase-mediated signalling is by removing phosphate moiety from the activated receptor by cellular phosphotyrosine phosphatase; the other mechanism is by ligand-induced removal of the cell surface receptors (receptor downregulation). Recently, a cytoplasmic protein, Cbl, that negatively regulates activated receptors has been identified (38). Cbl is a cellular homologue of a viral oncogene, v-*cbl*, product which is responsible for pre-B lymphomas and myeloid leukaemias in mice. Cbl is a multidomain, 120 kDa protein devoid of any catalytic activity (Fig. 2). The C-terminal part of Cbl (Cbl-C, aa 358–906) is composed of several functional motifs, including a RING finger (aa 365–428) which is thought to mediate protein–protein or protein–DNA inter-actions, a proline-rich region (aa 481–690) for interacting with SH3 domain-containing proteins, and a C-terminal leucine zipper involved in inter-molecular oligomerization; some of the tyrosine residues that are potential docking sites for SH2 domain-containing proteins are also located in this region. The N-terminal part of the molecule (Cbl-N, aa 1–357) contains a phosphotyrosine binding (PTB) domain which displays binding specificity different from that of other PTB domain-containing proteins, such as Shc and insulin receptor substrate (IRS) 1/2. Upon stimulation of cells by GFs, Cbl is associated, through its PTB domain, with the autophosphorylated receptor and is phosphorylated on multiple tyrosine residues. Overexpression of Cbl in mammalian cells results in decreased EGF receptor kinase activity and also reduced signalling by molecules downstream of the receptor. Conversely, expression of an oncogenic Cbl, generated either by deletion of Cbl-C or by small deletions N-terminal to the RING finger, results in hyperphosphory-lation and increased kinase activation of the PDGFRα and the EGFR. These results suggest that wild-type Cbl acts as a negative regulator of receptor kinase functions and the oncogenic Cbl reverses these effects. The molecular mechanism for Cbl-mediated kinase regulation becomes clear from recent studies with the PDGFR. The PDGFRα in cells overexpressing wild-type Cbl is ubiquitinated in a ligand-dependent fashion and the receptor in cells overexpressing Cbl undergoes a faster ligand-induced degradation compared with the control cells. Since ubiquitinated proteins are destined to be destroyed, ubiquitination may be one of the mechanisms by which activated PDGFR is regulated. It remains to be determined whether this is the general mechanism by which other receptor kinases are also regulated. It should be mentioned in this connection that Cbl itself is ubiquitinated and translocated to the plasma membrane in CSF-1-stimulated cells; however, it is not known whether CSF-1R is also ubiquitinated under these conditions.

4.2.6 Methodologies to identify signalling molecules that interact with the activated receptors

Protocol 8 describes a basic procedure to identify 120 kDa Ras–GAP as an intracellular protein that is associated with the activated PDGF receptor in intact cells (39). Different portions of the same blot can also be used to identify other receptor-associated proteins, such as PI3K (using an antibody directed to the p85 subunit), and 135 kDa PLC$_\gamma$. A slight variation of this method can be used to test the ability of a protein to interact with an activated receptor. For this purpose, a confluent culture of cells grown in DMEM–10% fetal bovine serum, is starved for 20 h by incubating at 37 °C with DMEM supplemented with 0.1% calf serum. The receptors from the lysed cells are incubated with an appropriate antibody, the immunocomplexes are isolated by protein A–Sepharose, and the receptors in the immunocomplexes are autophosphorylated with unlabelled ATP in the presence of the growth factor, or are left untreated. After washing, the immunoprecipitates are incubated with the protein under investigation, centrifuged to remove the unbound protein, and analysed by Western blot using a primary antibody directed to the interacting protein.

Protocol 8. Interaction of GTPase activating protein (GAP) with the activated β PDGF receptor

Reagents

- DMEM (Dulbecco's modified Eagle's medium)
- 10% fetal bovine serum
- 0.1% calf serum
- PDGF BB
- 20 mM Hepes, pH 7.4
- 0.15 M NaCl
- anti-β PDGF receptor antibody

- EB buffer: 1% Triton, 5 mM EDTA, 50 mM NaCl, 50 mM NaF, 2 mM sodium vanadate, 0.1% BSA, 1% aprotinin, 10 mM Tris, pH 7.6, 1 mM PMSF
- SDS–sample buffer: 1% SDS, 2.5% 2-mercaptoethanol, 10% glycerol, 100 mM Tris, pH 6.8, and 0.01% bromophenol blue
- anti-GAP antibody

Method

1. Grow cells in p100 dishes to confluency in DMEM supplemented with 10% fetal bovine serum.

2. Remove the medium and incubate the cells overnight in DMEM supplemented with 0.1% calf serum (cells become quiescent under these conditions).

3. Add 50 ng/ml PDGF BB to one dish and incubate at 37 °C for 5 min and use the second dish as a control.

4. Transfer the dishes immediately on to ice and wash the monolayer twice with ice-cold 20 mM Hepes, pH 7.4, containing 0.15 M NaCl.

5. Add 400 μl of EB buffer to each dish, scrape the cells with a rubber policeman rod, and transfer the solubilized cells to a microfuge tube on ice.

6. Rinse each of the culture dishes with an additional 400 μl of EB buffer and combine with the first extract.

7. Incubate the cell extracts on ice for 20 min with occasional vortexing and centrifuge at 14 000 r.p.m. for 15 min.

8. Incubate aliquots of both the stimulated and control cell lysates with an anti-β PDGF receptor antibody at 4 °C for 3–4 h.

9. Isolate the immunocomplex by incubating the reaction mixture with formaldehyde-fixed *S. aureus* at 4 °C for 60 min with constant vortexing.

10. Centrifuge and wash the bacterial pellet exhaustively to remove proteins which are non-specifically associated with the receptor.

11. Dissociate the receptor and the receptor-associated proteins from the immunocomplex by heating at 95 °C for 3 min with SDS–sample buffer.

12. Resolve the PDGF receptor and the associated proteins on SDS–PAGE

Protocol 8. *Continued*

> using 7.5% polyacrylamide gel, transfer the proteins electro-
> phoretically on to Immobilon (Millipore) and probe the blot with an
> anti-GAP antibody using a dilution of the antibody recommended by
> the manufacturer.
>
> 13. Identify the GAP–anti-GAP antibody complex by a secondary
>   antibody conjugated to alkaline phosphatase.

The presence of a 120 kDa band in the Western blot in PDGF-stimulated cells, and its absence
in the unstimulated cells, suggests the high affinity interaction between the activated PDGF
receptor and Ras–GAP.

5. Autophosphorylation results in conformational changes in the receptor

As phosphorylation introduces negative charges to a protein molecule, it is
highly likely that autophosphorylation should have a profound effect on the
receptor conformation. Our earlier studies with anti-receptor antibodies
suggested such autophosphorylation-induced structural alteration of the
growth factor receptors. While working with anti-peptide antibodies, we made
a serendipitous finding that allowed us to identify such a site. We have
previously reported the generation of a conformation-specific antibody to the
PDGFRβ. This anti-peptide antibody (AbP2), directed to amino acid residues
964–979(Glu–Gly–Tyr–Lys–Lys–Lys–Tyr–Gln–Gln–Val–Asp–Glu–Glu–Phe
–Leu–Arg) of the cytoplasmic domain of the human β receptor recognizes
specifically the phosphorylated receptor and not the unphosphorylated
receptor (40). Although the immunoprecipitation is inhibited by the peptide,
no such inhibition is seen with phenyl phosphate, an analogue of phosphotyrosine, suggesting that the antibody recognizes the phosphorylated
protein and not phosphotyrosine. A similar finding has also been reported by
another group with an anti-peptide antibody to PDGFRβ designated as Ab 83
(41). This indicates that receptor phosphorylation uncovers the antigenic
determinant and suggests a phosphorylation-induced conformational change
of the receptor. Subsequent to this finding, a number of anti-peptide
antibodies that specifically recognize activated receptors have been reported;
however, all of those antibodies are directed to phosphotyrosine-containing
peptides (42–44). In this respect, AbP2 (also Ab 83) is one of the only two
conformation-specific anti-receptor antibodies directed to an unphosphorylated peptide that recognizes the activated receptor; similar unmasking
of a peptide epitope following autophosphorylation has also been reported for
the insulin receptor (45). In addition to PDGFRβ, AbP2 also binds to the
EGFR and, interestingly, the recognition is also phosphorylation dependent
(37). In the P2 peptide, there are two tripeptide sequences, Asp–Glu–Glu and

Tyr–Gln–Gln, that are also present in the cytoplasmic domain of the EGFR at 979–981 and 1148–1150, respectively. Our earlier studies on antibody inhibition by P2-derived peptides (46) and our recent studies (12) conducted in collaboration with Dr Laura Beguinot (H. S. Rafaele, Milan, Italy), using EGFR C-terminal deletion mutants and point mutations (Tyr→Phe), suggest that Gln–Gln in combination with Asp–Glu–Glu forms a high affinity complex with AbP2 and such complex formation is dependent on tyrosine phosphorylation. Of the five phosphate-acceptor sites in the EGF receptor, clustered in the extreme C-terminal tail, phosphorylation of three tyrosine residues, 992, 1068, and 1086 located between Asp–Glu–Glu and Gln–Gln is necessary for AbP2 binding. In contrast, two more acceptor sites, Tyr-1173 and -1148, play no role in the conformation change. Asp–Glu–Glu and Gln–Gln are located 169 amino acids apart and it is highly likely that the interactions between three negatively charged phosphotyrosine residues in the receptor C-terminal may result in the bending of the peptide chain in such a way that these two peptides come close to each other to form an antibody-binding site. Such a possibility is also supported by our finding (12) that receptor dephosphorylation results in complete loss of AbP2 binding activity. This indicates that the conformation of a domain in the EGFR is positively regulated by phosphorylation of Tyr-992, -1068, and -1086.

Protocol 9. Reversibility of the phosphorylation-induced conformational change of the EGF receptor

Reagents

- methionine- and cysteine-free DMEM
- 2% dialysed fetal bovine serum
- Tran ^{35}S-label (ICN Pharmaceuticals Inc.) (100 μCi/ml, 1190 Ci/mmol)
- 20 mM Hepes, pH 7.4
- 0.15 M NaCl
- 20 μM unlabelled ATP
- 10 mM EDTA, pH 7.4

- solubilizing solution: 1% NP-40 in 20 mM Hepes, pH 7.4, 0.15 M NaCl, 10% glycerol and protease inhibitors (aprotinin, leupeptin, and PMSF)
- 40 mM phenyl phosphate
- wheat germ agglutinin–agarose
- 0.4 M *N*-acetyl glucosamine
- BSA

Method

1. Grow human epidermoid carcinoma cells, A431, for 20 h in p35 dishes (~10 cm^2 area).

2. Wash the cell monolayers with methionine- and cysteine-free DMEM containing 2% dialysed fetal bovine serum and pre-incubate in the same medium at 37 °C for 1 h.

3. Remove the medium and incubate the cells at 37 °C for 10 h with Tran ^{35}S-label in methionine- and cysteine-free DMEM containing 2% dialysed fetal bovine serum.

Protocol 9. *Continued*

4. Wash the cell monolayers three times with cold 20 mM Hepes, pH 7.4, containing 0.15 M NaCl and solubilize.

5. Centrifuge the lysed cells in a microfuge for 10 min in the cold and collect the clarified supernatant.

6. Incubate an aliquot of the supernatant with 1 μM EGF in 20 mM Hepes, pH 7.4, containing 10% glycerol, 0.15 M NaCl, 0.2% NP-40, 1 mM vanadate, 0.5 mM $MnCl_2$, and protease inhibitors at 4°C for 30 min.

7. Phosphorylate the activated receptor by incubating the reaction mixture at 4°C for 60 min with 20 μM unlabelled ATP.

8. Terminate the reaction by adding 10 mM EDTA, pH 7.4, and purify the tyrosine-phosphorylated EGF receptor by incubating the reaction mixture with 20 μl of monoclonal antibody to phosphotyrosine (such as 1G2) coupled to Sepharose (10 mg/ml antibody) at 4°C for 90 min.

9. Centrifuge and wash the gel beads three times (~200 μl per wash) to remove unbound radioactivity.

10. Elute the bound receptor by incubating the washed beads at 4°C for 30 min with 50 μl of 40 mM phenyl phosphate in 20 mM Hepes, pH 7.4, containing 0.15 M NaCl and 0.2% NP-40, and collect the dissociated receptor by centrifugation.

11. Repeat the elution one more time and pool both supernatants.

12. Incubate the purified receptor with 25 μl of wheat germ agglutinin–agarose at 4°C for 60 min, wash the beads three times, and elute the bound receptor with 0.4 M *N*-acetyl glucosamine in the above buffer.

13. Incubate an aliquot of the wheat germ agglutinin-purified receptor in the presence of 1 mg/ml BSA for 60 min at 4°C with (50 units/ml) or without alkaline phosphatase coupled to agarose.

14. Collect the supernatants after centrifugation and immunoprecipitate the samples with a conformation-specific antibody (such as AbP2).

15. Isolate the immunocomplex by protein A–Sepharose, dissociate the receptor by SDS–sample buffer, and subject to SDS–PAGE/fluorography.[a,b]

[a] Absence of the receptor band in the alkaline phosphatase-treated sample and its presence in the control sample suggest that the phosphorylation-induced conformational change is reversible.
[b] To demonstrate that the lack of antibody binding upon alkaline phosphatase treatment is due to the loss of phosphate groups, both the enzyme treated and untreated samples should also be immunoprecipitated with anti-phosphotyrosine antibody and analysed by SDS–PAGE.

6. Growth factor receptors in oncogenesis

Recent research has shown that perturbation of a growth factor/receptor system can lead to cell transformation and, in an animal, to tumorigenesis. For example, genetic deletion of portions of the EGFR and CSF-1R have generated the v-*erb-B* and v-*fms* oncogenes, respectively. The kinase activity of these oncogene-encoded truncated receptors are no longer regulated by their respective ligands and, as a result, they are constitutively active. Overexpression of a normal growth factor receptor gene can also produce transformation. For example, EGFR gene amplification has been associated with a number of neoplasms, including breast carcinoma, adenocarcinoma and squamous cell carcinoma of the lung, large cell carcinoma of the lung, gliomas, and a variety of bladder and gynaecological tumours (47). Several clinical and histopathological studies have shown that EGFR gene amplification is related to a shorter interval to relapse and poorer survival. Nearly 50% of grade IV gliomas (glioblastoma multiforme) have amplified EGFR genes. In the majority of such cases, the EGFR gene amplification is correlated with structural rearrangement of the gene, resulting in in-frame deletions that preserve the reading frame of the receptor message. To date, three truncated forms of EGFR have been identified. The type I deletion mutant lacks the majority of the extracellular domain and is unable to bind EGF. The type II deletion mutant contains an in-frame deletion of 83 amino acids (amino acid residues 520–603) and is capable of binding EGF. The type III deletion mutant, which occurs in 17% of the glioblastomas, appears to be the most prevalent. This deletion mutant is characterized by an 801 bp in-frame deletion resulting in the removal of NH_2-terminal amino acid residues 6–273 from the extracellular domain of the intact 170 kDa EGFR. This receptor (EGFRvIII), like the type I receptor, fails to bind EGF. However, it displays high tyrosine kinase activity and cells expressing this receptor are highly tumorigenic (48). In addition to EGFR, other members of the EGFR family are also found overexpressed in various tumour cells. *Erb-B-2* (*neu*) is frequently and specifically amplified in breast and ovarian cancers. In fact, amplification/overexpression of *neu* was found to correlate with poor clinical prognosis in cancer patients. Furthermore, a point mutation in the transmembrane domain ($Val^{664} \rightarrow Glu$) renders ErbB2, a protein of 185 kDa, constitutively active. This constitutively active kinase is capable of Shc phosphorylation and MAP kinase activation (35).

Overexpression of PDGF or its receptor has also been correlated with oncogenesis. Most of the human gliomas screened so far are known to express both types of PDGF; these cells also express PDGFα and PDGFβ receptors. Thus, the endogenously produced PDGF possibly leads to autocrine growth stimulation of tumour cells (5). Fusion of the PDGFRβ gene with other genes, such as Tel (a transcription factor), Cev14, and huntingtin interacting protein 1 (HIP1), has also been implicated in the pathogenesis of haematopoietic malignancies (49–51). In these fusion proteins, the receptor kinase is constitutively active.

References

1. Heldin, C-H. and Westermark, B. (1984). *Cell*, **37**, 9.
2. Cross, M. and Dexter, T.M. (1991). *Cell*, **64**, 271.
3. Fiore, P. P. D., Pierce, J. H., Kraus, M. H., Segatto, O., King, C. R., and Aaronson, S. A. (1987). *Science*, **237**, 178.
4. Massague, J. and Pandiella A. (1993). *Annual Review of Biochemistry*, **62**, 515.
5. Heldin, C-H. (1992). *EMBO Journal*, **11**, 4251.
6. Fantl, W. J., Johnson, D. E., and Williams, L. T. (1993). *Annual Review of Biochemistry*, **62**, 453.
7. Heldin, C-H. (1996). *Cancer Surveys*, **277** (Cell Signalling), 7.
8. Alroy, I. and Yarden, Y. (1997). *FEBS Letters*, **410**, 83.
9. Kanakaraj, P., Raj, S., Khan, S. A., and Bishayee, S. (1991). *Biochemistry*, **30**, 1761.
10. Heldin, C-H., Otsman, A., and Ronnstrand, L. (1998). *Biochemica et Biophysica Acta*, **1378**, F79.
11. Zhou, R. (1998). *Pharmacology and Therapeutics*, **77**, 151.
12. Bishayee, A., Beguinot, L., Bishayee, S. (1999). *Molecular Biology of the Cell*, **10**, 525–536.
13. He, W., Rose, D. W., Olefsky, J. M., and Gustafson, T. A. (1998). *Journal of Biological Chemistry*, **273**, 6860.
14. Hamilton, J. A. (1997). *Journal of Leckocyte Biology*, **62**, 145.
15. Vosseller, K., Stella, G., Yee, N. S., and Besmer, P. (1997). *Cell*, **8**, 909.
16. Mohammadi, M., Dikic, I., Sorokin, A., Burgess, W. H., Jaye, M., and Schlessinger, J. (1996). *Molecular and Cellular Biology*, **16**, 977.
17. Wolf, D. E., McKinnon-Thompson, C., Daou, M-C., Stephens, R. M., Kaplan, D. R., and Ross, A. H. (1998). *Biochemistry*, **37**, 3178.
18. Zisch, A. H., Kalo, M. S., Chong, L. D., and Pasquale, E. B. (1998). *Oncogene*, **16**, 2657.
19. Bennett, G. L. and Horuk, R. (1997). In *Methods in enzymology* (ed. ??), Vol. 288, p. 134. Academic Press, New York.
20. Scatchard, G. (1949). *Annals of New York Academy of Sciences*, **51**, 660.
21. Carpenter, G. (1985). In *Methods in enzymology*, Vol. 109, p. 101. Academic Press, New York.
22. Cohen, S., Ushiro, H., Stocheck, C., and Chinkers, M. (1982). *Journal of Biological Chemistry*, **257**, 1523.
23. Blechman, J. M., Lev, S., Barg, J., Eisenstein, M., Vaks, B., Vogel, Z., Givol, D., and Yarden, Y. (1995). *Cell*, **80**, 105.
24. Omura, T., Heldin, C-H., and Ostman, A. (1997). *Journal of Biological Chemistry*, **272**, 12676.
25. Lokker, N. A., O'Hare, J. P., Barsoumian, A., Tomlinson, J. E., Ramakrishnan, V., Fretto, L. J., and Giese, N. A. (1997). *Journal of Biological Chemistry*, **272**, 33037.
26. De Vos, A. M., Ultsch, M., and Kossiakoff, A. (1992). *Science*, **255**, 306.

27. Bishayee, S., Majumdar, S., Khire, J., and Das, M. (1989). *Journal of Biological Chemistry*, **264**, 11699.
28. Heldin, C-H., Ernlund, A., Rorsman, C., and Ronnstrand, L. (1989). *Journal of Biological Chemistry*, **264**, 8905.
29. Pawson, T. (1997). *Nature*, **385**, 582.
30. Qian, X., Vass, W. C., Papageorge, A. G., Anborgh, P. H., and Lowy, D. R. (1998). *Molecular and Cellular Biology*, **18**, 771.
31. Bonfini, L., Migliaccio, E., Pelicci, G., Lanfrancone, L., and Pelicci, P. (1996). *Trends in Biochemical Sciences*, **21**, 257.
32. Nantel, A., Mohammad-Ali, K., Sherk, J., Posner, B. I., and Thomas, D. Y. (1998). *Journal of Biological Chemistry*, **273**, 10475.
33. Liu, F. and Roth, R. (1995). *Proceedings of the National Academy of Sciences USA*, **92**, 10287.
34. O'Neill, T. J., Rose, T. W., Pillay, T. S., Hotta, K., Olefsky, J. M., and Gustafson, T. A. (1996). *Journal of Biological Chemistry*, **271**, 22506.
35. Porter, A. C. and Vaillancourt, R. R. (1998). *Oncogene*, **16**, 1343.
36. Alessi, D. R. and Cohen, P. (1998). *Current Opinion in Genetics & Development*, **8**, 55.
37. Ruggeri, B. A., Huang, L., Wood, M., Cheng, J. Q., and Testa, J. R. (1998). *Molecular Carcinogenesis*, **21**, 81.
38. Miyake, S., Lupher, M. L., Druker, B., and Band, H. (1998). *Proceedings of the National Academy of Sciences USA*, **95**, 7927.
39. Kazlauskas, A., Kashishian, A., Cooper, J. A., and Valius, M. (1992). *Molecular and Cellular Biology*, **12**, 2534.
40. Bishayee, S., Majumdar, S., Scher, C.D., and Khan, S. (1988). *Molecular and Cellular Biology* **8**, 3696.
41. Keating, M. T., Escobedo, J. A., and Williams, L. T. (1988). *Journal of Biological Chemistry*. **163**, 12805.
42. Bangalore, L., Tanner, A. J., Laudano, A. P., and Stern, D. (1992). *Proceedings of the National Academy of Sciences USA*, **89**, 11637.
43. Campos-Gonzalez, R. and Glenny, J. R. (1991). *Growth Factors*, **4**, 305.
44. Epstein, R. J., Druker, B. J., Roberts, T. M., and Stiles, C. D. (1992). *Proceedings of the National Academy of Sciences USA*, **89**, 10435.
45. Herrera, R. and Rosen, O. M. (1986). *Journal of Biological Chemistry*, **261**, 11980.
46. Panneerselvam, K., Reitz, H., Khan, S. A., and Bishayee, S. (1995). *Journal of Biological Chemistry*, **270**, 7975.
47. Harris, A. L., Nicholson, S., Sainsburg, R., Wright, C., and Fardon, C. (1992). *Journal of the National Institute Monograms*, **11**, 181.
48. Su Huang, H.-J., Nagane, M., Klingbeil, C. K., Lin, H., Nishikawa, R., Ji, X-D., Huang, C- M., Gill, G. N., Wiley, H. S., and Cavenee, W. K. (1997). *Journal of Biological Chemistry*, **272**, 2927.
49. Golub, T. R., Barker, G. F., Lovett, M., and Gilliland, D. G. (1994). *Cell*, **77**, 307.
50. Abe, A., Emi, N., Mitsune, T., Hiroshi, T., Marunouchi, T., and Hidehiko, S. (1997). *Blood*, **90**, 4271.
51. Ross, R. S., Bernard, O. A., Berger, R., and Gilliland, D. G. (1998). *Blood*, **91**, 4419.

6

Analysis of growth-signalling protein kinase cascades

ALESSANDRO ALESSANDRINI

1. Introduction

One of the main questions in signal transduction is how do cells take an extracellular signal and convert it into an intracellular one. The first step that enables this process involves receptors that bind to specific extracellular signals. Most receptors have intrinsic tyrosine kinase activity or have tyrosine kinase activity associated with them. This results in a series of phosphorylation events involving endogenous Ser/Thr kinases. The final steps in a cell's response to an extracellular signal may involve the reorganization of the cytoskeleton which may lead to motogenesis (cellular migrations) and the transcriptional activation of specific genes that may lead to mitogenesis and/or differentiation. This chapter will focus on the downstream components in signal transduction and, in particular, the MEK/ERK pathway.

2. Growth factors responsive pathways

Growth factors, such as platelet-derived growth factor (PDGF), epidermal-derived growth factor (EGF), fibroblast growth factor (FGF), insulin, insulin-like growth factor-1 (IGF-1), colony-stimulating factor-1 (CSF-1), nerve growth factor (NGF), etc., have been shown to activate signal transduction pathways, such as the MEK/ERK and Akt/PKB pathways. These growth factors may play an important role in cell growth and differentiation. Their receptors are described in Chapter 5. This chapter will deal specifically with the analysis of the MEK/ERK pathway, introducing techniques that can also be applied to the study of other pathways.

2.1 The MEK/ERK pathway (*Figure 1*)

MEK (<u>MAP</u>/<u>ERK</u> <u>k</u>inase) family members are key components in an intracellular signalling pathway called the ERK (<u>e</u>xtracellular-signal <u>r</u>egulated <u>k</u>inases)/MAP kinase pathway, implicated in the transition of cells from G_0 to

G_1 in the cell cycle (1–3). Members of this kinase cascade are highly conserved between species from yeast to mammals (3). In addition, proteins with sequence similarity to ERK/MAP kinases and MEK participate in other cellular signalling pathways; for instance those responding to stresses such as osmotic shock and UV-damaged DNA (4). The MAP/ERK kinase cascade is activated following stimulation of a wide variety of cell types with growth factors, hormones, or mitogens. Although there are seven known members of the MEK family of kinases at present, only two of them, MEK1 and MEK2, have been shown to play a role in the ERK/MAP kinase pathway (4–10). The binding of these various ligands to the appropriate cell surface receptor results in receptor activation, which in turn leads to GTP binding of Ras complexed to members of the Raf family of serine/threonine kinases. Translocation of Raf-1 to the plasma membrane by Ras is necessary for the activation of Raf-1 (11, 12). However, co-incubation of purified GTP-loaded Ras with c-Raf-1 fails to fully activate the latter, suggesting that a membrane-localized co-factor may be necessary. Raf family members then activate the kinases MEK1 and MEK2 by phosphorylating them on serines 218 and 222 for MEK1, or 222 and 226 in the case of MEK2. MEKs, as dual specificity kinases, subsequently activate their downstream targets, ERK-1 and ERK-2, by phosphorylating them on threonine and tyrosine. ERKs then phosphorylate both cytoplasmic substrates and nuclear transcription factors, which, thus modified, contribute to the early response of the cell after stimulation. The most likely candidates for such substrates are the 14-3-3 proteins (13, 14). The 14-3-3 proteins interact promiscuously with many different proteins, and the β isoform of 14-3-3 associates with the N-terminal regulatory domain of c-Raf-1 (15). Each 14-3-3 molecule can bind two c-Raf-1 molecules (reviewed in ref. 16). The 14-3-3 protein appears to allow c-Raf-1 molecules to interact more effectively when they are brought to the cell membrane by GTP-bound Ras. Oligomerization of c-Raf-1 is critical for activation of c-Raf-1 kinase activity (16). MEK1 and MEK2 are approximately 90% similar and 80% identical; the differences in these proteins being mainly in the amino-terminal region outside the kinase domain, and in a proline-rich region between conserved kinase domains IX and X (5, 9, 10). These differences between MEK1 and MEK2 may contribute to differences in interactions and specificity.

The contributions of MEK1 versus MEK2 in the ERK/MAP kinase pathway are not well defined. Recently, it has been shown that MEK2 is more highly expressed during mouse embryogenesis than MEK1, suggesting that the former may play a key role in development (17). However, relative MEK specificities towards these two known substrates in *in vivo* systems have not yet been rigorously examined. Differential activation of MEK1 and MEK2 by various Raf family members in HeLa cells has been described by Wu *et al.* (18), who find that MEK1 is activated *in vitro* by A-Raf, as well as Raf-1 and B-Raf; this is in contrast to MEK2, which is only phosphorylated by Raf-1 and B-Raf. Jelinek *et al.* (19) reported that immobilized Ras–Raf-1 and Ras–B-

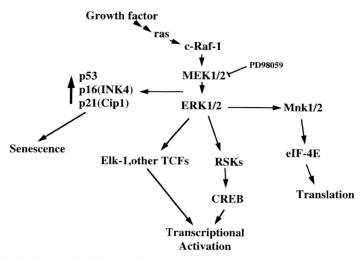

Figure 1. The MEK/ERK pathway. The MEK/ERK pathway is responsive to various growth factors and mitogens. GTP-loaded Ras recruits c-Raf-1 to the membrane. Activated c-Raf-1 then phosphorylates and activates MEK1 and MEK2, and, in turn, MEK1/2 phosphorylate and activate the ERKs. Substrates and pathways of the ERKs and the biological effects of their activation are shown. See text for details.

Raf complexes bind MEK1 but not MEK2. However, later studies indicated that both MEK1 and MEK2 interacted with Ras-bound Raf-1 through the proline-rich region in the kinase domain (20).

Another point that remains unclear is whether various external stimuli preferentially lead to activation of MEK1 versus MEK2 in cell lines and in tissue, or whether their functions are largely redundant. Stimulation of NIH-3T3, Rat1, and PC12 cells with TPA = (12-0-tetradecanoylphorbol-13-acetate), EGF, NGF, FGF, and PDGF failed to show a differential response between MEK1 and MEK2, suggesting that in some cell types, the functions of these two family members may be redundant. However, Downey *et al.* (21) have reported that MEK2 is more active than MEK1 after stimulation of neutrophils with chemotactic peptides and is sensitive to the PI-3 kinase inhibitor, wortmannin. Interestingly, this inhibitor, along with LY 294002, another PI-3 kinase inhibitor, has also been shown to inhibit the activation of proto-oncogene protein kinase B (PKB), also known as Akt/RAC. This family consists of Ser/Thr kinases that have also been been shown to respond to various growth factors. In addition, this pathway has been shown to be an anti-apoptotic pathway by phosphorylating Bad, resulting in the activation of Bcl2 (discussed below).

As mentioned previously, the activation of MEK1 and MEK2, leads to the phosphorylation and activation of the ERK/MAP kinases. ERKs then phosphorylate both cytoplasmic substrates, such as other kinases, and nuclear transcription factors, on specific amino acid sequences—(P/L)X(S/T)P. When modified, these proteins contribute to the early response of the cell after stimulation.

ERKs activate the p90 ribosomal S6 kinases, RSK1, RSK2, and RSK3, so named because their first identified substrate was the ribosomal S6 protein, although it has been shown that it is not an *in vivo* substrate. RSKs translocate into the nucleus and can also phosphorylate transcription factors (22, 23). RSK2 has been shown to play an important role in immediate early gene induction and mitogenesis since it phosphorylates the transcription factor, cAMP response element binding protein (CREB) at Ser-33, resulting in the increase expression of c-*fos* in response to some growth factors (24).

Recently, two additional ERK substrates have been identified and named MAP kinase-interacting serine/threonine kinase 1 and 2 (Mnk 1 and Mnk 2) (25, 26). Mnk1 phosphorylates eukaryotic initiation factor-4E (eIF-4E, also known as eIF-4a) at Ser-209. eIF-4E is a translation initiation factor which binds the 7-methyl-guanosine cap on all eukaryotic mRNAs. This protein plays an important role in the regulation of translation in mammalian cells (27).

Upon activation in response to growth factors, a portion of the ERK proteins translocate to the nucleus (28, 29). ERKs phosphorylate and activate a critical family of transcription factors, the ternary complex factors or TCFs (3, 30, 31). Elk-1 is a ternary complex factor that, when phosphorylated, forms a complex with SRF (serum response factor) and together bind to the promoter of a number of genes, including c-*fos*, that contain the serum response element, or SRE (32). These activated transcription factors play critical roles in the induction of immediate early genes and in the mitogenic response.

Expression of the D-type cyclins, the regulatory (activating) subunits for the cyclin-dependent kinase 4 and 6 (cdk4 and cdk6) catalytic subunits, increases in response to growth factor stimulation (reviewed in ref. 33). Expression of dominant negative mutants of MEK1 or ERK-1, or expression of the MAP kinase phosphatase, MKP-1, which dephosphorylates and inactivates the ERKs, inhibited growth factor-dependent expression of cyclin D1 (34–36). Activation of the MEK/ERK pathway also correlates with increased expression of cyclin E (the regulatory subunit of cdk2 which also promotes S phase progression) and decreased expression of the cyclin-dependent kinase inhibitor (cki) p27^{Kip1} (37). These data suggest that the Raf/ERK cascade may interact at multiple sites to promote cell cycle progression. Recently, it has also been shown that constitutively active MEK1 and oncogenic c-Raf-1 result in the upregulation of the tumour suppressor genes p53, p16(INK4), and p21(Cip1), in normal human fibroblasts, resulting in cell cycle arrest and senescence (38, 39).

2.2 The phosphoinositide-3 kinase pathways (*Figure 2*)

Many growth factors activate a kinase in cells that does not phosphorylate other proteins but phosphorylates phospholipids. This kinase, phospho-

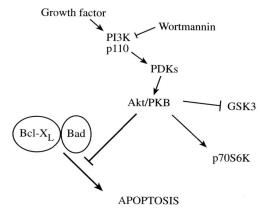

Figure 2. The PI3 kinase/Akt pathway. Various growth factors have been shown to activate the PI3 kinase/Akt pathway. Activation of Akt/PKB by phosphorylation by PKD1 and PKD2 results in inhibition of the Ser/Thr kinase, GSK3, activation of p70S6K, and the inhibiton of apoptosis by phosphorylating Bad and releasing Bcl2. See text for details.

inositide 3-kinase (PI3K), phosphorylates phosphoinositides, generating the second messengers, phosphatidylinositol-3,4-bisphosphate [PtdIns(3,4)P$_2$] and phosphatidylinositol-3,4,5-trisphosphate [PtdIns(3,4,5)P$_3$]. The activation of PI3K does result in the activation of the Ser/Thr kinase, PKB/Akt. For a more detailed description of PKB/Akt, see Chapter 5.

Once PKB/Akt is fully activated, it dissociates from the membrane to phosphorylate its cytoplasmic targets, such as glycogen synthase kinase-3 (GSK-3). Phosphorylation of GSK-3 inactivates it, allowing glycogen synthesis to proceed. PKB/Akt also phosphorylates and inactivates 4E-BP1, an eIF-4E-binding protein. Inactivation of 4E-BP1 releases eIF-4E, allowing mRNA translation to proceed (40).

Activation of the PKB/Akt pathway results in anti-apoptotic action (41–43). In fact, inhibiting PI3K blocks the ability of survival factors to protect cells from programmed cell death, and activated forms of PI3K and PKB/Akt protect cells from apoptosis induced by a number of factors. It is postulated that the anti-apoptotic effect of PKB/Akt may be explained by the phosphorylation of the pro-apoptotic protein, Bad (44, 45). Bad functions, in part, by binding to and inactivating the anti-apoptotic protein, Bcl-X$_L$. When Bad is phosphorylated, it dissociates from Bcl-X$_L$, allowing Bcl-X$_L$ to prevent cytochrome *c* release from mitochondria and thus prevent activation of the caspase cascade, leading to cell death .

This chapter will focus on laying out protocols that allow for the analysis of the MEK/ERK pathway, but modifications of these techniques can also be applied to the analysis of other pathways.

3. Techniques used to analyse protein kinases

In today's environment, the analysis of protein kinases has become routine. Various companies sell kits that allow one to test their kinase of choice. Here I outline an approach, though not the only one, whereby one begins to ask if a certain kinase is activated in response to a specific growth factor(s), and how to address the physiological significance of this activation.

3.1 Activation of MEK/ERK pathway in response to growth factors

3.1.1 *In vitro* kinase assay

To analyse if ERK, Akt, or whatever known kinase of interest is activated in response to certain growth factors, one must begin by starving the tissue culture cells for a period of time, then the cells are harvested, lysed in the appropriate buffer, the kinase is immunoprecipitated with a non-neutralizing antibody, and the kinase assay is performed (*Protocol 1*). Ordinarily, cells are grown in media containing 5–10% fetal bovine serum (FBS) or calf serum (CS). Starving cells in 0–0.1% serum for 24–48 hours is essential in order to obtain a significant kinase signal, since cells that are not starved exhibit basal level kinase activity and may mask the growth factor effect. By starving cells, they are synchronized at G_0. At this point, most growth factor-stimulated kinases are dormant.

Protocol 1. *In vitro* kinase assay—analysis of the MEK/ERK pathway

Equipment and reagents

- fetal bovine and calf serum
- PBS
- potassium phosphate buffer: 10 mM KPO_4, pH 7.05, 1 mM EDTA, 5mM EGTA, 10 mM $MgCl_2$
- 50 mM β-glycerophosphate
- 1 mM sodium vanadate
- 1 mM DTT
- 0.5% NP-40
- 0.1% Brij-35
- leupeptin

- 1 mM 4-(2-aminoethyl)benzenesulfonyl fluoride (Pefabloc®SC, Boehringer Mannheim)
- pepstatin A
- ST buffer: 100 mM NaCl, 10 mM Tris–HCl, pH 8.0
- $[\gamma^{32}P]ATP$
- sample buffer: 50 mM Tris–HCl, pH 6.8, 1% SDS, 10% glycerol, 0.05% bromophenol blue, 1% β PVDF = polyvinylidene difluoride 2-mercaptoethanol
- PVDF membrane
- X-OMAT AR film (Kodak)

Method

1. Grow cells in 100 mm tissue culture plates to 50% confluence in complete medium in the presence of 10% fetal bovine or calf serum (depending on the specific requirements of the cell line used).

---

2. Remove complete medium and replace it with medium containing 0–0.1% serum for 24–48 h. The amount of serum and the length of time should be determined for each cell line. I find that most cell lines can tolerate 0.1% serum for 48 h. Again, you should determine the conditions for each cell line used.

3. After 48 h, remove the medium and replace with complete medium for 5, 10, 15, 30, 60, and 120 min at 37 °C.[a]

4. Harvest cells after each time point by aspirating the medium with a Pasteur pipette and place the plates on ice and wash once with 1 ml ice-cold phosphate-buffered saline (PBS). Scrape the cells off in the 1 ml of PBS, transfer to an Eppendorf tube that is kept on ice, and spin in a microfuge at 1500 r.p.m. $\simeq$ 120 g at 4 °C.

5. Lyse the cell pellet in potassium phosphate buffer, 50 mM β-glycerophosphate, 1 mM sodium vanadate, 1 mM DTT, 0.5% NP-40, 0.1% Brij-35, 1 mM 4-(2-aminoethyl)benzenesulfonyl fluoride (Pefabloc®SC, Boehringer Mannheim), 10 μg per ml leupeptin, 10 μg per ml pepstatin A.

6. Centrifuge the lysates at 14 000 r.p.m. $\simeq$ 10 400 g for 10 min at 4 °C. Determine the protein concentration by the Bradford assay (Bio-Rad, 5000-006).

7. Perform immunoprecipitations by the addition of 10 μl anti-MEK1 antiserum (Santa Cruz), or 1 μg anti-MEK2 monoclonal antibody (UBI) to 400 μg of cell lysates. Incubate immunoprecipitates on ice for 1 h, and then add protein A–Sepharose beads (Zymed) rocking, for 30 min.

8. Wash immunoprecipitates three times with ST buffer.

9. Incubate the immunoprecipitates with 25 mM Tris–HCl, pH 7.5, 10 mM MgCl$_2$, 100 μM ATP, and 5 μCi [γ^{32}P]ATP, with 2 μg of GST-ERK1(K63M) (kinase-inactive) in a total volume of 30 μl, for 20 min at 30 °C.

10. Boil the samples in 1 sample buffer for 5 min, and then electrophorese on a 10% SDS–polyacrylamide gel. Transform the proteins to PVDF membrane, and then expose to X-OMAT AR film (Kodak).[b]

---

[a] In step 3, instead of adding complete medium, specific growth factors, such as PDGF and EGF, can be added directly to the plates, rocked by hand, and placed in a 37 °C incubator. One should also harvest cells that haven't been treated with growth factors or 10% serum for each time point.

[b] In step 10, the kinase assay samples were on an SDS–polyacrylamide gel and transferred to PVDF immobilon membrane (Millipore). This is done so that Western blot analysis can be performed (*Protocol 2*) to determine the relative amount of kinase that was immunoprecipitated per lane. One should obtain similar amounts precipitated in order for the experiment to be valid.

Protocol 2.   Western blot analysis[a]

Equipment and reagents
- sample buffer: 50 mM Tris–HCl, pH 6.8, 1% SDS, 10% glycerol, 0.05% bromophenol blue, 1% β-2-mercaptoethanol
- PBST: PBS plus 0.1% Tween-20
- BSA
- anti-MEK1/2 antibody
- anti-rabbit/anti-mouse HrPase-conjugated secondary antibody (Amersham)
- X-OMAT AR film (Kodak)

Method

1. Boil 30 μg of the lysate in 1 sample buffer for 5 min, and electrophorese on a 10% SDS–polyacrylamide gel. Transfer proteins to PVDF immobilon membrane (Millipore).

2. Block the membrane in PBS + 0.1% Tween-20 (PBST) containing 3% BSA for 1 h at room temperature.

3. Add anti-MEK1 and anti-MEK2 (depending on what you originally immunoprecipitated with) primary antibody at a dilution of 1:1000 in PBST and incubate for 1 h at room temperature.

4. Wash three times for 15, 5, and 5 min, respectively, with PBST at room temperature.

5. Incubate membranes with anti-rabbit or anti-mouse horseradish peroxidase-conjugated secondary antibody at a dilution of 1:2000 at room temperature for 25 min.

6. Wash membranes again as described in step 4.

7. Subject membranes to enhanced chemiluminescence (ECL, Amersham) and expose to X-OMAT AR film (Kodak).[b]

[a] This method works for most antibodies. The dilution of primary antibody may vary, especially if the antibody has never been tested in Westerns. In that case, appropriate dilutions to be tested are 1:100, 1:500, and 1:1000.

[b] With the advent of phosphospecific antibodies that recognize the phosphorylated and activated form of ERKs, performing radioactive kinase assays has become another option. The use of these antibodies (available from NEB), has allowed the analysis of phosphorylated ERK when one has a small amount of lysate available. The antibodies can be used at a dilution of 1:1000 in PBST+3%BSA overnight at 4°C. New England Biolabs also has available phosphospecific Akt, phosphospecific Bad (Akt substrate), phosphospecific Elk-1 (ERK substrate) antibodies, just to mention a few. One of the limiting factors in using these antibodies is the price. I found that you can save the 1:1000 dilution and that it can be used at least one more time. Store it at 4°C *without* sodium azide since this may interfere with the electrochemiluminescence reaction.

3.1.2 In-gel kinase assay

Another technique that is used to study the kinase activity of various kinases is the in-gel kinase assay (*Protocol 3*). This technique allows a snapshot picture of specific kinases that are activated at different time points during activation of signalling pathways with serum or a specific growth factor.

Protocol 3. In-gel kinase assay—analysis of myelin basic protein (MBP) kinases, such as ERKs

Equipment and reagents

- MBP
- isopropanol wash buffer: 20% isopropanol, 50 mM Tris, pH 8.0
- buffer B: 5 mM β-mercaptoethanol, 50 mM Tris, pH 8.0
- guanidine buffer: 6 M guanidine HCl, 5 mM β-mercaptoethanol, 50 mM Tris, pH 8.0
- renaturation buffer: 0.04% Tween-40 (Sigma), 5 mM β-mercaptoethanol, 50 mM Tris, pH 8.0

- kinase buffer: 20 mM Tris, pH 7.2, 10 mM MgCl$_2$, 15 mM β-glycerophosphate, 0.3 mM sodium vanadate
- [γ-^{32}P]ATP
- wash buffe: 5% trichloroacetic acid (TCA), 1% pyrophosphate
- rocker
- Geiger counter

Method

1. Make a mini SDS–polyacrylamide gel of the desired percentage containing 0.4 mg/ml myelin basic protein (MBP). Make sure the gel mix and MBP are well mixed before adding ammonium persulfate (i.e. before cross-linking the gel mix).[a]

2. Pour stacking gel (*without* MBP). Load gel samples and run the gel. To obtain a decent signal, use 30 μg of total cell lysate from culture cells or 100 μg of total lysate when analysing tissues.

3. After the run is complete, rinse gel briefly in 100 ml of isopropanol wash buffer. Wash two times for 30 min at room temperature with 250 ml isopropanol wash buffer. This and all subsequent incubations are carried out on a rocker with *gentle* rocking at room temperature, unless indicated otherwise. The tupperware container used measures 9.5 in. 6.5 in. 5.5 in.

4. Rinse briefly in 100 ml buffer B. Wash two times for a minimum of 30 min at room temperature with 250 ml buffer B. However, this can be carried out for 1–2 hours each.

5. Wash two times in 100 ml of guanidine buffer for 5 min each. Then wash for 1 h in 100 ml of guanidine buffer.

6. Rinse briefly in 200 ml renaturation buffer.[b] Wash two times for a minimum of 30 min at room temperature with 250 ml buffer B. However, this can be carried out for 1–2 hours each. Then wash the gel for 14–48 h in 1800 ml renaturation buffer at 4°C.

7. Carefully transfer the gel to a small container. The gel is very fragile at this point. Wash the gel in a small volume (10 ml) of kinase buffer for 1 h. The volume will depend on the number of gels and the container used.

Protocol 3. *Continued*

8. Incubate in kinase buffer containg 50 μM ATP and 20 μCi/ml [γ-^{32}P]ATP for 30 min. Use 5 ml per mini gel.[c]

9. Wash repeatedly in wash buffer until counts are no longer detectable in the wash solution. This can be monitored using a Geiger counter. For 1–5 gels, use a total of 2 litres of wash buffer with 6–7 changes. It should take at least 1.5–2 h.

10. Dry the gel and expose.[d]

[a] Though myelin basic protein was used as a substrate, my laboratory has also successfully used p27^{Kip1} as a substrate. One can also use GST fusion proteins (as described in *Protocol 4*). As controls, gels should also be prepared with no substrate added and/or GST alone.
[b] The longer one allows for renaturation in step 6, the better the signal. Also, it is important to use Tween-40, using Tween-20 does not work as well.
[c] Be careful when handling radioactivity. Designate containers that will be used for radioactivity.
[d] If the background signal is high after exposure of the gel, wash for longer in the subsequent experiment.

3.2 Is my protein an ERK substrate?

Like all kinases, ERK/MAP kinases recognize specific amino acid sequences which are referred to as consensus phosphorylation sites. In the case of ERKs, that sequence is (P/L)X(S/T)P, where serine or threonine is phosphorylated.

Analysis of the sequence of your protein should allow you to deduce if kinase consensus sites exist. Let us assume that your protein has an ERK phosphorylation site, how would you go about showing if your protein is phosphorylated by ERKs in an *in vitro* kinase assay and in culture cells.

3.2.1 Making a glutathione-*S*-transferase (GST) fusion with of your protein.

To make bacterially expressed GST fusion protein, the cDNA of the protein in question is cloned into pGEX vectors (Pharmacia) in the correct reading frame. There are several forms of this available that will allow the cloning of your gene in the correct reading frame. Once your gene is cloned in the appropriate vector, sequence the junction point before you proceed to making the fusion protein, to verify that the gene is in fact in the correct reading frame with GST. Transform the construct into LE392 bacteria, isolate a single colony, and proceed as described in *Protocol 4*.

Protocol 4. Preparation of a GST fusion protein

Equipment and reagents
- Luria Broth
- ampicillin

- IPTG (isopropylthio-β-D-galactoside)
- protease inhibitors (leupeptin, pepstatin)

- STE: 150 mM NaCl, 50 mM Tris, pH 7.2, 1 mM EDTA, 1 mM 4-(2-aminoethyl)-benzenesulfonyl fluoride (Pefabloc®SC, Boehringer Mannheim)
- 50 mg/ml lysozome
- 10% NP-40 (Sigma)
- 50% (v/v) glutathione agarose beads (Sigma)

- 5 mM reduced glutathione (Sigma)
- 50% ethylene glycol
- 50 mM Tris
- 50 mM NaCl
- 1mM DTT
- Falcon tube (50 ml)
- table-top centrifuge

Method

1. Grow up a single bacterial colony that has been transformed with your construct overnight at 37 °C in 100 ml of LB in the presence of 25 µg/ml ampicillin.

2. Next day, dilute the grown bacterial culture, 1:10 (total of 1 litre) in LB and shake at 37 °C for 1 h.

3. Add IPTG to a final concentration of 0.4 mM and shake for 3 h at 37 °C.

4. Spin the bacterial culture at 5000 g for 5 min to pellet cells.

5. Freeze pellet overnight at –70 °C.[a]

6. Resuspend the pellet in 1/50 of the original volume (20 mls) with ice-cold STE containing protease inhibitors (10 µg per ml leupeptin/10 µg per ml pepstatin A) and place on ice.

7. Add 200 µl of 50 mg/ml lysozyme. Final concentration should be 5 mg.

8. Incubate for 10 min on ice.

9. Add 100 µl 10% NP-40. Final concentration should be 0.05%.

10. Incubate for 10 min on ice.

11. Spin at 12 000 r.p.m. ≃ 17 210 g for 20 min at 4 °C using an SS-34 rotor.

12. Collect the supernatant and transfer to a 50 ml conical Falcon tube.

13. Add 400 µl of a 50% (v/v) glutathione agarose bead suspension in STE.

14. Rock in the cold room for 1 h.

15. Spin down in a table-top centrifuge at setting 5 for 2 min in the cold room.

16. Wash four times with ice-cold STE plus protease inhibitors—two 10 ml washes in a 50 ml conical tube; the next two, using 5 mls in a 15 ml conical tube. In between washes, the tubes are spun as described in step 15.

17. Elute fusion protein from beads using 1.5 ml of 5 mM reduced glutathione in 50 mM Tris, pH 8.0.[b]

18. Incubate on ice for 2 min and spin down as described in step 15.

Protocol 4. *Continued*

19. Save the supernatant and dialyse overnight in the cold room against 1 litre of 50% ethylene glycol, 50 mM Tris, pH 7.5, 50 mM NaCl, 1 mM DTT.[c]

20. Remove dialysed protein, tranfer to an Eppendorf tube, and store at −20 °C.[d–g]

[a] It is important to freeze the pellet before lysis, otherwise there will be less lysis of bacteria and a reduction in the yield of fusion protein.
[b] Increasing the amount of glutathione, even to 10 mM, results in reduction of the pH and less efficient elution of the fusion protein off the beads.
[c] A rule of thumb: when dialysing against 50% ethylene glycol, the final volume of your sample is roughly halved.
[d] To estimate the yield of your fusion protein, run a mini SDS–polyacrylamide gel, fix the gel and stain with Coomassie blue. Analyse 5 and 10 μl samples of your fusion protein. Also run known BSA amounts (ranging from 100 to 1000 ng) so as to roughly estimate the concentration of your fusion protein.
[e] One should also prepare GST protein alone to be used as a negative control in *in vitro* kinase assays, as well as in in-gel kinase assays (see *Protocol 3*).
[f] If the yield of your fusion protein is low, one can prepare 4–6 litres of bacterial culture; one can also add another 0.4 mM IPTG at the last hour of the 3 h incubation.
[g] If the concentration of fusion protein is too dilute (i.e. 100 ng/10 μl volume), elute with a smaller volume of glutathione—again, remembering that the final volume is halved after dialysis. This is important because a final concentration of ethylene glycol of greater than 10% may inhibit kinase activity.

One can proceed with an *in vitro* kinase assay using at least 100 ng of fusion protein following *Protocol 1*. Instead of anti-MEK antibodies, one can use antibodies made to the C-terminus of ERKs (available from Santa Cruz) to immunoprecipitate active ERK from lysates described in *Protocol 1*, or one can use GST–ERK (available from various laboratories, including mine), if the size of your GST fusion protein is not roughly the same size as GST–ERK, which itself exhibits auotphosphorylation and therefore may mask the signal of the fusion protein of interest. As a negative control, use GST–ERK (kinase inactive) instead of GST–ERK (wild-type); another negative control is to use GST alone as a substrate.

Showing that a protein is a substrate for a kinase, such as ERK/MAP kinases, is the first step in determining the physiological significance of this phosphorylation. In the next section I shall outline experiments that will aid in determing if a protein is indeed a substrate for a specific kinase in cells.

4. Determining the physiological significance of a protein phosphorylated by a specific kinase: the use of specific techniques

In this section, I will outline a series of experiments that will determine if the phosphorylation of a protein by a specific kinase in an *in vitro* kinase assay is physiologically relevant. I will outline (a) the use of mammalian expression vectors, (b) *in vivo* labelling to determine the phosphorylation of the substrate, (c) two-dimensional (2D) tryptic phosphopeptide analysis of the phosphorylated substrate to determine if phosphorylation *in vitro* gives a similar tryptic peptide map to phosphorylation in culture cells, and (d) co-localization of the kinase and the substrate using indirect immunofluroscence analysis.

4.1 Expression in mammalian cells

Once it is established that a protein is a substrate for a specific kinase in the *in vitro* kinase assay, it is important to establish if the protein is also phosphorylated in culture cells. One can begin to answer this question by cloning the gene in a mammalian expression vector that also 'epitope tags' the protein with a specific amino acid sequence. This allows one to distinguish the ectopically expressed gene product from its endogenous counterpart, which may be expressed at lower levels. In addition, having an epitope-tagged protein allows one to use available antibodies to these tags, antibodies that may not be available or good enough for immunoprecipitation, indirect immunofluorescence, and/or Western blot analysis.

Two vectors that are very useful for tagging are the FLAG vectors (Kodak, IBI), which add the FLAG amino acid sequence, DYKDDDDK, to the protein, and pJ3H (46), which adds the haemagglutinin (HA) amino acid sequence, YPYDVPDYA, to the protein. Monoclonal antibodies are available that recognize each tag: monoclonal M2 antibodies (IBI and Sigma) which recognize the FLAG sequence may be used in Western blot analysis, immunoprecipitations, and indirect immunofluorescence; monoclonal antibodies to the HA tag are also available, 12CA5 (Babco), which may be used in Western blot analysis and immunoprecipitations.

After cloning the cDNA into one of these vectors, the construct is transfected into a mammalian cell line. Some easily transfectable cell lines are COS7 and 293. Both are widely used and available. There are a number of transfection techniques that yield different results with different cell lines. My laboratory uses a transfection protocol that is described in Pear *et al.* (*Protocol 5*). At 48 h post-transfection, the level of expression of the the tagged protein should be determined by Western blot analysis (*Protocol 2*) before proceeding to *in vivo* ^{32}P-labelling (*Protocol 6*) of cells to determine phosphorylation of the substrate in cells and to obtain enough phosphorylated material for 2D tryptic phosphopeptide analysis (*Protocol 7*) (47).

4.2 *In vivo* ^{32}P-labelling of culture cells and two-dimensional tryptic phosphopeptide analysis

In vivo ^{32}P-labelling of cells to evaluate if a protein is phosphorylated as a result of activation of the MEK/ERK pathway or the PI3 kinase pathway can be facilitated by the use of drugs that specifically inhibit each pathway. PD98059 (NEB) is a compound that specifically inhibits MEK1 (48, 49). Wortmannin (Sigma) is a compound that specifically inhibits PI3 kinase (43).

Protocol 5. Transfection of mammalian cells

Equipment and reagents
- HBS: 50 mM *N*-2-hydroxyethylpiperazine-*N'*-2'-ethanesulfonic acid (Hepes), pH 7.05, 10 mM KCl, 12 mM dextrose, 280 mM NaCl, 1.5 mM Na_2HPO_4

Method

1. Grow cells to subconfluency (60–70%) on 100 mm tissue culture plates in complete medium.

2. Just prior to transfection, replace the medium with medium containing 25 μM chloroquine.

3. Transfection should be carried out by mixing equal volumes of 2 HBS (50 mM *N*-2-hydroxyethylpiperazine-*N'*-2'-ethanesulfonic acid (Hepes), pH 7.05, 10 mM KCl, 12 mM dextrose, 280 mM NaCl, 1.5 mM Na_2HPO_4 and 250 mM calcium chloride solution containing 10 μg DNA.[a,b]

4. Add the mixture to cells within 1–2 min.

5. Incubate for 10 h at 37°C.

6. Aspirate medium and replace with complete medium without chloroquine.

7. Incubate cells for an additional 38 h at 37°C.

8. Perform Western blot analysis as described in *Protocol 2* to determine the level of expression of the transfected gene.

[a] Chloroquine can inhibit lysosome function and thus reduce the breakdown of the transfected DNA. Incubating the cells for more than 12 h with chloroquine may result in cellular toxicity.
[b] It is crucial that the pH of the 2 HBS be 7.05, otherwise tranfection efficiency is reduced.

Protocol 6. *In vivo* ^{32}P-labelling of transfected culture cells

Equipment and reagents
- phosphate-free medium
- 40 μM PD98059 (NEB)
- 100 nM wortmannin (Sigma)
- [^{32}P]orthophosphate (NEN)
- PBS
- monoclonal antibodies (mAb): M2 or 12CA5
- lysis buffer
- sample buffer: 100 mM TrisHCl, pH 6.8, 2% SDS, 20% glycerol, 0.1% bromophenol blue, 2% β-2-mercaptoethanol
- X-OMAT AR film (Kodak)
- PVDF membrane

Method

1. At 48 h post-transfection (see *Protocol 5*), starve the cells for 24 h in medium containing 0.1% serum.

2. At the 23rd hour, replace the medium with phosphate-free medium containing 0.1% dialysed serum ± 40 μM PD98059 or 100 nM wortmannin.

3. Aspirate medium, and add to each 100 mm plate, phosphate-free medium containing 10% serum (or a specific growth factor), 0.5 mCi/ml [^{32}P]orthophosphate (NEN) ± 40 μM PD98059 or 100 nM wortmannin.[a–c]

4. Incubate plates for 15, 30, and 60 min in a 37 °C incubator. **Important**: place cells in a radioactive-proof container when placing in the incubator.

5. Wash plates once with ice-cold PBS.

6. Scrape the cells in 0.5 ml of ice-cold PBS, transfer to an Eppendorf tube, and spin at 1500 rpm ≃ 120 g in a microfuge.

7. Carefully remove supernatant and lyse and process the pellet as described in *Protocol 1*.

8. Add 1:100 dilution of the appropriate mAb—M2 mAb for FLAG-tagged proteins or 12CA5 for HA-tagged proteins.

9. Incubate on ice for 1 h.

10. Add 30 μl of protein G–agarose beads (Santa Cruz) to each tube and rock for 1 h at 4 °C.

11. Wash beads three times with lysis buffer and then resuspend in 30 μl of 2 sample buffer. Boil the samples for 5 min, and then electrophorese on an SDS–polyacrylamide gel of the appropriate percentage.

12. Transfer proteins to PVDF membrane, and then expose to X-OMAT AR film.

13. If you obtain a reasonable signal (overnight exposure or less), proceed to *Protocol 7*.

---

[a] Be extremely careful when working with [^{32}P]orthophosphate. Use a shield and constantly monitor yourself and your surroundings for contamination. Use disposable pipettes and discard them, as well as the washes, in the appropriate radioactive waste containers.

[b] The use of the drugs will elucidate if either of the pathways is responsible for the phosphorylation of your protein.

[c] It is important to titre the concentration of PD98059 to avoid any toxicity resulting from the drug. Test from 10 to 50 μM PD98059 for each cell line and monitor for cell death.

The use of PD98059 or wortmannin in a ^{32}P-labelling *in vivo* experiment will aid in discerning if the MEK/ERK or PI3 kinase pathways play any role in the phosphorylation of your protein. The 2D tryptic phosphopeptide analysis experiment is important in determining if the specific kinase used in the *in vitro* kinase assay is involved in the phosphorylation of your protein *in vivo*, thus establishing the first steps in elucidating physiological relevance.

Protocol 7. 2D tryptic phosphopeptide analysis

Equipment and reagents

- 0.5% polyvinylpyrrolidone
- 100 mM acetic acid
- 50 mM ammonium bicarbonate
- 1 mM DTT
- trypsin

- 10 μg of L-1-tosylamido-2-phenylethyl-chloromethyl ketone (TCPK)
- TLC plates (Chromagram plates, Kodak)
- HLTE-7000 apparatus (C.B.S. Scientific)

Method

1. Excise the ^{32}P-labelled band from the PVDF membrane (from *Protocol 6*) and incubate at 37 °C for 30 min in 500 μl of 0.5% polyvinyl-pyrrolidone and 100 mM acetic acid. This is a blocking step so that no non-specific binding occurs of trypsin on to the membrane.

2. Remove supernatant and wash the membrane slices once with 500 μl of distilled water and once with 500 μl of 50 mM ammonium bicarbonate.

3. Add 200 μl of 50 mM NH$_4$HCO$_3$ and 1 mM DTT containing 10 μg of L-1-tosylamido-2-phenylethylchloromethyl ketone (TCPK)–trypsin and incubate for 16 h at 37 °C.[a]

4. Add an additional 10 μg of trypsin and incubate for an additional 3 h at 37 °C.[b]

5. Transfer the supernatant to a freash Eppendorf tube and vacuum-dry.

6. Wash the pellet three times with 50 μl of distilled water, vacuum-drying in between washes.

7. Dissolve the pellet in distilled water at 1000 cpm (Cerenkov) per 5 μl.[c]

8. Spot the 5 μl on to thin layer chromatography cellulose plates and electrophorese in the first dimension from cathode (−) to anode (+) at 1000 V for 35 min in 0.1% ammonium carbonate (pH 8.9) using a HLTE-7000 apparatus.[d]

9. The second dimension consists of ascending chromatography in buffer containing 1-butanol/pyridine/acetic acid/water, 375:250:75:300 (vol/vol), for 3 h at room temperature.

10. Air-dry the plates in a fume hood and the detect the ^{32}P-labelled phosphopeptides by autoradiography.[e]

[a] TPCK–trypsin is treated so as to inhibit any contaminating chymotrypsin.

[b] Try not to overtreat with trypsin because the presence of too much trypsin will result in increased smearing of the signal due to the presence of too much protein.

[c] 500–1000 cpm samples should yield a reasonable signal.

[d] When spotting the sample on to chromatography plates, add 1 μl at a time, drying with a hair dryer in between loadings. Set the hair dryer on cool; using the hot setting will bake the sample on to the plate, resulting in very little migration of the phosphopeptides.

[e] To compare the phosphopeptide pattern of your protein phosphorylated in an *in vitro* kinase assay to that of your protein labelled *in vivo*, run the following plates: (1) 1000 cpm of the protein sample labelled *in vitro*; (2) 1000 cpm of the protein sample labelled *in vivo*; and (3) 500 cpm of the protein sample labelled *in vitro* plus 500 cpm of the protein sample labelled *in vivo*. This will allow you to discern if the phosphopeptide pattern between *in vtro* and *in vivo* labelling is occuring at the same sites.

Another step in evaluating physiological relevance is to analyse if in fact the kinase of interest and its putative substrate co-localize in the cell, especially under conditions of activation of the specific pathway. Indirect immuno-fluorescence (*Protocol 8*) is used to establish co-localization of the proteins in question. With the advent of phosphospecific antibodies to ERK and Akt, one can begin to analyse the co-localization of the activated forms of these kinases and their putative substrates.

Protocol 8. Indirect immunofluorescence and the analysis of co-localization of proteins

Equipment and reagents
- 4% paraformaldehyde[a]
- 5% sucrose[a]
- PBS
- 10% fetal bovine serum
- TBST: Tris-buffered saline containing 0.1% Triton X-100
- M2 mAb and phosphospecific ERK/Akt (NEB)
- 3% BSA
- Cy3™-conjugated anti-rabbit antibodies (Jackson ImmunoResearch)
- fluorescein isothiocyanate (FITC)-conjugated anti-mouse antibodies (Jackson ImmunoResearch)
- fluorescence microscope

Method

1. Transfect cells as described in *Protocol 6*. After 24 h, split the cells into 6-well plates. Split the cells at various dilutions so that in 24 h the cells are 50% confluent.

2. Aspirate medium and replace with medium containing 0.1% serum and incubate for 24 h at 37 °C.

3. Start a time course by adding medium with 10% serum or a specific growth factor; I would take 15, 30, and 60 min time points. Incubate at 37 °C.

Protocol 8. *Continued*

4. At each time point, quickly aspirate the medium and fix cells with 4% paraformaldehyde and 5% sucrose in 1 PBS.

5. Incubate for 20 min at room temperature.

6. Wash three times with PBS for 5 min each at room temperature.

7. Block for 10 min at room temperature with PBS containing 10% fetal bovine serum.

8. Wash once in TBST for 5 min at room temperature.

9. Add primary antibodies [1:100 dilution for phosphospecific ERK or Akt (polyclonal) (NEB); 1:100 M2 mAb (if using FLAG-tagged protein)] in TBST containing 3% bovine serum albumin (BSA) and incubate for 90 min at room temperature.

10. Wash three times with TBST for 5 min each at room temperature.

11. Add secondary antibody [1:500 for Cy3™-conjugated anti-rabbit antibodies; 1:100 for fluorescein isothiocyanate (FITC)-conjugated anti-mouse antibodies] in TBST containing 3% BSA and incubate for 60 min at room temperature. During this step, cover the plate in aluminum foil so as not to bleach the fluorescent antibodies.

11. Wash three times with TBST for 5 min each at room temperature.

12. Visualize using a fluorescence microscope.

[a]Preparing the 4% paraformaldehyde and 5% sucrose solution, mix all the ingredients in a fume hood:
- heat 60 ml of distilled water to 60°C
- add 4 g of paraformaldehyde and mix
- add one drop of 1 N NaOH to allow the paraformaldehyde to go into solution
- add 5 g of sucrose and mix
- add 10 ml of 10 PBS
- adjust volume to 100 ml with distilled water
- the pH of the solution should be around 7.2

5. Concluding remarks

In this chapter, I have introduced techniques used to analyse the phosphorylation of proteins by kinases, specifically those involved in the MEK/ERK and Akt pathways. Showing that a protein is phosphorylated in an *in vitro* kinase assay is one thing, establishing a physiological relevance for such a phosphorylation event is another. That is why this chapter also included an outline of determining the physiological significance of the phosphorylation of a protein by a specific kinase.

Both the MEK/ERK and Akt pathways have been implicated in the growth and differentiation of cells. Their roles in many processes have been outlined at the beginning of this chapter. The expression of constitutively active MEK1 results in the transformation of established cell lines, such as NIH-3T3 and Swiss 3T3 cells (50, 51); overexpression of the same MEK1 mutant in primary human fibroblast results in cell cycle arrest and senescence (38, 39). Recently, we have shown that inhibition of MEK1 results in the protection of hippocampal cells from damage resulting from seizure activity (52). The Akt pathway has already been established as an anti-apoptotic pathway (43).

What roles the various growth factor-dependent kinases play in other processes, whether normal or pathophysiological, remain to be determined. The identification of their physiological substrates will aid in the elucidation of these roles, and will help shed light on our understanding of cell growth and differentiation.

Acknowledgements

This work was supported in part by grants from the American Heart Association, Massachusetts Affiliate, Inc., and Massachusetts General Hospital Interdepartmental Stroke Project Grant.

References

1. Blenis, J. *Proc. Natl. Acad. Sci.* (1993) **90**: 5889–5892.
2. Davis, R. J. *Biol. Chem.* (1993) **268**: 14553–14556.
3. Seger, R., Krebs, E.G. *FASEB J.* (1995) **9**: 726–735.
4. Davis, R.J. *TIBS* (1994) **19**: 470–473.
5. Brott, B.K., Alessandrini, A., Largaespada, D.A., Copeland, N.G., Jenkins, N.A., Crews, C.M., Erikson, R.L. *Cell Growth and Diff.* (1993) **4**: 921–929.
6. Crews, C.M., Alessandrini, A., Erikson, R.L. *Science* (1992) **258**: 478–480.
7. Seger, R, Ahn, N.G., Posada, J., Munar, E.S., Jensen, A.M., Cooper, J.A., Cobb, M.H., Krebs, E.G. *J. Biol. Chem.* (1992) **267**: 14373–14381.
8. Wu, J., Dent, P., Jelinek, T., Wolfman, A., Weber, M.J., Sturgill, T.W. *Science* (1993) **262**: 1065–1069.
9. Wu, J., Harrison, J.K., Dent, P., Lynch, K.R., Weber, M.J., Sturgill, T.W. *Mol. Cell. Biol.* (1993) **13**: 4539–4548.
10. Zheng, C.-F., Guan, K.-L. *J. Biol. Chem.* (1993) **268**: 11435–11439.
11. Leevers, S.J., Paterson, H.F., Marshall, C.J. *Nature* (1994) **369**: 411–414.
12. Stokoe, D., Macdonald, S.G., Cadwallader, K., Symons, M., Hancock, J.F. *Science* (1994) **264**: 1463–1467.
13. Liu, D., Bienkowska, J., Petosa, C., Collier, R.J., Fu, H., Liddington, R. *Nature* (1995) 376: 191–194.
14. Xiao, B., Smerdon, S.J., Jones, D.H., Dodson, G.G., Soneji, Y., Aitken, A., Gamblin, S.J. *Nature* (1995) **376**: 188–191.
15. Li, S., Janosch, P., Tanji, M., Rosenfeld, G., Waymire, J.C., Mischak, H., Kolch, W., Sedivy, J.M. *EMBO J.* (1995) **14**: 685–696.
16. Marshall, C.J. *Nature* (1996) **383**: 127–128.
17. Alessandrini, A., Brott, B.K., Erikson, R.L. *Cell Growth Differ* (1997) **8**: 505–11.
18. Wu, X., Noh, S.J., Zhou, G., Dixon, J.E., Guan, K.-L. *J. Biol. Chem.* (1996) **271**: 3265–3271.
19. Jelinek, T., Catling, A.D., Reuter, C.W., Moodie, S.A., Wolfman, A., Weber, M.J. *Mol. Cell. Biol.* (1994) **14**: 8212–8218.
20. Catling, A.D. *Mol. Cell. Biol.* (1995) **15**: 5214–5225.

21. Downey, G.P., Butler, J.R., Brumell, J., Borregaard, N., Kjeldsen, L., Sue-A-Quan, A., Grinstein, S. *J. Biol. Chem.* (1996) **271**: 21005–21011.
22. Sturgill, T.W., Ray, L.B., Erikson, E., Maller, J.L. *Nature* (1988) **334**: 715–718.
23. Zhao, Y., Bjorbaek, C., Weremowicz, S., Morton, C., Moller, D.E. *Mol. Cell. Biol.* (1995) **15**: 4353–4363.
24. Xing, J., Ginty, D.D., Greenberg, M.E. *Science* (1996) 273: 959–960.
25. Fukunaga, R., Hunter., T. *EMBO J.* (1997) **16**: 1221–1933.
26. Waskiewicz, A.J., Flynn, A., Proud, C.G., Cooper, J.A. *EMBO J.* (1997) **16**: 1909–1920.
27. Flynn, A., Proud, C.G. *Cancer Surv* (1996) **27**: 293–310.
28. Lenormand, P., Sardet, C., Pages, G., L'Allemain, G., Brunet, A., Pouyssegur, J. *J. Cell Biol* (1993) **122**: 1079–1088.
29. Gonzalez, F, Seth, A., Raden, D., Bowman, D., Fay, F., Davis, R. *J. Cell Biol.* (1993) **122**: 1089–1101.
30. Treisman, R. *Curr. Opin. Cell. Biol.* (1996) **8**: 205–215.
31. Treisman, R. *EMBO J.* (1995) **14**: 4905–4913.
32. Gille, H., Sharrocks, A.D., Shaw, P.E. *Nature* (1992) **358**: 414–416.
33. Grana, X., Reddy, E.P. *Oncogene* (1995) 11: 211–219.
34. Lavoie, J.N., L'Allemain, G., Brunet, A., Muller, R., Pouyssegur, J. *J. Biol. Chem.* (1996) **271**: 20608–20616.
35. Woods, D., Parry, D., Cherwinski, H., Bosch E., Lees, E., McMahon, M. *Mol. Cell. Biol.* (1997) **17**: 5598–611.
36. Sewing, A., Wiseman, B., Lloyd, A.C., Land, H. *Mol Cell Biol* (1997) **17**: 5588–97.
37. Alessandrini A., Chiaur D.S., Pagano M. *Leukemia* (1997) **11**: 342–5.
38. Zhu, J., Woods, D., McMahon, M., Bishop, J.M. *Genes Dev* (1998) **12**: 2997–3007.
39. Lin, AW, Barradas, M, Stone, JC, van AL, Serrano, M, Lowe, SW. *Genes Dev* (1998) **12**: 3008–19.
40. Gingras, A.C., Kennedy, G., O'Leary, M.A., Sonenberg, N., Hay, N. *Genes Dev.* (1998) **15**: 502–513.
41. Kauffmann-Zeh, A., Rodriguez-Viciana, P., Ulrich, E., Gilbert, C., Coffer, P., Downward, J, Evan, G. *Nature* (1997) **385**: 544–548.
42. Kennedy, S.G., Wagner, A.J., Conzen, S.D., Jordan, J., Bellacosa, A., Tsichlis, P.N., Hay N. *Genes and Devel.* (1997) **11**: 701–713.
43. Franke, T.F., Kaplan, D.R., Cantley, .LC. *Cell* (1997) **88**: 435–7.
44. Datta, S.R. *Cell* (1997) 91: 231–241.
45. del Peso, L., Gonzales-Garcia, M., Page, C., Herrera, R., Nune, G. *Science* (1997) **278**: 786–689.
46. Chou, M., Blenis, J. *Cell (*1996) **85**: 573–583.
47. Pear, W.S., Nolan, G.P., Scott, M.L., Baltimore, D. *Proc. Natl. Acad. Sci. USA* (1993) **90**: 8392–6.
48. Alessandrini, A., Crews, C.M., Erikson, R.L.. *Proc. Natl. Acad. Sci. USA* (1992) **89**: 8200–8204.
49. Alessi, D.R., Cuenda, A., Cohen, P., Dudley, D.T., Saltiel, A.R. *J Biol Chem* (1995) **270**: 27489–94.
50. Klemke, R.L, Cai, S., Giannini, A.L., Gallagher, P.J., de L.P., Cheresh, D.A. *J Cell Biol* (1997) **137**: 481–92.
51. Cowley, S., Paterson, H., Kemp, P., Marshall, .C. *Cell* (1994) **77**: 841–852.
52. Alessandrini, A., Greulich, H., Huang, W., Erikson, R.L. *J. Biol. Chem.* (1996) **271**: 31612–31618.
53. Murray, B., Alessandrini, A., Cole, A.J., Yee, A.G., Furshpan, E.J. *Proc. Natl. Acad. USA* (1998) **95**: 11975–11980.

7

Cytometric analysis of the pRB pathway

GLORIA JUAN, FRANK TRAGANOS, and ZBIGNIEW
DARZYNKIEWICZ

1. Introduction

The term retinoblastoma protein (pRB) pathway ('pRB pathway') has recently been introduced to define the components and function of the cell cycle regulatory machinery directly associated with phosphorylation of the retinoblastoma susceptibility gene product (1–6). In addition to the central element, pRB, the pathway includes the cell cycle-dependent kinases which phosphorylate this protein (Cdks), activators of these kinases (several types of cyclins), their inhibitors from both the Ink and Kip families (Ckis), and the products whose activation is controlled by pRB phosphorylation, the E2F family of transcription factors (7–9). The pRB pathway serves as the master regulatory switch triggering the cell's entrance to and progression through the S phase. In non-tumour cells, the pRB pathway responds to changes in cell environment, primarily to growth factors, hormones, and cell to cell interactions, through the respective receptors and signal transduction pathways.

A common feature of most cancer cell types is dysregulation of the pRB pathway (3, 10). Dysregulation may result from defects in the components of the pathway itself or from changes upstream of pRB, at the level of growth factor receptors and/or signal transduction. In both instances, unscheduled pRB phosphorylation, unrelated to normal regulatory signals, takes place. Because E2F is not sequestered at any time during the cycle, its perpetual availability means that, following mitosis, the cells are already committed to enter the S phase of the next cycle, regardless of the presence or absence of growth factors or other stimuli from the environment that normally provide stimulatory or restrictive signals for cell proliferation.

Immunocytochemical detection and semi-quantitative analysis of the expression of particular components of the pRB pathway in relation to the cell's position in the cycle, as offered by multiparameter cytometry (reviewed in refs 11 and 12), opened new possibilities in exploring the mechanism of cell cycle regulation by the pathway. There is already a wealth of data in the literature resulting from multivariate analysis of cyclins D, E, and A, or Ckis, in relation to the cell cycle position as is discussed in more detail later in this chapter. The recently developed methodology that allows one to monitor the state of pRB phosphorylation in individual cells (13, 14) is the newest tool in cytometry to probe the differences in the mechanisms of cell cycle regulation between normal and tumour cells, to screen the activity of new anti-tumour drugs in terms of their ability to suppress pRB phosphorylation, and to investigate whether constitutive pRB phosphorylation in human tumours is of prognostic value. Cytometry of the pRB pathway provides new information which complements data obtained by classical methods of molecular biology.

2. Cyclins D, E, A, and B1 and Cdk inhibitors (Ckis)

Cyclins are the key components of the cell cycle regulatory machinery. They activate their respective partner Cdks and target them to specific protein substrates. Expression of cyclins D, E, A, or B1 is periodic and occurs at specific and well-defined points in the cycle (*Figure 1*). D-type cyclins are G_1 cyclins and in normal cells they are maximally expressed in response to mitogenic stimulation, e.g. by growth factors. Expression of particular D cyclins is tissue- and cell-type specific. For example, cyclin D1 is predominant in fibroblasts and epithelial cells, while cyclins D2 and D3 are expressed in cells of lymphocytic lineage; the latter cells are cyclin D1-negative (15, 16). Cyclin D1 activates Cdk4/6 and targets pRB for phosphorylation by these kinases.

Cyclin E also is a G_1 cyclin. The onset of its accumulation is in mid-G_1 and its maximal expression is at the time of cell entrance to S. Cyclin E undergoes degradation by a ubiquitin–proteasome pathway as the cell progresses through S (17). The kinase partner of cyclin E is Cdk2. The enzymatic activity of this complex is essential for cell entrance to the S phase. pRB is one of the substrates phosphorylated by cyclin E–Cdk2, but the phosphorylation sites are different than those phosphorylated by cyclin D–Cdk4.

Accumulation of cyclin A starts early in the S phase, is maximal at the end of G_2, and the protein is rapidly degraded, also by the proteasome system, in the prometaphase. Cyclin A may associate with either Cdk2 or Cdk1; the kinase activity of the complex drives the cell through S and G_2, respectively (18). The rate of DNA replication is correlated with the cellular level of cyclin A (19).

Cyclin B1 activates Cdk1 (alternatively denoted as Cdc2) forming maturation promoting factor (MPF), whose kinase activity is needed to prepare cells to undergo mitosis (20). Cyclin B1 begins accumulating late in S phase, reaches maximal levels as the cell enters mitosis, and is destroyed during the transition to anaphase. In contrast to cyclins D, E, and A, which

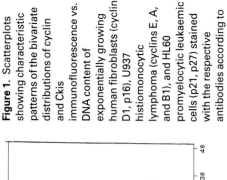

Figure 1. Scatterplots showing characteristic patterns of the bivariate distributions of cyclin and Ckis immunofluorescence vs. DNA content of exponentially growing human fibroblasts (cyclin D1, p16), U937 histomonocytic lymphoma (cyclins E, A, and B1), and HL60 promyelocytic leukaemic cells (p21, p27) stained with the respective antibodies according to *Protocol 1*.

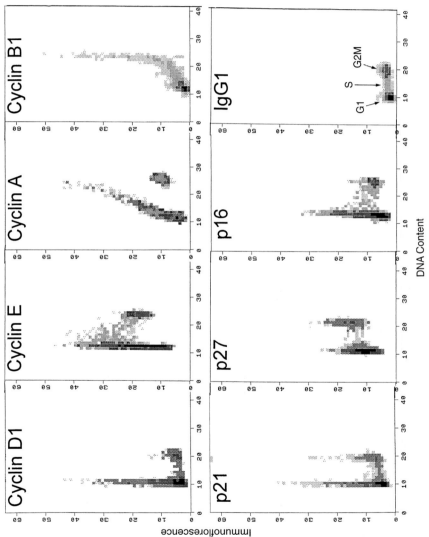

are localized in the nucleus, cyclin B1 is present in the cytoplasm during G_2 but is translocated into the nucleus at prophase (20).

The cellular content of particular cyclins is regulated not only at the transcriptional and translational level but also by the altered rate of their degradation, which, as mentioned, is through the ubiquitin–proteasome system. Thus, for example, the relatively long half-life of cyclin B1 during G_2 is shortened during G_1; overexpression of cyclin E, on the other hand, prolongs its half-life (21). Thus, at particular phases of the cycle, the message level may not always correlate with the amount of the respective cyclin protein.

Regardless of the mechanism regulating intracellular level of cyclins D, E, A, and B1, the periodic and timely expression of these proteins, which takes place during unperturbed growth of normal cells, represent landmarks of the cell cycle which complement the traditionally recognized ones, such as DNA replication and mitosis. The timing of particular events that occur during the cycle, or the point of action of anti-tumour agents can therefore be related to these landmarks (11).

Immunocytochemical detection of cyclins, when combined with analysis of cellular DNA content, made it possible, by multiparameter flow cytometry, to correlate the presence of a particular cyclin with the cell's position in the cycle (13, 19, 22–32). The expression of cyclins *vis-a-vis* the cell cycle phase, therefore, can be studied in asynchronous cell populations in a manner that does not perturb cell cycle progression or induce growth imbalance, which otherwise occurs when cells are synchronized in the cycle, e.g. by DNA polymerase inhibitors (28, 30). Furthermore, the intercellular variability in cyclin expression is revealed, rare cells and cell subpopulations with distinct features can be detected, and the presence of thresholds (27) in expression of these proteins at particular phases of the cycle determined.

A new form of cell cycle regulation emerged with the discovery of Ckis that directly or indirectly inhibit the kinase activity of the complex. These inhibitory proteins fall into two classes based on structural homology: (1) the Ink4 (underline{in}hibitors of cd\underline{k}4) family consisting of $p15^{INK4b}$, $p16^{INK4a}$, $p18^{INK4c}$, and $p19^{INK4d}$; and (2) the Cip/Kip (\underline{c}yclin/\underline{k}inase \underline{i}nhibitory \underline{p}roteins) family consisting of $p21^{CIP1}$, $p27^{KIP1}$, and $p57^{KIP2}$ (reviewed in refs 33 and 34).

INK4 family members act exclusively on cyclin D/Cdk4 or Cdk6 complexes which, in turn, are intimately related to activation of pRB. Members of this family are composed primarily of tandemly repeated ankyrin motifs. Human p16 and p15 map in tandem to the short arm of human chromosome 9 (9p21), while p18 has been assigned to chromosome 1, and p19 to chromosome 19. All members of this family of Ckis appear to be capable of inactivating Cdks in two ways: (1) by preventing complex formation between cyclin D and Cdk 4 or 6; and/or (2) by binding to the already formed cyclin D–Cdk complex (34).

CIP/KIP family members are potent inhibitors of the G_1 and S phase cyclin–Cdk complexes (e.g. cyclin D2–Cdk4, cyclin E–Cdk2, cyclin A–Cdk2) and, to a lesser extent, of the mitotic cyclin, cyclin B1–Cdk 1. These molecules have no structural similarity to the INK4 family of Ckis, nor do they directly inhibit cyclin-activating kinase (Cak). CIP/KIP family members inhibit Cdks by binding to the complex and preventing Cak from phosphorylating Thr-161. Overexpression of p21 and p27 causes a G_1 arrest, suggesting that, although they have broad specificity *in vitro*, these Ckis may act only on G_1 Cdks *in vivo*. Interestingly, binding of one molecule of p21 to a Cdk complex stimulates kinase activity—the inhibitory effect only occurs upon binding of a second p21 molecule (34).

Both p21 and p27 participate in numerous regulatory responses. DNA damage is mediated through the tumour suppressor gene product p53, which elevates p21 levels transcriptionally, causing inhibition of various G_1 Cdks. Mitogen activation increases p21 expression and stimulates Cdk activation, although, as noted above, p21 also acts to inhibit this activity when a second molecule binds. p27 expression increases in contact-inhibited or mitogen-depleted cells, and, contrary to p21, decreases with mitogen-induced exit from quiescence. p57 has much the same activity as p27 although it has different tissue distribution (33, 34).

Protocol 1.

Reagents

- phosphate-buffered saline (PBS)
- *fixative:* 80% ethanol or absolute methanol, keep at –20°C
- solution of 0.25% (v/v) Triton X-100 in PBS, pH 7.4 (store at 4°C)
- *cyclin antibodies:* The antibodies listed below were tested in our laboratory and were found to be satisfactory for use by flow cytometry: mouse monoclonal antibodies to human cyclin B1 (clone GNS-1), cyclin A (clone BF-683), cyclin D1 (clone G124-326), cyclin D3 (clone G107-565), and to cyclin E (clone HE12) (PharMingen, San Diego, CA); cyclin D1 (clone SD 4) (Immunotech/Coulter Corporation, Miami. FL) was also satisfactory

- *rinsing buffer:* 1% (w/v) bovine serum albumin (BSA) in PBS, pH 7.4 (store at 4°C)
- *Ckis antibodies:* anti-p16^{INK4a} (clone G175-405); anti-p21^{CIP1} (clone SX 118) and anti-p27^{KIP1} (clone G173-524) (PharMingen)
- mouse IgG1 (isotypic control)
- FITC-conjugated goat anti-mouse IgG
- *PI staining buffer:* 5 μg/ml PI (Molecular Probes, Eugene, OR) and 200 μg/ml DNase-free RNase A (Sigma Chemical Co., St. Louis, MO) in PBS, made fresh
- silanized or polypropylene 15 ml conical tube
- flow cytometer equipped with 488 nm argon laser

Procedures

A. Cell harvesting

1. *Cells growing in suspension, haematological samples.* Rinse once with PBS (300 *g*, 5 min room temperature) and resuspend in PBS to have $\sim10^6$ cells per 1 ml.

2. *Cells growing attached to tissue culture dishes.* Collect cells from flasks or Petri dishes by trypsinization and pool the trypsinized cells

with the cells floating in the medium (the latter consist of detached mitotic, apoptotic, and dead cells). Centrifuge at 300 *g* at room temperature for 5 min. Resuspend in medium containing 10% serum (to inactivate trypsin) and centrifuge again as before. Suspend cells in PBS at ~10^6 cells/ml.

Note. Other means of trypsin inactivation, such as addition of protease inhibitors may also be used

3. *Cells isolated from tissues (e.g. tumours).* Rinse free of any enzyme used for cell dissociation and resuspend in PBS as above.

Note. In the final suspension in PBS, the cells should be well dispersed (not in aggregates) and should not exceed the density of 5 × 10^6 cells/ml.

B. *Cell fixation*

1. With a Pasteur pipette, transfer 1 ml of the cell suspension in PBS into 15 ml tubes containing 10 ml of 80% ethanol or 100% methanol on ice. Time of fixation (storage) at 4°C may vary from 4 h to several days.

Note. To minimize cell loss, all the subsequent steps should be done in the same tube.

C. *Incubation with anti-cyclin or anti-Cki antibody*

1. Centrifuge for 5 min at 300 *g*, room temperature, remove alcohol, resuspend cells in 5 ml PBS, and centrifuge as before.
2. Remove supernatant and resuspend the cell pellet (10^6 cells) in 1 ml of 0.25% Triton X-100 in PBS. Keep on ice for 5 min, add 5 ml of PBS and centrifuge at 300 *g* for 5 min at room temperature.
3. Suspend the cell pellet in 100 μl of 1% BSA in PBS (rinsing buffer), containing the primary antibody at the appropriate dilution to obtain 0.5 μg of antibody per sample. Incubate for 60 min at room temperature with gentle agitation or at 4°C overnight. Treat control cells as above, except incubate them with the appropriate isotypic antibody at the same titre.

D. *Labelling with FITC-conjugated secondary antibody*

1. Add 5 ml of rinsing buffer, centrifuge for 5 min at 300 *g* at room temperature.
2. Suspend the cell pellet in 100 μl of FITC-conjugated anti-mouse IgG antibody diluted 1:30 in rinsing buffer. Incubate for 30 min in the dark at room temperature with gentle agitation.

Note. If the cyclin or Cki antibody directly conjugated to FITC is used, the basic protocol can be simplified by omitting these steps.

E. *Staining cellular DNA with PI*

1. Add 5 ml of rinsing buffer, centrifuge for 5 min at 300 *g* at room temperature.

2. Suspend the cell pellet in the PI staining solution. Incubate for 20 min at room temperature in the dark before measurement.

F. *Measuring cell fluorescence by flow cytometry*

1. Set up and adjust the flow cytometer for excitation with blue light (488 nm laser line). Use a 530±20 nm band pass filter for detection of FITC emission and a 620 nm long pass filter for PI emission (e.g. FL1 and FL3 on a FACScan, Becton–Dickinson, San Jose, CA).

2. Measure the cyclin- or Cki-associated green fluorescence of FITC and DNA-associated red fluorescence of PI (*Figure 1*).

3. Expression of cyclins D, E, A, and B1 and Ckis detected by flow cytometry

The periodic expression of cyclins D, E, A, and B1 in relation to the major phases of the cell cycle, as discussed in the Introduction, is reflected by a very characteristic bivariate pattern of cyclin vs. cellular DNA content distribution (*Figure 1*). The presence of cyclin D1 in exponentially growing normal fibroblasts is limited to cells in $G_{0/1}$; most S and G_2/M cells are cyclin D1-negative, with the exception of a very few cells with a G_2/M DNA content. The latter may be G_1 cell doublets. Such a pattern suggests that only early- to mid-G_1 cells are cyclin D1-positive. Because G_2/M cells are cyclin D-negative, immediately post-mitotic cells do not inherit this protein, which then has to be transiently synthesized early in G_1 and degraded prior to entrance to S phase.

The maximal expression of cyclin E is detected in cells undergoing transition from G_1 to S. The level of this protein decreases during cell progression through S, with the result that G_2/M cells are essentially cyclin E-negative. A distinct threshold in cyclin E expression is also apparent at the G_1/S transition: To enter S phase the cells have to accumulate cyclin E above a threshold level.

The pattern of expression of cyclin A indicates that its onset of accumulation is in early S phase. The level of this protein progressively increases during S, reaching a maximum in the G_2 phase. Because cyclin A is rapidly degraded in prometaphase, metaphase cells as well as post-mitotic, G_1 cells show no expression of this protein. In contrast to cyclin A which is present in all S phase cells, the expression of cyclin B1 is essentially limited to the cells with a G_2/M content of DNA although the onset of its accumulation occurs at the very end of S phase (19). Because this protein is degraded at the entrance to the anaphase, all mitotic cells prior to this stage are cyclin B1-positive. The 'scheduled' patterns of expression of cyclins D, E, A, and B1 with respect to the cell cycle, as shown in *Figure 1*, are characteristic of normal fibroblasts or mitogen-stimulated proliferating lymphocytes which grow asynchronously and exponentially (11, 12). Such patterns were also observed in several tumour cell lines.

There are situations, however, when the pattern of cyclin expression *vis-a-vis* cell cycle position is 'unscheduled', i.e. much different to that shown in *Figure 1*. The 'unscheduled' pattern defines the appearance of G_1 cyclins (i.e. cyclin E) in G_2/M phase, and/or G_2/M cyclins (cyclins A and/or B1) in G_1 phase cells. The unscheduled expression of cyclins E, A, and B1 was observed either as a result of perturbation of cell growth in an attempt to synchronize cells in the cycle (28, 30), or manifested as a phenotype of some tumour cell lines (22, 27). In the first instance, the suppression of cell cycle progression during cell synchronization by DNA polymerase inhibitors resulted in gross overexpression of cyclin E in G_1/S cells (fivefold increase; 28) in comparison with their unperturbed counterparts. Considering that the cells were being held arrested at G_1/S, i. e. in a phase when cyclin E is normally synthesized and accumulates in the cell, the overexpression of cyclin E might be expected. Quite unexpected, however, was the pattern of expression of cyclins B1 and A in the synchronized cells. Namely, while in the unperturbed cultures the cells at the G_1/S boundary are cyclin B1- and A-negative, while in the synchronized cultures they have almost as high a level of these cyclins as do G_2 cells in exponentially growing cultures. Synchronization by double thymidine block (or by aphidicolin, mimosine, or hydroxyurea, not shown) thus induces untimely (unscheduled) expression of cyclins B1 and A. The observed unscheduled expression of cyclins B1 and A in G_1 may be a result of the prolongation of their half-life by cyclin E (21). The latter, which is known to stabilize cyclins B1 and A (21), when overexpressed in cells arrested by inhibitors of DNA replication, may prevent degradation of these cyclins.

As mentioned, while many tumour cell lines exhibit patterns of expression of cyclins B1, A, and E similar to the expression seen for normal fibroblasts or lymphocytes as shown in *Figure 1*, some lines demonstrate distinctly 'unscheduled' expression of these cyclins (26, 27). Thus, for example, cyclin B1 was detected in G_1 and in early S phase of human promyelocytic leukaemic HL60, breast carcinoma Hs578T, and T-47D cells (26). Expression of cyclin E in G_2/M cells was observed in other tumour lines (26). It is possible that the unscheduled expression of cyclins B1 or E observed in these tumour lines is associated with dysregulation of the cell cycle drive machinery. Namely, the presence of G_1 cyclins during G_2/M and vice versa suggests that their partner Cdks may remain persistently active throughout the cell cycle. This may result in a loss of the regulatory control mechanisms at particular checkpoints of the cycle. If this is the case, the evidence of unscheduled expression of cyclins may be of prognostic value in oncology.

Cyclin D is the most frequently expressed in an 'unscheduled' fashion by tumour cells (27, 35). While the presence of D-type cyclins in normal cells is restricted to a fraction of G_1 cells and S, and G_2/M cells are cyclin D-negative (*Figure 1*), these cyclins appear to be expressed in all phases of the cycle in tumour cell lines, regardless of the overall level of their expression (35). This may be a reflection of the dysregulation of the pRB pathway, which as mentioned in the Introduction, appears to be a common occurrence in cancer. Such a pattern of cyclin D expression suggests that phosphorylation of pRB by Cdk4 is not restricted to G_1 phase in these cells, but continues throughout most of the cell cycle, including mitosis and the immediately post-mitotic phase.

In exponentially growing cells the expression of some Ckis, in analogy to that of cyclins D, E, A, and B, has a very characteristic pattern *vis-a-vis* the cell cycle position (*Figure 1*). For example, the cell cycle changes in level of p21 resemble those of cyclin D1. The expression of p21 is maximal in a fraction of G_1 cells, then declines in the cells entering S phase and is minimal in S phase cells. In contrast to cyclin D1, however, the level of p21 in some G_2/M cells is relatively high. The respective patterns of expression of cyclins D1 and E and p21 indicate that in the cells progressing through the early to mid-section of G_1, the level of cyclin D1 is high, of cyclin E is low, and expression of p21 is maximal. Just prior to S phase the levels of p21 and cyclin D1 drop to a minimum, but the expression of cyclin E rises to a maximum. This pattern of changes reflects the regulatory function of these proteins during cell transit through G_1 and entrance to S phase.

4. Critical steps in detection of cyclins or Ckis

The critical steps are cell fixation, permeabilization, and choice of the proper antibody. The fixative stabilizes the antigen *in situ* and preserves its epitope in a state where it remains accessible and reactive with the antibody. In most studies on cyclins, precipitating fixatives such as 70–80 % ethanol (22–26), absolute methanol (29, 30), or a 1:1 mixture of methanol and acetone (31), cooled to –20 to –40°C, have been used. Brief treatment with 1% formaldehyde followed by 70% cold ethanol appears to be a preferred procedure for fixation of D-type cyclins (12), although this cyclin can also be detected following fixation with cold methanol (22). Addition of detergent (e.g. 0.1% Triton X-100) to the solutions used after cell fixation with formaldehyde increases cell permeability to the antibody. The choice of fixative, thus, appears not to be a critical factor for cyclin detection and, although the absolute level of the immunofluorescence may vary, various fixation protocols yield essentially similar cyclin distributions with respect to cell cycle position. Each fixative has some undesirable effects (e.g. increased cell clumping in the case of ethanol/acetone mixture, or cell autofluorescence and poor DNA stainability when formaldehyde is used) and one often has to compromise between these effects and the optimal detection of a particular cyclin.

More critical for the detection of cyclins is the choice of a proper antibody. Very often the antibody, while applicable to Western blotting, fails in immunocytochemical applications, and vice versa. This may be due to differences in accessibility of the epitope or differences in the degree of denaturation of the antigen on the immunoblots compared with its *in situ* location. Some epitopes may not be accessible *in situ* at all. Since there is strong homology between different cyclin types, cross-reactivity may also be a problem. Because commercially available mAbs may differ in specificity, degree of cross-reactivity, etc., it is essential for the authors to provide information (the vendor and hybridoma clone number) of the reagent used in their study. We have noticed that concentrations of mAbs generally lower than those advised by the vendors provide better cyclin stainability due to decreased background fluorescence. It is advisable to titrate antibody on the actual cells used in each experiment to find the lowest concentration that saturates the binding sites (approaches plateau).

The abundance of a particular cyclin within the cell plays a role in its detection. The signal to noise ratio (ratio of fluorescence intensity of the cyclin-positive cells to the control cells, stained with the isotype immunoglobin) is generally higher for cyclin B1 than for cyclins E or A. This indicates that cyclin B1 is more abundant than the latter cyclins. The level of expression of D-type cyclins varies markedly depending on the cell type and the phase of cell growth. High sensitivity of the instrument and a low level of cell autofluorescence, therefore, are of greater importance for the detection of the less abundant cyclins E or A compared with cyclin B1 or D-type cyclins.

5. Detection of pRB phosphorylation

As mentioned, the activity of pRB is modulated by its phosphorylation (1–6). pRB is hypophosphorylated in quiescent cells and in early G_1 cells. Its phosphorylation on serine and threonine residues by Cdk4/6 during G_1 commits cells to enter S phase via release of transcription factors such as E2F, which are sequestered by hypophosphorylated pRB. The genes activated by E2F are essential for DNA replication and cell proliferation (7, 8). In cells progressing through S, G_2, and mitosis (until cell division), pRB remains in a hyperphosphorylated state. During terminal differentiation, when cells exit the cycle, expression of pRB is upregulated, but this protein remains hypophosphorylated (14).

Some questions that relate to the mechanism by which pRB controls the cell cycle cannot be resolved using classic methods of molecular biology that involve probing proteins by Western, or RNA by Northern, blotting. For example, is phosphorylation of pRB within the cell an 'all or nothing' phenomenon or is there a mixture of hypophosphorylated and hyperphosphorylated pRB molecules at varying proportions at all times during the cycle? What proportion of pRB molecules are phosphorylated within the cell during G_1 prior to entrance to the S phase? Is there a critical threshold in the ratio of hypophosphorylated to hyperphosphorylated pRB molecules that determines the transition of cells to quiescence or commitment to enter S? Is it the ratio of hypo- to hyper-phosphorylated pRB or the total level of the latter that is critical for cell commitment to enter S? The answers to these questions are important for the better understanding of the mechanism of cell cycle regulation in individual cells and, in particular, for explaining the heterogeneity in cell kinetics.

Some of the queries listed above may be answered by using a new approach to probe the status of pRB phosphorylation in individual cells (13, 14). The assay is based on the use of two anti-pRB mAbs, one of which specifically detects the hypophosphorylated form of this protein (pRB^{P-}) and another of which reacts with total pRB, regardless of its phosphorylation (pRB^{T}). Conjugation of each of these mAbs with fluorochromes of different colour has made it possible to immunocytochemically estimate the ratio of pRB^{P-}/pRB^{T} on a cell by cell basis. Utilizing multiparameter flow cytometry, in which cyclin expression and cellular DNA content are detected, the state of pRB phosphorylation can be correlated with the cell cycle position without the need for cell synchronization. Using this method we have studied phosphorylation of pRB during mitogenic stimulation of human lymphocytes and correlated it with induction of D cyclins and initiation of DNA replication (13), while human leukaemic HL60 cells were compared during normal proliferation and following differentiation in culture (14).

Protocol 2.

Reagents

- *fixative:* 1% formaldehyde in PBS, keep at 4°C ,and 80% ethanol, keep at –20°C.
- solution of 0.25% (v/v) Triton X-100 in PBS, pH 7.4 (store at 4°C)
- *rinsing buffer:* 1% (w/v) bovine serum albumin (BSA) in PBS, pH 7.4 (store at 4°C)
- *pRB antibodies:* anti-pRB^{T} mAb (Phar-Mingen, clone G3-245) conjugated with Cy-Chrome and anti-pRb^{P-} mAb (PharMingen, clone G99-549) conjugated with FITC
- phosphate-buffered saline (PBS)

- *PI staining buffer:* 5 μg/ml PI (Molecular Probes) and 200 μg/ml DNase-free RNase A (Sigma) in PBS, made fresh
- *DAPI staining buffer:* 1 μg/ml 4,6-diamidino-2-phenyl indole (DAPI) (Molecular Probes)
- silanized or polypropylene 15 ml conical tube
- flow cytometer equipped with 488 nm argon laser or 488 nm and UV light laser

Procedures

A. *Cell harvesting*
Follow the steps described in *Protocol 1*.

B. *Cell fixation*

1. With a Pasteur pipette, transfer 1 ml of the cell suspension into 15 ml tubes and add 1% formaldehyde in PBS for 15 min on ice, then wash

Protocol 2. *Continued*

with PBS and resuspend in ice-cold 80% ethanol on ice. Time of fixation (storage) at 4°C may vary from 4 h to several days.

Note. To minimize cell loss, all the subsequent steps should be done in the same tube.

C. *Incubation with anti-pRB antibodies*
1. Centrifuge for 5 min at 300 *g* at room temperature, remove alcohol, suspend cells in 5 ml PBS, and centrifuge as before.
2. Remove supernatant and resuspend the cell pellet ($\leqslant 10^6$ cells) in 1 ml of 0.25% solution of Triton X-100 in PBS. Keep on ice for 5 min, add 5 ml of PBS, and centrifuge at 300 *g* for 5 min at room temperature.
3. Suspend the cell pellet in 100 μl of 1% BSA in PBS (rinsing buffer), containing the pRb antibodies at the appropiate dilution to obtain 0.5 μg of pRB^T antibody and 1 μg of pRB^{P-} antibody per sample. Incubate for 2 h at room temperature with gentle agitation or at 4°C overnight.

D. *Staining cellular DNA with PI or DAPI*
1. Add 5 ml of rinsing buffer, centrifuge for 5 min at 300 *g* at room temperature.
2. Suspend the cell pellet in a PI staining solution (two colours: pRB^{P-}–FITC/PI) or in a DAPI staining solution (three colours: pRB^{P-}–FITC, pRB^T–Cychrome and DAPI). Incubate for 20 min at room temperature in the dark before measurement.

E. *Measuring cell fluorescence by flow cytometry*
1. Set up and adjust the flow cytometer for excitation with blue light (488 nm laser line). Use a 530±20 nm band pass filter for detection of FITC emission and a 620 nm long pass filter for PI emission or Cychrome. Alternatively, set up an additional UV light laser (340 nm lasing line) and use a filter/dichroic mirror combination to collect light at 480±20 nm wavelength for DAPI emission.

Figure 2 presents the bivariate distributions of pRB^T, pRB^{P-}, and pRB^{P-}/pRB^T vs. cellular DNA content of exponentially growing HL60 cells. Because the measurement of DNA content reveals the cell cycle position, the bivariate analysis as shown in this figure makes it possible to correlate expression of total pRB, as well as its phosphorylation state (pRB^{P-}, pRB^{P-}/pRB^T), with the cell cycle phase of individual cells. As is evident, the G$_1$ cell population is variable in expression of pRB^T. Also, a threshold in pRB^T during G$_1$ is apparent: nearly all cells entering S, as well as the cells in S and G$_2$/M, express pRB^T above the threshold level (*Figure 2*, arrow).

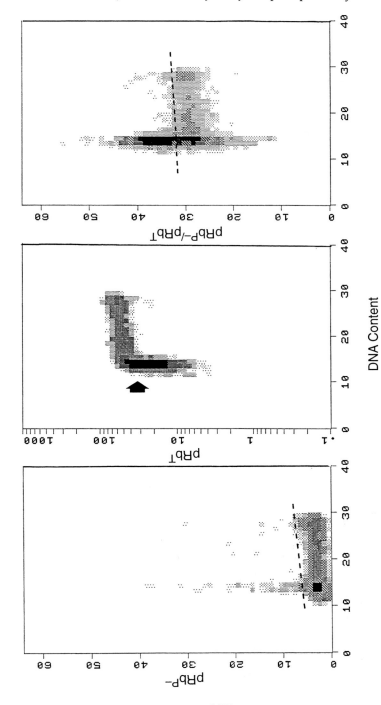

Figure 2. Scatterplots representing bivariate distributions of pRB^{P-}, pRB^T, and pRB^{P-}/pRB^T ratio vs. DNA content of HL60 cells stained with the respective antibodies and DAPI, as described in *Protocol 2*.

In the exponentially growing, untreated cultures, the expression of pRB^{P-} is confined to a rather small (~10 %) fraction of G_1 cells (*Figure 2*). This indicates that during G_1, pRB remains underphosphorylated for a relatively short period of time, equivalent to approximately 10% of the duration of G_1. The cell cycle phase differences in pRB^T and pRB^{P-} are reflected in the change of pRB^{P-}/pRB^T across the cell cycle (*Figure 2*). The intercellular variability of the G_1 population with respect to pRB^{P-}/pRB^T, and the presence of a pRB^{P-}/pRB^T threshold prior to entrance to S, are both evident. The cells in S and G_2/M are much more uniform in their pRB^{P-}/pRB^T ratio compared with G_1 cells.

Acknowledgements

Supported by NCI Grant RO1 28704, the 'This Close' Foundation for Cancer Research, and the Robert Welke Foundation for Cancer Research.

References

1. Dowdy, S.F., Hinds, P.W., Loule, K., Reed, S.I., Arnold, A., and Weinberg, R.A. (1993). *Cell*, **73**, 499.
2. Krek, W., Livingston, D.M., and Shirodkar, S. (1993). *Genes Dev.*, **7**, 1850.
3. Weinberg, R. A. (1995). *Cell*, **81**, 323.
4. Wu, C.-L., Classon, M., Dyson, N., and Harlow, E. (1996). *Mol. Cell Biol.*, **26**, 3698.
5. Bartek, J., Bartkova, J., and Lukas, J. (1996). *Curr. Opin. Cell Biol.*, **8**, 805.
6. Huang, Y., Ishiko, T., Nakada, S., Utsugisawa, T., Kato, T., and Yuan, Z.-M. (1997). *Cancer Res.*, **57**, 3640.
7. Sladek,T.L. (1997). *Cell Prolif.*, **30**, 97.
8. Farnham, P.J., Slansky, J.E., and Kollmar, R. (1993). *Biochim. Biophys. Acta*, **1155**, 125.
9. Xu, G., Livingston, D.M., and Krek, W. (1995). *Proc. Natl. Acad. Sci. USA.*, **92**, 1357.
10. Bartek, J., Bartkova, J., and Lukas, J. (1997). *Exp. Cell Res.*, **237**, 1.
11. Darzynkiewicz, Z., Gong, J., Juan, G., Ardelt, B., and Traganos, F. (1996). *Cytometry*, **25**, 1.
12. Darzynkiewicz, Z., Traganos, F., and Gong (1994). *Meth. Cell Biol.*,**41**, 421.
13. Juan, G., Gruenwald, S., and Darzynkiewicz, Z. (1998). *Exp. Cell Res.*, **239**, 104.
14. Juan, G., Li, X., and Darzynkiewicz, Z., (1998). *Exp. Cell Res.*, **244**, 83.
15. Ajchenbaum, F., Ando, K., DeCaprio, J.A., and Griffin, J.D. (1993). *J. Biol. Chem.*, **268**, 4113.
16. Tam, S.W., Theodoras, A.M., Shay, J.W., Draetta, G.F., and Pagano, M. (1994). *Oncogene*, **9**, 2663.
17. Dulic, V., Lees, E., and Reed, S.I. (1992). *Science*, **257**, 1958.
18. Krek, W., Xu, G., and Livingston, D.M. (1995). *Cell*, **83**, 1149.
19. Juan, G., Li, X., and Darzynkiewicz, Z. (1997). *Cancer Res.*, **57**, 803.

20. Pines, J. and Hunter, T. (1991). *J. Cell Biol.*, **115**, 1.
21. Amon, A., Irniger, S., and Nasmyth, K. (1994). *Cell*, **77**, 1037.
22. Gong, J., Ardelt, B., Traganos, F., and Darzynkiewicz, Z. (1994). *Cancer Res.*, **54**, 4285.
23. Gong, J., Bhatia, U., Traganos, F., and Darzynkiewicz, Z. (1995). *Leukemia*, **9**, 983.
24. Gong, J., Li, X., Traganos, F., and Darzynkiewicz, Z. (1994). *Cell Prolif.*, **27**, 357.
25. Gong, J., Traganos, F., and Darzynkiewicz, Z. (1993). *Cancer Res.*, **53**, 5096.
26. Gong, J., Traganos, F., and Darzynkiewicz, Z. (1994). *Cancer Res.*, **54**, 3136.
27. Gong, J., Traganos, F., and Darzynkiewicz, Z. (1995). *Cell Prolif.*, **28**, 337.
28. Gong, J., Traganos, F., and Darzynkiewicz, Z. (1995). *Cell Growth Differ.*, **6**, 1485.
29. Sherwood, S.W., Rush, D.P., Kung, A.L., and Schimke, R.T. (1994). *Exp. Cell Res.*, **211**, 275.
30. Urbani, L., Sherwood, S.W., and Schimke, R.T. (1995). *Exp. Cell Res.*, **219**, 159.
31. Bartkova, J., Lukas, J., Strauss, M., and Bartek, J. (1995). *Oncogene*, **10**, 775.
32. Lukas, J., Bartkova, J., Welcker, M., Petersen, O.W., Peters, G., Strauss, M., and Bartek, J. (1995). *Oncogene*, **10**, 2125.
33. Morgan, D.O. (1995). *Nature*, **374**, 131.
34. Sherr, C. J., and Roberts, J.M. (1995). *Genes Dev.*, **9**, 1149.
35. Juan, J., Gong, J., Traganos, F., and Darzynkiewicz, Z. (1996). *Cell Prolif.*, **29**, 256.

8

Methods for oncogene detection

STEPHEN C. COSENZA, STACEY J. BAKER, and
E. PREMKUMAR REDDY

1. Introduction

It is now widely agreed that cancer results from mutations in genes that regulate a cell's ability to proliferate, differentiate, and die. Despite the fact that the human genome contains, on average $1-3 \times 10^8$ base pairs of DNA, relatively few mutated genes have been implicated as causative agents in the induction of cancer. As a result, the identification of these genes, termed oncogenes, within the past 10–15 years has allowed the field of molecular biology to make vast contributions to the field of cancer research.

Although relatively few human cancers are caused by viruses, viruses are the causative agents of tumours in many animal species such as mouse and chicken. In 1911, Peyton Rous isolated the first tumour-forming retrovirus (termed Rous sarcoma virus, or RSV) from a naturally occurring tumour of a chicken (1). The virus was later proven to be transforming due to the acquisition of an additional sequence from the host cell. This sequence, termed v-*src*, is the viral homologue of the normal c-*src* cellular proto-oncogene and is transforming due to the fact that DNA sequences that encode key amino acid residues that regulate the activity of the Src protein were lost during transduction by RSV (2). While the significance of these studies initially seems irrelevant to the study of human tumours, the notion that mutation of a cellular gene can lead to deregulated cell growth was the first key to our understanding of cancer in humans.

The ability to introduce foreign DNA into mammalian cells discovered during the 1970s (3, 4) revolutionized the techniques used to investigate genes which could be implicated in human cancers. The data provided by the retroviral systems suggested that cellular genes, called proto-oncogenes, could become oncogenic when they are mutated and thus are converted to oncogenes, or genes that promote cancer. The ability to confer the trans-forming phenotype from transformed human cells to the non-transformed murine recipient cell line NIH-3T3 in a focus-forming assay, led to discovery of dominant-acting human transforming genes (5–8). Subsequent cloning and identification of the genes responsible for the transformation event proved that small changes in the proto-oncogene sequence can result in its conversion to an oncogene. One of the first experiments to demonstrate this phenomenon was performed using the T24 bladder carcinoma (9). While the initial transfection of T24-derived genomic DNA into NIH-3T3 cells induced transformed foci, sequence analysis of the cloned transforming sequences revealed that a single point mutation in the cellular *ras* gene at codons 12 or 61 was responsible for the conversion of the normal cellular proto-oncogene into a transforming oncogene. Therefore, like the oncogenes of many retroviruses, those present in human tumours are mutated versions of a normal gene that promote cancerous growth.

Since *ras*, approximately 200 oncogenes have been identified (10). Activation of cellular genes resulting in dominant transforming genes can occur by point mutations, amplifications, and translocations. This chapter describes the NIH-3T3 cell transformation assay and the subsequent co-transfection/nude mouse assays which can be performed for the detection and identification of oncogenes.

2. Detection of oncogenes using NIH-3T3 cell focus formation assays

To determine whether a tumour sample contains dominant transforming sequences or to determine whether a cloned gene is oncogenic, the DNA is introduced into non-transformed cells and its ability to induce a transformed phenotype is studied. One of the most widely used model systems to measure transformation potential is the NIH-3T3 cell system, an immortalized, but non-transformed cell line derived from mouse fibroblasts. Under normal circumstances, these adherent cells grow as a monolayer and cease to divide when they are contact inhibited upon reaching confluence. When contact inhibition occurs, the cells enter a quiescent state and remain viable for long periods of time without proliferating. However, when these cells become transformed they are not contact inhibited, fail to enter quiescence, and therefore continue to proliferate and form foci. The ability of a particular gene or DNA isolated from a tumour sample to induce foci in NIH-3T3 *in vitro* focus formation assays is a good indication of whether it encodes transforming sequences; in general, genes with the highest oncogenic potential induce the formation of more foci per microgram of DNA than those with lesser oncogenic potential. Generally, two types of DNA are transfected into NIH-3T3 cells when inducing foci: (1) cellular genomic DNA derived from a tumour, cell line, or other source, and (2) plasmid DNA in which constitutive expression of a cloned gene is driven by an exogenous promoter (i.e. RSV or CMV) or retroviral LTR (when using a retroviral vector).

3. Transfection using cellular genomic DNA

Roughly 20% of the DNAs derived from tumour samples induce foci in NIH-3T3 cell transformation assays. This is in contrast to DNAs derived from normal tissues, which do not induce transformation. The NIH-3T3 cell transformation assay is depicted in *Figure 1*. Briefly, high molecular weight genomic DNA is isolated from a tumour tissue culture sample, sheared (to make smaller DNA fragments that can be easily introduced into cells), and combined with calcium chloride. The addition of the DNA/CaCl$_2$ solution to a sodium phosphate-based buffer forms a precipitate that 'sits' on the cells. Because the calcium phosphate precipitate that contains the DNA makes the cellular membranes porous, the DNA can enter the cells and will eventually be stably integrated as part of the cell's genome.

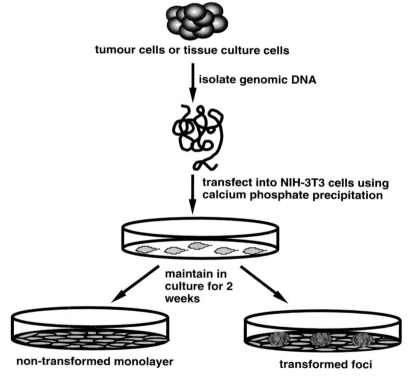

Figure 1. Schematic representation of NIH-3T3 cell transfection using calcium phosphate precipitation.

163

4. Isolation of high molecular weight genomic DNA from tumour samples and tissue culture cells and transfection into NIH-3T3 cells using calcium phosphate precipitation

Protocol 1. Isolation of genomic DNA from tissue or tissue culture cells (11)

Reagents

- lysis buffer: 100 mM NaCl, 10 mM Tris–HCl, pH 8.0, 25 mM EDTA, pH 8.0, 0.5% SDS, 0.1 mg/ml proteinase K
- PBS
- phenol/chloroform (1:1)
- chloroform/isoamyl alcohol (24:1)
- 7.5 M ammonium acetate

Method

1. Mince tissue samples immediately after excision. Freeze the minced sample in liquid nitrogen.

2. Using a mortar and pestle, or a hammer, break up the sample until it reaches a sand-like consistency.

3. Add 1.0–1.5 ml lysis buffer (freshly added prior to use) per 100 mg of tissue. The sample will not go into solution at this step. Do not forcibly resuspend, as shearing of the DNA can occur.

4. If the DNA is to be isolated from tissue culture cells, trypsinize the cells and collect them by centrifugation.

5. Discard the supernatant and wash the pellet twice with PBS.

6. Resuspend the sample in 1 ml DNA lysis buffer (see above) per 1×10^8 cells.

7. Incubate the samples overnight (minimum of 12 h) at 50°C. The samples will be very viscous. Do not vigorously pipette, as shearing of the DNA can occur.

8. Extract the samples twice with an equal volume of phenol/chloroform mixture (1:1).

9. Perform a final extraction using chloroform/isoamyl alcohol (24:1).

10. Precipitate the supernatant (containing the genomic DNA) using 1/2 vol. of 7.5 M amonium acetate (pH 7.5) and 2 original vols of ethanol. One original vol of isopropanol can also be used to minimize the total volume. Mix thoroughly; the precipitate should form immediately. If it does not, the samples can be precipitated overnight at –20°C.

11. Pellet the DNA by centrifugation at 1500 g for 15–30 min.

12. Wash the pellet in 70% ethanol.

13. Decant the ethanol and allow the pellet to air-dry for 10–15 min. Drying under vacuum is not recommended.

14. Resuspend the pellet in 10 mM Tris–HCl, pH 8.0, overnight at 50 °C. Again, do not vigorously pipette to resuspend, as shearing of the DNA can occur.

15. In order to use the DNA in transfection assays, it must be sheared. To shear the DNA, pass it through a syringe attached to a 21 gauge needle. (It is recommended that only an aliquot be sheared at this point, as the DNA may be needed for future experiments that require intact high molecular weight genomic DNA.)

Protocol 2. Transfection of NIH-3T3 cells using calcium phosphate precipitation (3, 9)

Reagents

- DMEM (Dulbecco's modified Eagle medium)
- calf serum
- penicillin/streptomycin
- 2 × HEBS (500 ml): 8 g NaCl, 0.37 g KCl, 0.125 g Na_2HPO_4, 99.4 mg Na_2PO_4, 1 g dextrose, 5 g Hepes

Method

1. Plate 1–2 × 10^5 cells/100 mm dish. The cells should be obtained from healthy cultures, actively proliferating in DMEM supplemented with 10% calf serum and penicillin/streptomycin.

2. The following day, combine 30 μg sheared DNA with sterile water, bringing the final volume to 450 μl.

3. Add the mixture to 50 μl of 2.5 M $CaCl_2$.

4. Add the DNA/$CaCl_2$ mixture dropwise into 500 μl of 2 × HEBS (Hepes-buffered saline) (pH 7.05–7.10) while bubbling with a pipette-aid, and incubate at room temperature for 30 min (not more than 1 h).

   Adjust the final pH to 7.05–7.10 with 10 N NaOH. Filter sterilize and keep frozen. The pH of the buffer must be checked prior to each use as it can dramatically affect transfection efficiency.

5. Add one precipitated sample to one dish of NIH-3T3 cells (plated in step 1). The precipitate may not be visible at this point. It should be easily visible the following day, appearing as sand-like particles on top of the cells.

6. Twenty-four hours post-transfection, change the medium, using DMEM that is supplemented with 5% calf serum and penicillin/streptomycin. Replace the medium weekly, or when it becomes acidic.

7. Foci are scored approximately 2 weeks post-transfection and are normally reported as the number of foci per microgram of DNA.

5. Transfection of NIH-3T3 cells by calcium phosphate precipitation using plasmid DNA (3, 12)

As previously stated, approximately 20% of the DNAs derived from tumour samples will induce focus formation in NIH-3T3 cells. While this number seems high with respect to normal cells that yield zero transformants, it actually represents a limitation of the assay itself. Because the genomic DNA must be sheared to facilitate uptake into the cells, the possibility exists that a gene with transforming potential may be cleaved by the shearing process. This cleavage may destroy coding and/or splicing sequences and render the gene inactive. As a result, a tumour that contains oncogenic sequences that are encoded in large loci may phenotypically resemble a normal cell and produce zero foci. Therefore, it is advantageous to sub-clone the cDNA that encodes a suspected oncoprotein into an expression or retroviral vector and introduce the plasmid into the cells, especially when the starting material that was used to isolate the oncogene is available in limited quantity. In addition, sequencing and manipulation of the gene can be easily performed on plasmid DNAs, which are easily purified in large quantities. Two types of plasmids work well in NIH-3T3 cells. The first type is an eukaryotic expression vector that places the gene of interest under the control of a strong viral promoter, such as that of CMV or RSV. The second is a retroviral vector, a plasmid that places the gene of interest under the control of the LTRs (Long Terminal Repeat) of a retrovirus, such as MSV or RSV. In general, retroviral vectors such as pMV7 (13) drive higher levels of gene expression in NIH-3T3 cells than do more general eukaryotic expression vectors; this is most likely due to the fact that the 5'- and 3'-LTRs contain some of the strongest promoter/ enhancer sequences identified thus far. However, since retroviral vectors tend to be large (most are greater than 9.0 kb), they can be more difficult to purify in large quantities. In addition to concerns with vector types, it has also been shown that for some very strong transforming genes, such as v-*abl*, only small amounts of linear DNA should be used for transfection procedures to ensure optimal focus formation. This is presumably due to the fact that there is an upper limit to the amount of a transfected oncogene that a cell can 'tolerate', and beyond this amount the gene is toxic to the cells (12).

Protocol 3. Transfection of NIH-3T3 cells using calcium phosphate precipitation using plasmid DNA

Reagents

As for *Protocol 2.*

Method

1. Plate 1–2 × 10^5 cells/100 mm dish. The cells should be obtained from healthy cultures, actively proliferating in DMEM supplemented with 10% calf serum and penicillin/streptomycin.

2. The following day, combine 10–100 ng linearized plasmid DNA and 20 μg calf thymus high molecular weight DNA and bring the final volume to 450 μl with sterile water. Both the plasmid DNA and the calf thymus DNA (which acts as a carrier) should be clean, without protein or RNA contamination.

   Note. While linearization is not required, it can increase the likelihood that the gene of interest will be stably integrated into the cellular genome. In order for stable integration to occur, the DNA must be nicked. By linearizing the DNA [in a part of the plasmid that is non-essential for the cell (i.e. in the ampicillin resistance gene), the DNA is cut in a region that lies outside the cDNA sequences, the promoter sequences that drive its expression, and any drug resistance markers that may be used for selection (see below) at a later time]. If whole plasmid DNA is used, the cell's machinery will randomly nick the DNA, possibly disrupting the coding and/or regulatory sequences.

3. Add the mixture to 50 μl of 2.5 M CaCl$_2$.

4. Add the DNA/CaCl$_2$ mixture dropwise into 500 μl of 2 × HEBS (Hepes-buffered saline) (pH 7.05–7.10) while bubbling with a pipette-aid, and incubate at room temperature for 30 min (not more than 1 h).

5. Add one sample to one dish of NIH-3T3 cells (plated in step 1). The precipitate may not be visible at this point. It should be easily visible the following day, appearing as sand-like particles on top of the cells.

6. Twenty-four hours post-transfection, change the medium, using DMEM that is supplemented with 5% calf serum and penicillin/streptomycin. Replace the medium weekly, or when it becomes acidic.

7. Foci are scored approximately 2 weeks post-transfection and are normally reported as the number of foci per microgram of DNA. Several oncogenes, such as *ras*, routinely and rapidly induce the formation of 10μ000 foci per microgram of plasmid (9, 14), whereas others, such as v-*abl*, induce roughly 400–500 (12).

6. NIH-3T3 co-transfection/nude mouse tumorigenicity assay

Although the NIH-3T3 cell focus formation assay system was originally very sensitive for identifying dominant-acting oncogenes, it became obvious that this system was biased to members of the *ras* family. Analysis of all the cloned genes derived from this system demonstrated that the gene products were activated forms of their cellular counterparts. Taken together with the observation that most DNA from tumour samples fails to produce foci in transfected NIH-3T3 cells, it was speculated that gene products consisting of other forms of genetic mutations or structural abnormalities such as amplifications would be missed by this system. In addition to the above theoretical concerns, the NIH-3T3 cell focus formation assay system has a few technical problems which makes the interpretation subjective when scoring for the number of foci. This is mainly due to culture conditions and the formation of spontaneously formed foci. One of the major problems is that the identification of positive foci relies on a visual comparison of the morphological outgrowth of transformed foci compared with a normal confluent monolayer and spontaneously formed foci. We now know that the morphological phenotype of a focus is directly related to the activity of a particular oncogene. Although some foci, such as those produced by v-*abl* transformation of NIH-3T3 cells, are readily distinguished from spontaneously transformed cells, other foci derived by transformation by genes such as v-*fgr* are a little more difficult to distinguish from background. The time required for the formation of the foci in a monolayer is an experimental factor which must not be taken lightly. Spontaneously derived foci from an NIH-3T3 monolayer will appear in the culture beyond 24 days. Therefore, as mentioned above, foci produced by the transfected DNA should appear before this time.

This entire issue is dependent upon the origin of the NIH-3T3 cells and the culturing conditions of the cells before and after transfection. This is the one aspect of the assay system that could make this protocol problematic. The development of alternative assay systems by Blair *et al.* (15) and Fasano *et al.* (16) addressed these issues. The assay system is now routinely called the NIH-3T3 co-transfection/nude mouse system. The major changes included in these modified protocols relies on the use of short-term transfected NIH-3T3 cells followed by injection into nude mice to look for the development of tumours (15). The athymic nude (nu/nu) mouse model system was found to be a very sensitive system for transformation potential (17) and therefore provided a direct *in vivo* assay for solid and lymphocytic tumour development. A further improvement on this system was developed by Fasano *et al.* (16) in which they co-transfected a selectable marker along with the genomic DNA in order to select for cells that were competent for DNA uptake and, as a result, increase the assay's sensitivity. Secondly, the drug selection step greatly reduces, if not eliminates, the problem of spontaneously transformed NIH-3T3 cells. A schematic of this system is shown in *Figure 2*. This system has proved to be useful in identifying oncogenes which include not only activating mutations but also genomic rearrangements (18) and gene amplifications (15, 19). Important points will be addressed during each step of the protocol.

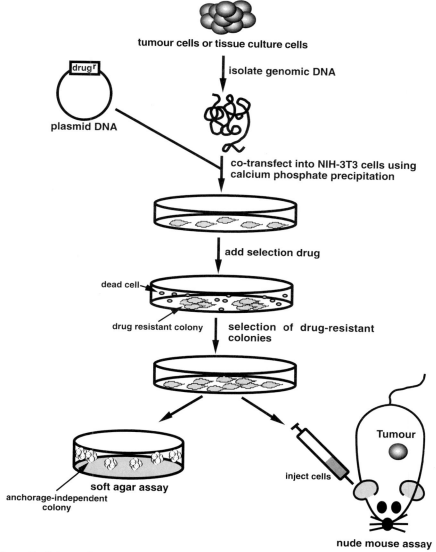

Figure 2. Schematic representation of the NIH-3T3 co-transfection/nude mouse tumorigenicity assay.

Protocol 4. Nude Mouse Assays using Transfected NIH-3T3 cells

This assay is performed essentially the same previously described in *Protocols 2* and *3* with the only modification being the addition of plasmid-containing bacterial genes expressed from a mammalian promoter, conferring selection to antibiotics such as neomycin (neo) and hygromycin phosphotransferase (hygro), which confer resistance to G418 and hygromycin, respectively, or the bacterial *gpt* gene, which allows cells to grow in medium containing mycophenolic acid, xanthine, and aminopterin. These plasmids are commercially available from commercial sources such as Stratagene, Clontech, and Invitrogen.

Method

1. Plate 1–2 × 10^5 cells/100 mm dish. The cells should be obtained from healthy cultures, actively proliferating in DMEM supplemented with 10% calf serum and penicillin/streptomycin.

2. The following day, combine 30 µg sheared DNA and 0.3–5 µg of selection plasmid with sterile water, bringing the final volume to 450 µl.

   Note. The amount of selection marker used will influence the number of resistant cells obtained. However, this ultimately depends on the vector itself and should be determined empirically.

3. Add the mixture to 50 µl of 2.5 M CaCl$_2$.

4. Add the DNA/CaCl$_2$ mixture dropwise into 500 µl of 2 × HEBS (Hepes-buffered saline) (pH 7.05–7.10) while bubbling with a pipette-aid, and incubate at room temperature for 30 min (not more than 1 h).

5. Add one precipitated sample to one dish of NIH-3T3 cells (plated in step 1). The precipitate may not be visible at this point. It should be easily visible the following day, appearing as sand-like particles on top of the cells.

6. Twenty-four hours post-transfection, change the medium, using DMEM that is supplemented with 5% calf serum and penicillin/streptomycin. The cells are permitted to grow for another 24 h.

7. The transfected cells are trypsinized and re-seeded in DMEM supplemented with 5% calf serum, penicillin/streptomycin, and the appropriate antibiotic or drug. The concentration of selection agent should be determined empirically or by literature review, but, for example, NIH-3T3 would require 250 µg/ml active G418 or 150 µg/ml hygromycin. The cells should be plated such that you achieve a 1:4 split. That is, each 100 mm transfection plate should be trypsinized and divided equally into four new plates. This is important since all

selection processes require cell division and therefore low cell density is required.

8. The plates are re-fed every 3–4 days with the same selection medium.

9. During the first 3–5 days you should observe the death of the majority of cells. If you do not observe death then there is a problem with the selection process. Equally important is that the death is not too rapid (less than 3 days), indicating that the genomic DNA sequences have not had a chance to integrate into the NIH-3T3 cell genome, which again requires cell division. It is therefore important to always include as control plates non-transfected or simply genomic DNA-transfected cells.

10. The plates should eventually contain 200–400 colonies and should be maintained until they just reach confluency (14–18 days post splitting).

11. At this time the cells from each four-plate group are trypsinized, pooled, washed three times with phosphate-buffered saline to remove all traces of selection agent, and counted using traditional methods.

12. For injections, there are normally enough pooled cells to inject 0.5–1 10^7 cells per transfection for each athymic nude mouse. Alternatively, the cells can be used for the soft agar assay (see below).

Protocol 5. Nude mouse injections and tumour volume determinations

Method

1. Investigators should follow all rules and guidelines set forth by the institution's Institutional Animal Care and Use Committee (IACUC) or equivalent.

2. Athymic nude mice (nu/nu) can be purchased from a variety of animal suppliers with a few different genetic backgrounds. Female mice are generally easier to handle since you can house five at a time without much chance of fighting.

3. The mice are housed in barrier facilities in Hep-filtered top cages with sterile food and water.

4. The mice are injected with the pooled resistant cells (volume should not exceed 0.2 ml) using a 1 ml tuberculin syringe containing a 27 gauge needle into the flank of the hind leg of each mouse. Be careful not to inject into the muscle tissue just below the skin.

5. The mice are checked weekly for tumour development. When the

Protocol 5. *Continued*

tumours become visible, tumour growth will be monitored by weekly volume determinations.

6. The tumours are measured in three diameters with vernier calipers.

7. Tumour volume is calculated by the following formula: $V = \pi(XYZ)/6$ where X, Y, Z are the three measured diameters minus the folded skin thickness (1 mm).

8. In order to analyse the genomic DNA from the primary tumours and to perform secondary foci and tumour assays, the tumours can be excised, explanted for expansion, and genomic DNA isolated as described above.

Note. Tumour development with this assay system should occur within 21–80 days post-implantation. If one uses DNA from the *ras*-transformed T24 cells, foci develop between 21 and 35 days. This is compared with tumours that have been shown to arise from transfections using normal NIH-3T3 or human lung diploid fibroblast DNA with a latency of 63–112 days (20).

7. Secondary NIH-3T3 foci and tumour induction assays

The primary foci and tumours derived from both methods are subsequently re-tested by simply repeating the entire experimental sequence. Genomic DNA is isolated from primary focus or a tumour sample and analysed by routine methods in order to determine the presence of human genomic repetitive (*alu*) sequences. This powerful internal sequence tag allows a comparison between primary and secondary tumours without actually cloning the DNA sequences. This is performed by Southern-blotting digested genomic DNA and probing with alu sequence probes such as blur-8. (This probe can be obtained from the American Type Culture Collection.) The first Southern blot will show multiple bands that hybridize to the probe. However, after secondary and tertiary foci are isolated, only one band will hybridize to the probe. *Figure 3* shows a schematic representation of this technique. The techniques for the analysis and cloning of the transforming genes are beyond the scope of this chapter. Readers should refer to the original articles and recombinant DNA technology method books.

8. Soft agar assays

Transformation of anchorage-dependent cells such as primary fibroblasts results from a progression of genetic events. Experimentally, the first plateau

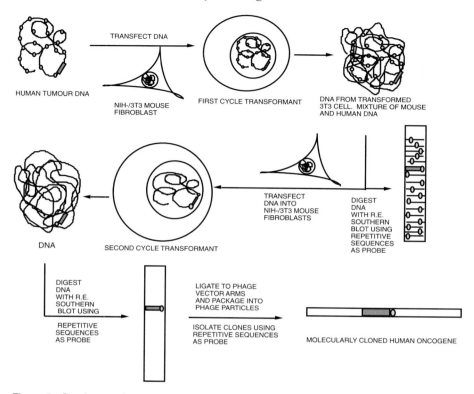

Figure 3. Cloning *ras* from the T24 bladder carcinoma.

that a cell needs to reach to achieve the fully transformed state is immortalization. These cells (NIH-3T3 cells, for example), as described in Chapter 14, no longer senesce but still retain the ability to contact-inhibit and enter into quiescence. Cells which are fully transformed will not only lose the ability to enter into a typical contact-dependent quiescence and form foci, but also gain the advantage of growth in semi-solid supports or soft agar. This anchorage-independent growth in soft agar is thought to be characteristic of a more highly transformed phenotype. Transformed cells derived by the NIH-3T3 transformation assay can be tested for their ability to grew in semi-solid medium by the soft agar assay. This assay can replace the nude mouse assay if an animal facility is not readily available and can be used to compare the transforming potentials of tumour-derived DNA and cloned cDNAs. The following protocol will describe an assay which can be used directly following step 12 in *Protocol 4* in the co-transfection, but simple adjustments of cell numbers and volumes, while maintaining the appropriate soft agar concentration, can be done for other experimental assays. This would include anchorage-independent growth of isolated foci and the possible cloned oncogenes transfected into NIH-3T3 cells.

Protocol 6.

Method

1. Prepare 2.4% stock solution of noble agar in sterile tissue culture quality water. This stock solution is melted and placed in a 60 °C water bath until needed.

2. In order to prepare bottom agar, aliquot 22 ml of DMEM supplemented with 5% calf serum and penicillin/streptomysin into a sterile tube and add 11 ml of the 2.4% noble agar solution. The final concentration of noble agar should be approximately 0.8%

3. Quickly add 10 ml of the bottom agar to pre-warmed 100 mm dishes and allow to harden in the hood. The use of gridded plates can facilitate counting, if necessary.

4. The cells pooled from the transfection assay (*Protocol 4*, step 12 section) should be mixed into the standard growth medium to a final volume of 5.8 ml. To this tube add 1.2 ml of the 2.4% warmed noble agar, mix, and quickly add to the bottom agar. The final agar concentration should be 0.4–0.5%.

5. The top agar is permitted to harden and the plates are returned to the incubator.

6. The plates can be fed with 3–5 ml fresh top agar/complete growth medium mixture every 5 days, or as needed. Do not allow the plates to dry out.

7. Colonies should become visible by 7 (for genes with high transformation potential) to 14 days.

8. Colonies can be counted microscopically as they are or following staining with a 0.05% (w/v) nitroblue tetrazolium solution (NBT), prepared under sterile conditions in phosphate-buffered saline (PBS). Place enough 0.05% NBT solution to just cover the top agar. Incubate overnight in the incubator and count the next day. The cells are still viable following this staining procedure such that colonies can be isolated and expanded if necessary.

---

9. Summary

The above protocols describe classical methods that have withstood the test of time in both sensitivity and relative ease for the purpose of identifying oncogenes. As stated in previous sections, both the focus-forming assay and the co-transfection/nude mouse assay can be used separately or in conjunction with each other for experiments aimed at identifying potential oncogenes from both clinical and experimental samples. The ease and sensitivity of these experimental systems has led many investigators to identify the activation of proto-oncogenes in many systems which deal with the effects of various carcinogens (21–23). For example, oncogenes such as *ras*, *met*, *trk*, *dbl*, *vav*, *raf*, and *ret* have been identified using the NIH-3T3 transformation assay (reviewed in ref. 24), while others, such as *mas*, *axl*, and *bcd* have been identified using the NIH-3T3 co-transfection/nude mouse assay (18, 19, 25). It should also be noted that there are a number of other methods that have been used in addition to the above protocols to increase the sensitivity of the assay itself. These modifications have mainly been aimed at optimizing the transfection systems in combination with subtractive library expression systems (26, 27). These newer protocols may prove to be useful in cases when NIH-3T3 cells are found to be insensitive to the particular activity of a potential oncogene found within a differing genetic background.

References

1. Rous, P. (1911). *J. Exp. Med.* **13**, 397.
2. Brugge, J.S. and Erikson, R.L. (1977). *Nature* **269**, 346.
3. Graham, F.L. and Van der Eb, A.J. (1973). *Virology* **52**, 456.
4. Wigler, M., Sweet, R., Sim, G.K., Wold, B., Pellicer, A., Lacy, E., Maniatis, T., Silverstein, S. and Axel, R. (1979). *Cell* **16**, 777.
5. Shih, C., Shilo, B., Goldfarb, M., Dannenberg, A. and Weinber, R. (1979). *Proc. Natl. Acad. Sci. USA* **76**, 5714.
6. Krontiris, T.G. and Cooper, G.M. (1981). *Proc. Natl. Acad. Sci. USA* **78**, 1181.
7. Murray, M.J., Shilo, B.Z., Shih, C., Cowing, D., Hsu, H.W. and Weinberg, R.A. (1981). *Cell* **25**, 355.
8. Perucho, M. Goldfarb, M., Shimizu, K., Lama, C., Fogh, J. and Wigler, M. (1981). *Cell* **27**, 467.
9. Reddy, E.P., Reynolds, R.K., Santos, E., and Barbacid, M. (1982). *Nature* **300**, 149.
10. Baker, S.J. and Reddy, E.P. (1999). *The encyclopedia of molecular biology*, in press.
11. Ausubel, F., Brent, R., Kingston, R.E., Moore, R.E., Seidman, J.G., Smith, J.A. and Struhl, K. (ed.) (1991). *Current protocols in molecular biology*. John Wiley & Sons, New York.
12. Shore, S.K. and Reddy, E.P. (1989). *Oncogene* **4**, 1411.
13. Kirchmeier, P.T., Housey, G.M., Johnson, M.D., Perkins, A.S., and Weinstein, I.B. (1988). *DNA* **7**, 219.
14. Yuasa, Y., Srivastava, S.K., Dunn, C.Y., Rhim, J.S., Reddy, E.P. and Aaronson, S.A. (1983). *Nature* **303**, 775.
15. Blair, D.G., Cooper, C.S., Eader, L.A. and Vande Woude, G.F. (1982). *Science* **218**, 1122.

16. Fasano, O., Birnbaum, D., Edlund, L., Fogh, J. and Wigler, M. (1984). *Mol. Cell. Biol.*, **4**, 1695.
17. Fogh, J. and Giovanella, B.C. (ed.) (1978). *The nude mouse in experimental and clinical research.* Academic Press, New York.
18. El Rouby, S. and Newcomb, E.W. (1996). *Oncogene* **13**, 2523.
19. O'Bryan, J.P., Frye, R.A., Cogswell, P.C., Neubauer, A., Kitch, B., Prokop, C., Espinosa III, R., Le Beau, M.M., Earp, H.S. and Liu, E.T. (1991). *Mol. Cell. Biol.* **11**, 5016.
20. Reynolds, S.H., Anna, C.K., Brown, K.C., Wiest, J.S., Beattie, E.J., Pero, R.W., Iglehart, J.D. and Anderson, M.W. (1991). *Proc. Natl. Acad. Sci. USA* **88**, 1085.
21. Reynolds, S.H., Stowers, S.J., Maronpot, R.R., Anderson, M.W. and Aaronson, S.A. (1986). *Proc. Natl. Acad. Sci. USA* **83**, 33.
22. Newcomb, E.W., Steinberg, J.J. and Pellicer, A. (1988). *Cancer Res.* **48**, 5514.
23. Hegi, M.E., Fox, T.R., Belinsky, S.A., Devereux, T.R. and Anderson, M.W. (1993). *Carcinogenesis* **14**, 145.
24. Patterson, H. (1992). *Eur. J. Cancer* **1**, 258.
25. Young, D., Waitches, G., Birchmeier, C., Fasano, O. and Wigler, M. (1986). *Cell* **45**, 711.
26. Miki, T., Fleming, T.P., Crescenzi, M., Molloy, C.J., Blam, S.B., Reynolds, S.H. and Aaronson, S.A. (1991). *Proc. Natl. Acad. Sci. USA* **88**, 5167.
27. Shen, R., Su, Z., Olson, C.A. and Fisher, P.B. (1995). *Proc. Natl. Acad. Sci. USA* **92**, 6778.

Isolation and visualization of the principal components of nuclear architecture

GARY S. STEIN, MARTIN MONTECINO, SANDRA MCNEIL,
SHIRWIN POCKWINSE, ANDRÉ J. VAN WIJNEN,
JANET L. STEIN, and JANE B. LIAN

1. Introduction

It is becoming increasingly evident that gene expression *in vivo* is functionally interrelated with nuclear architecture. The linear ordering of protein coding sequences in genes, as well as the representation of flanking and intragenic regulatory elements, only partially accounts for transcriptional responsiveness to the broad spectrum of cues from physiological signalling pathways that modulate gene expression under diverse biological circumstances. Equally important, cataloguing protein–DNA and protein–protein interactions at gene regulatory sequences is necessary but insufficient to explain the transcriptional mechanisms that are operative in intact cells and tissues. Rather, there is growing awareness that gene expression must be understood from the perspective of contributions by multiple levels of nuclear organization.

The packaging of DNA as chromatin reduces distances between gene promoter elements, supporting the integration of regulatory activities. Regulatory mechanisms mediating remodelling of chromatin organization that render promoter sequences competent to interact with cognate transcription factors and support protein–protein interactions are required for fidelity of gene expression. The spatial distribution of genes and regulatory factors within the nucleus facilitates the formation of sites that support replication, transcription, and processing of gene transcripts. Mechanisms that direct factors to intranuclear domains, where replication and gene expression occur, are the principal parameters of control that interrelate biochemical components of regulation with nuclear morphology.

Consequently, there is a necessity to employ biochemical and *in situ* analyses to experimentally address control of gene expression within the three-dimensional context of nuclear architecture. The power of combining these approaches is the ability to identify and characterize regulatory factors and functional activities, as well as to visualize spatial components of gene regulatory mechanisms in intact cells. As the resolution of 'probes' and detection systems advance, we can anticipate further insight into linkages between nuclear structure and function.

In this chapter, we will restrict our consideration to *in vitro* and *in vivo* methods for identification, characterization, and visualization of nuclear proteins and nucleic acids that regulate or are regulated by nuclear structure—gene expression interrelationships. In addition, powerful approaches are available for pursuing the characterization and activities of nuclear domains where replication occurs.

2. *In vitro* experimental approaches

2.1 Chromatin organization

Regulation of chromatin structure plays an essential role in the modulation of transcription in eukaryotic cells (1). The organization of genomic DNA as chromatin allows the DNA to be partially protected in nucleosomes, but more readily accessible in linker regions. Thus, various enzymes and reagents can be used to cleave and detect DNA where it is accessible. We will focus on two of the most widely used enzymes to study changes in chromatin structure associated with transcriptional activation, micrococcal nuclease (MNase) and deoxyribonuclease I (DNase I).

Micrococcal nuclease (MNase) cuts double-stranded DNA, leaving fragments with a free hydroxyl group at their 5'-ends. The enzymatic activity shows marked preference for single-stranded DNA and an absolute requirement for Ca^{2+}. As MNase cuts DNA preferentially in the linker regions between the nucleosomes (2), its effect on bulk chromatin is to free oligomers corresponding to 1, 2, 3 nucleosomes, etc. This pattern is reflected by a ladder of bands when digested samples are analysed electrophoretically (3). Their sizes vary according to the size of the segment, which corresponds to the minimal nucleosomal repeat unit. Variations in this value might be found between different cell types or tissues that reflect changes in the linker length (4).

Deoxyribonuclease I (DNase I) from bovine pancreas is an endonuclease that cleaves phosphodiester linkages yielding 5'-phosphate-terminated polynucleotides with a free hydroxyl group at the 3'-position. This enzyme shows preference for double-stranded DNA and has some sequence specificity for AT-rich regions. It requires divalent cations (Ca^{2+}, Mg^{2+}, Mn^{2+}) for activity, cleaves the DNA in the minor groove, and cuts one of the two strands. A double-stranded break is observed only if a second cut occurs on the other strand in close proximity to the first cut (4). Extensive analysis of chromatin structure has suggested that most active genes contain DNase I hypersensitive regions. These domains generally reflect alterations in the classical nucleosomal organization and the binding of specific nuclear factors. Thus, DNase I digestion has been widely used to probe structures *in vivo* and *in vitro* based on the premise that accessibility to DNase I reflects chromatin accessibility to nuclear regulatory molecules (5, 6).

2.1.1 Analysis of chromatin structure by low resolution methods

Analysis of the chromatin structure can be initially carried out by examining extensive areas surrounding genes using low resolution approaches like the indirect end-labelling method (7). This approach may allow detection of hypersensitive domains associated with transcriptional activity and the nucleosome translational positioning at regulatory sequences. On average, this method will allow mapping of cleavage sites generated by DNase I and MNase with an accuracy of ±20 bp.

The sequential experimental steps that are normally followed while carrying on chromatin structure analysis by the indirect end-labelling method are summarized in *Protocols 1–4*.

Protocol 1. Nuclei isolation

Reagents

- nuclei isolation buffer: 60 mM KCl, 15 mM NaCl, 2 mM EDTA, 0.5 mM EGTA, 15 mM Tris–HCl, pH 7.4, 0.15 mM spermine, 0.5 mM spermidine, 0.5 mM DTT (dithio-threitol), 0.2 % (v/v) Nonidet NP-40
- digestion buffer: 60 mM KCl, 15 mM NaCl, 15 mM Tris–HCl, pH 7.4, 3 mM MgCl$_2$, 0.5 mM CaCl$_2$
- PBS

Method

1. Wash plated cells with cold phosphate-buffered saline (PBS), scrape in 2 ml of PBS per 100 mm plate, and pool into an ice-cold polypropylene conical tube. Pellet the cells at 800 *g* for 5 min at 4°C.

2. Resuspend the cell pellets in 8 vols of nuclei isolation buffer and lyse by dounce homogenization (pestle B). The composition of the nuclei isolation buffer may vary depending on the type of cell being studied. We have successfully used the nuclei isolation buffer in the analysis of several bone-derived cell lines as well as normal diploid primary rat osteoblasts (5).

3. Collect nuclei by low speed centrifugation (500 *g*) for 5 min at 4°C and gently resuspend in ice-cold digestion buffer.

Protocol 2. Nuclei digestion

Reagents

- DNase I and Mnase (Worthington Biochemicals)
- EDTA
- EGTA

Protocol 2. *Continued*

Method

1. Digest isolated nuclei (20 units per ml of absorbance at 260 nm) to a limited extent with DNase I or MNase. To perform this step two alternative approaches can be taken: incubate the nuclei in the presence of increasing concentrations of enzyme for a fixed period of time or incubate with a fixed amount of enzyme during increasing periods of time. Although both approaches have been utilized successfully, we prefer the former since it eliminates an increasing potential contribution of endogenous nuclease activity to the digestion pattern generated. Our digestion reactions are carried out at 18°C for 10 min in the case of DNase I (0, 1, 3, 5, 7, 10, 15 units per ml) and for 5 min with MNase (0, 1, 3, 5, 7, 10, 15, and 20 units per ml).

2. The digestion reaction is stopped by the addition of EDTA and EGTA to a final concentration of 25 mM and 10 mM, respectively.

Protocol 3. Genomic DNA purification

Reagents

- proteinase K (100 μg/ml)
- 0.5% SDS
- 3 M sodium acetate
- TE

Method

1. Purify partially digested genomic DNA by incubating overnight with proteinase K (100 μg/ml) in the presence of 0.5% (w/v) SDS at 37°C.

2. Extract the samples with phenol/chloroform/isoamyl alcohol (25:24:1) and precipitate the DNA from the aqueous phase by adding 0.1 vols of 3 M sodium acetate pH 7.5 and 2.5 vols of 95% (v/v) ethanol (–20°C).

3. Collect purified DNA samples by centrifugation at 4°C, wash with 70% (v/v) ethanol, dry, and resuspend in 100 μl of TE (Tris-EDTA buffer).

Protocol 4. Detection of nucleosome positioning and DNase I hypersensitivity

Method

1. Cut the partially digested purified genomic DNA to completion with a restriction enzyme using the conditions provided by the supplier. This enzyme should cleave near the region of interest (0.5–3 kb), providing a reference end.

2. Fractionate the DNA fragments by electrophoresis in agarose gels, denature, transfer to a membrane, and hybridize to a random primer-labelled probe that specifically recognizes the end generated by the restriction enzyme.

3. Autoradiography reveals a series of bands in the nuclei samples. Each band represents a DNA fragment with a reference restriction site at one end and the nuclease cleavage site at the other end. The size of the band allows determination of the distance of the nuclease cleavage site from the reference end (see *Figure 1*).

A critical limitation in the analysis of nucleosome positioning by MNase is the sequence preference shown by the MNase enzyme, cutting pA and pT faster than pC or pG (4, 8, 9). This sequence preference leads to a series of defined cuts in deproteinized DNA. This is why the strict comparison of chromatin-digested patterns with those obtained after limited digestion of naked genomic DNA is an absolute requirement in this type of analysis. Although sequence preference may occasionally restrict the interpretation of the results, it provides the advantage of allowing the measurement of cutting sites on naked DNA and of measuring those sites that are accessible in chromatin and those sites that are protected.

2.1.2 Analysis of chromatin structure by high resolution methods

The chromatin digestion patterns obtained following incubation of isolated nuclei with MNase or DNase I can be further analysed at the single nucleotide resolution level by using ligation-mediated PCR (LMPCR) (10). This may allow determination of precise nucleosome positioning as well as protein–DNA interactions occurring within DNase I hypersensitive domains.

The procedure described below in *Protocols 5–8* is based on previously published detailed protocols (11) and considers the analysis by LMPCR of the same nuclease-digested samples studied by the indirect end-labelling method (6, 12).

Protocol 5. First strand synthesis

Equipment and reagents

- first strand buffer mix (prepare just before use): 40 mM NaCl, 10 mM Tris–HCl, pH 8.9, 5 mM MgSO$_4$, 0.01% (w/v) gelatin, 200 μM of each dNTPs, 0.3 pmol gene-specific primer 1, 0.5 units of Vent DNA polymerase (New England Biolabs)
- MNase and DNase I
- TE
- thermal cycler

Method

1. The first step defines the fixed end of the DNA sequence ladder by denaturing the nuclease-cleaved genomic DNA and annealing a gene-specific primer. For this purpose dilute 2–3 μg of the MNase or DNase I

Protocol 5. *Continued*

partially digested DNA (in 5 μl TE, pH 7.5) in 25 μl of first strand synthesis buffer containing the gene-specific primer. Denature the mix for 5 min at 95°C, anneal the primer for 30 min, and carry out the synthesis reaction at 76°C for 10 min. All these steps can best be carried out in a thermal cycler.

2. The annealing temperature for the gene-specific primer 1, as well as for the other two primers to be used in the subsequent steps (see below), will depend on the percentage of CG present in the oligonucleotide and will have to be determined experimentally.

Protocol 6. Ligation of a unidirectional linker primer

Reagents

- T4 polynucleotide kinase
- MNase and DNase I
- dilution solution (prepare just before use): 110 mM Tris–HCl, pH 7.5, 17.5 mM MgCl$_2$, 50 mM DTT, 125 μg/ml purified BSA
- 2.7 M sodium acetate, pH 7.0

- ligase mix (prepare just before use): 10 mM MgCl$_2$, 20 mM DTT, 3 mM ATP, pH 7.0, 50 μg/ml purified BSA, 4 μM linker, 50 mM Tris–HCl, pH 7.7, 3 Weiss units of T4 DNA ligase
- 1 mg/ml yeast tRNA

Method

1. Extend the gene-specific primer 1 to the variable nuclease (MNase or DNase I) cleavage site to generate a family of blunt-ended duplex molecules, which can be a substrate for the T4 DNA ligase-catalysed addition of a specially designed linker (10) to each member of the sequence ladder, thereby providing each with a common, defined end.

2. As explained above, MNase activity produces DNA fragments containing 3′-phosphate and 5′-hydroxyl groups which will work well in LMPCR following phosphorylation by T4 polynucleotide kinase. On the other hand, DNase I digestion of DNA provides the 5′-phosphorylated end required for this ligation step.

3. Dilute samples by the addition of 20 μl of ice-cold ligase dilution solution, kept in an ice bath, and then add 25 μl of ice-cold ligase mix. Incubate the samples at 17°C overnight.

4. The specificity in this step is provided by the fact that the bulk of genomic DNA does not serve as a substrate in the ligation reaction because it lacks a blunt, double-stranded end (10).

5. Stop the ligation reaction by addition of 9.4 μl of 2.7 M sodium acetate, pH 7.0, with 1mg/ml yeast tRNA.

6. Precipitate the DNA samples with 220 μl of ice-cold 100% ethanol and kept at −20°C.

(a)

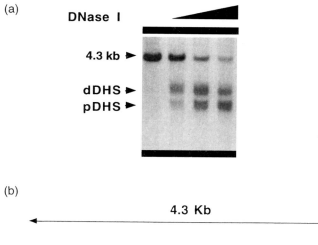

(b)

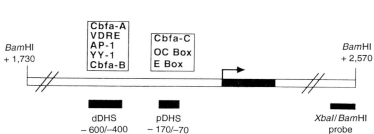

Figure 1. DNase I hypersensitivity at the osteocalcin (*OC*) gene promoter in bone-derived cells. Confluent rat osteosarcoma ROS 17/2.8 cells, which continuously express the OC gene, were collected, the nuclei isolated, and then incubated with increasing concentrations of DNase I for 10 min at 18°C. Following purification, the partially digested genomic DNA was cleaved to completion with *Bam*HI. The samples were then electrophoresed in 1.2% agarose gel, blotted, and hybridized with *Xba*I/*Bam*HI probe. (a) Two DNase I hypersensitive domains, a proximal (pDHS) and a distal (dDHS), are detected in the promoter of the *OC* gene only in bone cells expressing this gene (5, 60). (b) Schematic representation of the *OC* gene and flanking sequences. The diagram shows the position of the *Bam*HI restriction sites at both extremes of the 4.3 kb genomic DNA segment and the *Xba*I/*Bam*HI probe used in the indirect end-labelling analysis. The black box represents the *OC* gene coding region (including exons and introns) and the horizontal arrow over the gene marks the direction of transcription. The position of both DNase I hypersensitive sites, as well as the location of key transcription regulatory elements with these two sites, are indicated.

Protocol 7. Amplification

Reagents

- 5 × amplification buffer: 200 mM NaCl, 100 mM Tris–HCl, pH 8.9, 25 mM $MgSO_4$, 0.05% (w/v) gelatin, 0.5% (v/v) Triton X-100
- Vent polymerase mix (prepare just before use): 0.6 μl 5 × amplification buffer, 1.9 μl H_2O, 0.5 μl Vent polymerase
- amplification mix (prepared just before use): 20 μl 5 × amplification buffer, 1 μl 10 pmol/μl gene-specific primer 2, × 1 μl 10 pmol/μl linker primer (10), 0.8 μl 25 mM 4 dNTP mix, pH 7.0, 7.2 μl H_2O

Method

1. Carry out amplification of the DNA ladder by PCR, using a second gene-specific primer and the linker primer. To increase specificity, the second gene-specific primer should be positioned so that its extending end is 3′ to that of the first primer. Use the original genomic DNA again in this reaction, although now it has a longer strand provided by the linker covalently attached to it, and the extension products read through this added sequence. Each member of the sequence ladder now has two defined ends (the common linker primer and the second gene-specific primer), and is suitable for PCR.

2. Collect the DNA samples by centrifugation at 4°C, wash with 500 μl of 70% ethanol, dry, and resuspend in 70 μl of double-distilled water.

3. Add 30 μl of ice-cold amplification mix and 3 μl of Vent polymerase mix to each sample.

4. Perform 18 cycles of PCR, each involving denaturation for 1 min at 96°C and elongation for 3 min at 76°C. The annealing temperature will have to be estimated according to the sequence composition of the gene-specific primer 2 and confirmed experimentally. The first denaturation should be longer than 1 minute (3–4 minutes) and the last extension should be allowed to proceed for 10 minutes.

Protocol 8. Labelling

Reagents

- end-labelling mix (prepare just before use): 1 μl 5 × amplification buffer, 2.3 μl 1 pmol/μl end-labelled gene-specific primer 3, 0.4 μl 25 mM 4dNTP mix, pH 7.0, 0.5 μl Vent polymerase
- Vent stop solution (prepare just before use): 260 mM sodium acetate, pH 7.0, 10 mM Tris–HCl, pH 7.5, 4 mM EDTA, 68 μg/ml yeast tRNA

Method

1. After 18 cycles of PCR, the amplified DNA ladder can be visualized by primer extension of an end-labelled, third, gene-specific primer. This

primer should overlap the gene-specific primer 2, and also be positioned so that its extending end is 3′ to that of the second primer.

2. Transfer the samples to ice, spin down, and quickly return them to ice. Add 5 μl of end-labelling mix.

3. Carry out two more cycles of PCR to label the DNA. The first denaturation should be 4 min and the second 1 min at 95°C. Extension should be 10 min at 76°C. Again, the annealing temperature should be calculated according to the theoretical melting temperature and confirmed experimentally.

4. Stop the reaction by transferring the samples to ice and by adding Vent stop solution.

5. Extract vigorously with 1 vol of phenol/chloroform/isoamyl alcohol (25:24:1), microcentrifuge, divide the aqueous phase into three, and precipitate them with 2.5 vols of cold absolute ethanol at –20°C.

6. Collect the samples by centrifugation, wash with 70% (v/v) ethanol, dry, and resuspend in 7 μl of sequencing gel loading buffer.

7. Electrophorese the samples in a 6% sequencing gel, fix, and dry the gel, and autoradiograph overnight.

8. The appearance will be that of the corresponding sequence ladder, except that it will be uniformly longer by the additional length of the linker (see *Figure 2*).

Because nucleosome core DNA is resistant to MNase digestion, a precisely positioned nucleosome is expected to protect approximately 146 base pairs of DNA against double-strand cutting by MNase (3). However, because the linker length between individual nucleosomes is variable, and because MNase has sequence specificity, the distance between cutting sites in the linkers flanking a positioned nucleosome could vary between 146 and 200 base pairs.

Although the use of MNase digestion has provided a powerful tool to study nucleosome positioning, there are limitations. Among these are several reports of intranucleosomal cleavage (13–15), producing nicks on the nucleosome surface. This will not significantly affect the mapping of double-stranded cuts by low resolution methods, but it must be considered when individual strands are analysed by high resolution approaches such as LMPCR.

It is also necessary to consider that nucleosomes are dynamic structures that may associate and reassemble or move along the DNA sequence (16, 17). Hence, cleavage sites might be protected in only a fraction of the population. In addition, the chromatin organization and transcription factor binding in certain genes may be heterogeneous with respect to the cell cycle (18) and therefore produce a complex pattern that might be interpreted as non-positioned nucleosomes or absence of binding.

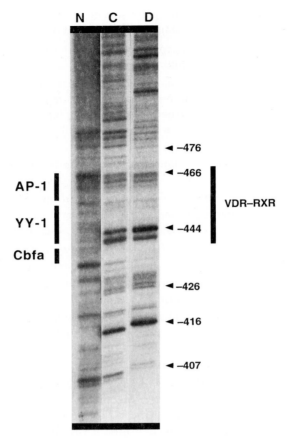

Figure 2. High resolution analysis of the distal DNase I hypersensitive site in the *OC* gene. Nuclei isolated from ROS 17/2.8 cells, control, and treated with vitamin D_3 for 4 h were digested with DNase I. Following purification, the partially digested genomic DNA samples were analysed by LMPCR as described (6, 61). Footprints encompassing binding elements for the transcription factors YY-1, Cbfa, AP-1, and the VDR–RXR heterodimers are shown by vertical bars. The nucleotide positions are indicated at both sides. N = protein-free DNA, C = control, and D = vitamin D_3.

2.2 Isolation and characterization of nuclear matrix components

2.2.1 *In vitro* biochemical isolation of nuclear matrix components

The three-dimensional organization of the nucleus plays a central role in the regulation of gene expression. Evidence includes the preferential association of actively transcribing genes and regulatory components of transcription, as well as processing of gene transcripts, with the nuclear matrix. DNA and regulatory factors are brought into juxtaposition by the nuclear matrix, a non-

Table 1. Biochemical isolation of nuclear matrix components

Extraction step	Outcome
Cytoskeleton buffer	Release of soluble cytoplasmic and nuclear proteins and phospholipids
RSB-Majik buffer	Extraction of cytoskeletal proteins and polyribosomes
Digestion buffer	Release of DNA and associated histones
Disassembly, reassembly buffers	Release of nuclear matrix proteins

chromatin scaffold within the nucleus. By regulating these interactions, the nuclear matrix plays a fundamental role in transcriptional control (19).

To determine the spatial relationships and functional interactions of nuclear matrix-associated proteins, the nuclear matrix is isolated by releasing soluble nuclear and cytoplasmic proteins, DNA, phospholipids, and chromatin (20). Protein interactions are evaluated by Western blot analysis, protein–protein interactions are assayed by immunoprecipitations, and protein–DNA interactions are defined by electrophoretic mobility shift assays (EMSA). As this fractionation proceeds, non-matrix proteins can be analysed at each step to show relationships to the nuclear matrix proteins.

Table 1 shows the results of the biochemical isolation of nuclear matrix components.

Protocol 9. Isolation of nuclear matrix and non-matrix proteins from cell cultures

Reagents

- PBS
- 1.2 mM PMSF
- 0.5% Triton X-100
- Dnase I (100 μg/ml)
- RNase A (50 μg/ml)
- cytoskeletal (CSK) buffer: 100 mM NaCl, 300 mM sucrose, 10 mM 1,4 piperazine-diethane sulfonic acid (Pipes) (pH 6.8), 3 mM MgCl$_2$, 1 mM EGTA
- disassembly buffer: 8 M urea, 20 mM MES [2-(N-morpholino) ethanesulfonic acid], pH 6.6, 1 mM EGTA, 0.1 mM MgCl$_2$, 1.0% 2-mercaptoethanol, 1.2 mM PMSF

- RSB–Majik buffer: 100 mM NaCl, 10 mM Tris, 3 mM MgCl$_2$, 1.0% Tween-40, 0.5% deoxycholate (Na salt), 1.2 mM PMSF
- digestion buffer: 50 mM NaCl, 300mM sucrose, 10 mM Pipes (pH 6.8), 3 mM MgCl$_2$, 1 mM EGTA
- assembly buffer: 150 mM KCl, 25 mM imidazole hydrochloride (pH 7.1), 5 mM MgCl$_2$, 0.125 mM EGTA, 2 mM dithiothreitol, 0.2 mM PMSF
- buffer D: 100 mM KCl, 20 mM Hepes (pH 7.9), 0.2 mM EDTA, 0.5 mM dithiothreitol, 0.5 mM PMSF

Method

1. Wash plated cells with cold phosphate-buffered saline (PBS), scrape in 1.0 ml of PBS per 100 mm plate, and pool into an ice-cold polypropylene centrifuge tube (21). Pellet cells by centrifugation at 880 *g* for 5 min at 4°C.

Protocol 9. *Continued*

2. Resuspend pellets in cytoskeletal (CSK) buffer at 0 °C for 10 min and then pellet as above. Freeze in aliquots at –20 °C. Before use, add Triton X-100 to a final concentration of 0.5% (v/v) from a 20 × stock solution, and phenylmethanesulfonyl fluoride (PMSF) to a final concentration of 1.2 mM from a 100 × stock solution. PMSF stock solution is 20 mg/ml in isopropanol stored at room temperature and is added immediately before extraction.

3. Extract cytoskeletal proteins by resuspending the pellet in RSB–Majik buffer. Following centrifugation, as above, remove the supernatant, which contains polyribosomes.

4. Digest chromatin with RNase-free DNase in digestion buffer for 20 min at room temperature on a tilt shaker. Freeze in aliquots at –20 °C. Add Triton X-100, PMSF, DNase I, and RNase A just before extraction. After digestion, add ammonium sulfate to a final concentration of 250 mM to release the DNA and associated histones from the nucleus. Recover the nuclear matrix-intermediate filament (NM-IF) fraction, consisting of <5% of the total cellular protein, by centrifugation at 880 *g* for 10 min at 4 °C.

5. Nuclear matrix proteins. Resuspend the NM-IF pellet in disassembly buffer. Carry out dialysis in bags (2000 MW cut-off) at room temperature against assembly buffer. Carry out for two 1 hour periods in 100-fold excess of assembly buffer followed by an overnight dialysis against 200-fold excess of assembly buffer. Centrifuge the resulting suspension at 150 000 *g* for 95 min at 20 °C using a fixed-angle rotor to pellet the reassembled intermediate filaments (IF). Concentrate the supernatant containing the nuclear matrix using Centricon tubes (Amicon). Dialyse the concentrate twice in the Centricon tubes with buffer D lacking glycerol.

6. After dialysis, add glycerol to a concentration of 20% (v/v), and analyse nuclear matrix proteins by electrophoretic mobility shift assay (EMSA) or Western blots. Non-matrix proteins are prepared by the 0.42 M KCl extraction method of Dignam *et al.* (1983) (22).

2.2.2 Assays

Western blot analysis. Nuclear matrix and non-matrix proteins (25 μg) are resolved in a 10% SDS–PAGE mini gel, electroblotted on to membrane, and visualized using ECL reagents (Amersham) (21, 23).

Immunoprecipitation. Protein–protein interactions are revealed by immunoprecipitation which involves lysing the cells; binding a specific antigen to an antibody; precipitating the antibody–antigen complex; washing the precipitate and dissociating the antigen from the immune complex. The dissociated antigen is then analysed by electrophoretic methods (11).

Electrophoretic mobility shift assay (EMSA). Protein–DNA interactions are analysed by EMSA. The binding reaction mixture (20 μl) includes 1–2 μg of nuclear protein, 75 mM KCl, 0.1 mM dithiothreitol (except when antibody was present), 200 ng of poly (dI–dC), and 50 fM probe. Probes are prepared by 5′-end-labelling using T4 polynucleotide kinase. Electrophoresis is performed at 4°C in a 5% polyacrylamide gel in TGE buffer consisting of 50 mM Tris, 380 mM glycine, 2.1 mM EDTA, pH 8.5 (21).

3. *In vivo* experimental approaches

3.1 Subcellular fractionation of components of nuclear architecture for *in situ* immunofluorescence analysis

3.1.1 *In situ* immunofluorescence of endogenous proteins

In situ immunofluorescence microscopy of fractionated cells is widely used to study the subnuclear distribution of proteins and their association with functional sites in the nucleus. This high resolution analysis permits visualization of the biochemical machinery for gene replication and transcription. Evidence has accumulated that rather than a subcellular compartment with a random distribution of DNA, RNA, and proteins, the nucleus is a highly organized structure that changes throughout the cell cycle and development (24–27). Nuclear foci that may represent the sites of DNA replication and RNA processing are retained throughout sequential fractionation of the cell and are associated with the nuclear matrix. Insight into regulatory mechanisms that operate within the context of nuclear architecture requires identification of proteins that are present at nuclear matrix-associated subnuclear sites. It is necessary to experimentally define which regulatory proteins co-localize and how often, and to determine the sequences that are required to establish and sustain these associations.

Many proteins are associated with the nuclear matrix. Specific proteins are localized to regions of the nucleus and have a reproducible pattern of localization. The AML family of transcription factors provides a paradigm for tissue-restricted transcriptional control. These regulatory proteins, additionally, are an example of factors that are associated with the nuclear matrix. AML1B and AML3 are observed in a punctate pattern throughout the nucleus and this distribution is retained *in situ* in nuclear matrix intermediate filament preparations. AML1B is associated with sites of active transcription where it co-localizes with RNA polymerase II (*Figure 3*). The non-overlapping immunostaining observed with antibodies that distinguish AML1B from SC35 splicing factors, indicate the spatial independence of sites for transcription and transcript processing (28, 29). YY1 is also associated with the nuclear matrix but exhibits a very different subnuclear distribution,

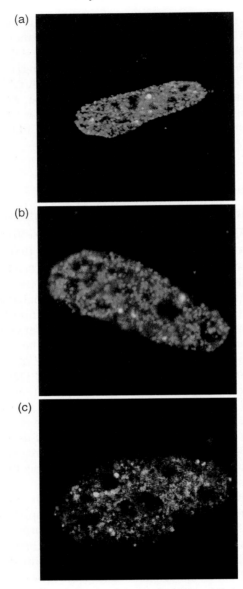

Figure 3. Co-localization of a subset of AML1B with RNA pol II_0 in the nuclear matrix of human Saos-2 osteosarcoma cells shows that CBFA2/AML1B is directed to transcriptionally active nuclear foci that contain the hyperphosphorylated form of RNA polymerase II (pol II_0). The images were obtained by immunofluorescence microscopy using antibodies against AML1B (green) and RNA pol II_0 (red), while co-localization is reflected by yellow signals. Immunofluorescence signals were recorded using (a) standard 35 mm slide photography, (b) a CCD camera interfaced with a digital microscope system, and (c) a confocal laser-scanning microscope.

predominantly in the nucleoli where it co-localizes with another nuclear matrix protein, B23 (30). The steroid hormone receptors, vitamin D receptor (VDR) and glucocorticoid receptor (GR) are also nuclear matrix associated and have specific distribution patterns.

There are three points to consider in the methodology of *in situ* immuno-fluorescence microscopy experiments. First, the fixation of cells to preserve structure and to maintain epitope recognition is critical to accurately represent the protein distribution *in vivo*. Second, permeabilization of cells must be sufficient to allow antibody access while maintaining cellular structure. And third, primary and secondary antibodies must be chosen carefully.

Glutaraldehyde is the most effective fixative for preserving fine structure; however, it is harsh and can often destroy epitope sites. Cold methanol/acetone (1:1) provides rapid fixation and permeabilization of cells while maintaining antibody recognition sites; however, some alterations can occur in protein distribution. Formaldehyde or paraformaldehyde fixation preserves structure as well as retaining antigen recognition. Formaldehyde is the preferred fixative for most *in situ* immunofluorescence applications. It is routine to use 2–4% (v/v) formaldehyde in phosphate-buffered saline (PBS, pH 7.4) for 10 min. The temperature may vary but fixation typically begins on ice and is allowed to warm to room temperature during the fixation period.

Permeabilization of cells is essential to allow antibody access to epitopes. Many detergents are commonly used for permeabilization. Triton X-100 (0.1–0.5%, v/v) and NP-40 (0.1–0.5%, v/v) are used to permeabilize the plasma and nuclear membranes, both before and after fixation of the cells. Saponin (0.5%, v/v) is used to temporarily permeabilize membranes but must be included throughout antibody incubations and washes to be effective. Digitonin (50 mg/ml, 5 min at 20 °C) selectively permeabilizes the plasma membrane while leaving other membranes intact. Cold methanol/acetone will permeabilize the membranes during fixation.

Non-immunological labelling of the nucleus is routinely accomplished through staining with DAPI, a fluorescent stain that labels DNA, detecting the nucleus in interphase cells and the chromosomes in mitotic cells. Other nuclear stains include Hoechst and propidium iodide.

Immunological labelling of nuclear proteins requires antibodies that recognize specific molecules. The specificity of antibodies can be determined by Western blot analysis to establish recognition of only the appropriate protein. Comparison of immune and pre-immune sera *in situ* is also important to demonstrate antibody specificity. Alternatively, when analysing proteins expressed in cells of transfected cultures, the untransfected cells can be used as controls for background labelling. Incubation conditions can vary, with temperatures of 4 to 37 °C. The length of antibody incubation can also vary from 1 to 24 h. Routinely, primary antibody incubations are carried out for

1–2 h at 37°C, or overnight at 4°C. Secondary antibody incubations are routinely carried out for 30–60 min at 37°C, or 1–2 h at room temperature. Longer secondary antibody incubation times may result in high background staining. Optimal conditions must be determined for each antibody. Secondary antibodies must be chosen to recognize the host species of the primary antibody. For example, antisera raised in rabbit against human PML protein must be detected by an anti-rabbit secondary antibody. Use of multiple antibodies requires consideration of species compatibility for all primary and secondary antibodies used. In addition, each antibody concentration must be determined by titrating different dilutions of the antibody to determine the optimal concentration for the experiment. The signal must be high and the background low. With the use of multiple antibodies, the relative strength of the signals must be considered and optimized. Blocking conditions are also important for maintaining low background by reducing non-specific antibody binding. This is routinely done using either BSA (0.5%, w/v) or normal serum (10% v/v horse serum). When using normal serum as a blocking agent, it is critical that the blocking antibody not be the same species as the specific antibody.

3.1.2 Flow chart

Prepare cells

- grow adherent cells on glass, gelatin-coated coverslips
- cytospin suspension cells on to glass slides that have been treated to form a highly adherent surface

Process cells

- whole cell (WC) preparation → intact cell
- cytoskeletal (CSK) preparation → extraction of soluble cytoplasmic and nuclear proteins
- nuclear matrix intermediate filament (NM-IF) preparation → extraction of soluble cytoplasmic and nuclear proteins and chromatin

Antibody staining

- primary antibody: labels antigen: 37°C for 1 h, or 4°C overnight
- secondary antibody: fluorochrome labels primary antibody: 37°C for 30–60 min or 1–2 h at room temperature.
- mount in an anti-photo bleach medium which prevents the rapid loss of fluorescence (fading or bleaching) during microscopic examination. In general, mounting medium can be made in a glycerol base with a buffer and an anti-fade reagent. Mounting mediums such as Vectorshield (Vector Laboratories) are available commercially.

3.1.3 Protocol for immunofluorescence analysis of cells on coverslips

Protocol 10. Whole cell (WC), cytoskeletal (CSK), or nuclear matrix-intermediate filament (NM-IF) preparations

Equipment and reagents

- 10× phosphate-buffered saline (PBS): 80 g NaCl, 2 g KCl, 14.4 g, Na_2HPO_4, 2.4 g KH_2PO_4.
- Add water to almost 1 litre, adjust pH to 7.4, adjust volume to 1 litre, filter using a 0.2 μM filter, and dilute 1:10 for use. [It is optional to add 0.02% (v/v) sodium azide (a preservative) to this solution. If you add sodium azide to PBS, do not use on cells until they are fixed *as it is toxic to cells*.] If there is no azide in the PBS, store in the refrigerator to retard bacterial growth.
- phosphate-buffered saline with bovine serum albumin (PBSA): PBS with 0.5% (w/v) bovine serum albumin (Sigma)
- cytoskeletal buffer (CSK): 10 mM Pipes, 300 mM sucrose, 100 mM NaCl, 3 mM $MgCl_2$, 1 mM EGTA. To make 1 litre of stock solution use: 3.024 g Pipes, 102.69 g sucrose, 5.844 g NaCl, 0.6099 g $MgCl_2$.6 H_2O, 0.3804 g EGTA. Titrate to pH 6.8 with 1

M NaOH. Filter using a 0.2 μm filter. Freeze in aliquots at −20°C. Immediately before use, Triton X-100 is added to a final concentration of 0.5% (v/v) from a 20 × stock solution and Vanadyl ribonucleoside complex (VRC) is added to a final concentration of 2 mM.
- digestion buffer (DB): 10 mM Pipes, 300 mM sucrose, 50 mM NaCl, 3 mM $MgCl_2$, 1 mM EGTA. To make 1 litre of stock solution, use: 3.024 g Pipes, 102.69 g sucrose, 2.922 g NaCl, 0.6099 g $MgCl_2$.6 H_2O, 0.3804 g EGTA. Titrate to pH 6.8 with 1 M NaOH. Filter using a 0.2 μm filter. Freeze in aliquots at −20°C. Before use, Triton X-100 is added to a final concentration of 0.5% (v/v) from 20 × stock solution. VRC is added to a final concentration of 2 mM and PMSF to a final concentration of 1.2 mM.

Methods of cell attachment

A. *Adhering cells*

1. Use glass, gelatin-coated coverslips since glass will not autofluoresce and gelatin enhances the adherence of cells to coverslips.

2. Microwave 0.5 g gelatin in 100 ml of distilled H_2O until dissolved.

3. Make fresh solution for each experiment.

4. Put coverslips in a beaker with 0.5% (w/v) gelatin solution and autoclave for 30 min.

5. Allow coverslips to dry in six-well tissue culture plates in a laminar flow hood for 1–2 h and then plate cells.

6. Cells will not grow well if coverslips are not dry.

B. *Suspension cells*

1. Cytospin 70000 cells/slide on to Vectabond (Vector Laboratories) or Cell-Tak (Collaborative Biomedical Products) treated slides (800 r.p.m., 4 min). Vectabond chemically modifies glass surfaces to form a highly adherent surface. Cell Tak is a protein solution designed to coat slides to increase adherence.

Protocol 10. *Continued*

Process cells

This method is a modification of the method described by Fey *et al.* (20).

1. *Adherent cells.* Rinse all wells (on ice) twice with 2 ml of filtered phosphate-buffered saline (PBS).

2. *Suspension cells.* When cytospun monolayer of cells is dry, immerse directly in fixative for whole cell preparation or in CSK buffer for CSK and NM-IF preparations.

A. *Whole cell preparation*

This method fixes or preserves cellular structures *in situ*. The cellular membrane is then permeabilized with Triton X-100 for antibody penetration.

1. Fix with 3.7% (v/v) formaldehyde in PBS for 10 min at 0°C warming to room temperature.

2. Wash with PBS.

3. Extract with 0.5% (v/v) Triton X-100 in PBS for 5 min at room temperature.

4. Wash twice with PBS.

5. Wash twice with PBS with 0.5% (w/v) bovine serum albumin (PBSA). BSA is a blocking agent that reduces non-specific antibody binding.

6. Stain with antibodies.

B. *Cytoskeletal (CSK) preparation*

Soluble cytoplasmic and nuclear proteins are extracted and the cells are then fixed. Triton X-100 is added to both the CSK buffer and the fixative for antibody penetration.

1. Extract twice for 5 min each with cytoskeletal buffer (CSK) with 0.5% (v/v) Triton + 2 mM vanadyl ribonucleoside complex (VRC) (5′–3′) + 1.2 mM PMSF (Sigma) at 0°C or alternatively, extract transfected cells twice for 15 min each with CSK at 0°C.

2. VRC is included as a ribonuclease inhibitor. Because the nuclear matrix contains both protein and RNA, it is important to inhibit ribonucleases at all stages of nuclear matrix preparation.

3. Phenylmethylsulfonyl fluoride (PMSF) is included as a protease inhibitor. A stock solution (20 mg/ml in isopropanol) is stored at room temperature and used at a final concentration of 0.2 mg/ml.

4. These inhibitors are added immediately before extraction.

5. Add fixative [CSK buffer + 0.5% (v/v) Triton + 3.7% formaldehyde (v/v) at 0°C].

6. Fix for 10 min, gradually warming to room temperature during the incubation period.

7. Wash twice with PBS.

8. Wash twice with PBSA.

C. *Nuclear matrix-intermediate filament (NM-IF) preparation*

After extraction with CSK buffer to remove the soluble nuclear and cytoplasmic proteins, the NM-IF preparation undergoes two chromatin digestions with RNase free DNase I {Boehringer Mannheim Biochemicals) in digestion buffer at 25–30°C. The NM-IF then undergoes two ammonium sulfate extractions to remove additional cytoskeletal proteins and any undissociated chromatin. DNase I cleaves the chromatin but it may not be completely extracted. The ammonium sulfate completes the extraction of chromatin. The cells are then fixed. As in the CSK preparation, 0.5% (v/v) Triton is included in the CSK buffer, in the digestion buffer, and in the fixative to increase membrane permeabilization.

11. Extract as in the CSK preparation.

12. To conserve DNase I, use a 50 μl drop on a parafilm-covered glass plate. Invert the coverslip on to the drop with the cells facing the DNase I drop. Cover the inverted slides/coverslips with parafilm and put the slides/coverslips in a humidity chamber (layer the bottom of a tupperware container with wet paper towels) and cover tightly to avoid evaporation.

13. Incubate the cells in digestion buffer with 0.5% (v/v) Triton X-100, 2 mM VRC, 1.2 mM PMSF (Sigma) + DNase I (Boehringer) (400 units/ml) for 15–30 min at 25–30°C. Prepare fresh solution for each experiment.

14. Incubate cells with a fresh aliquot of DNase in digestion buffer for 15–30 min at 25–30°C. Replace coverslips, cell side up, in wells of the six-well plate.

15. Add 1–2 ml/well ammonium sulfate extraction buffer at 0°C (digestion buffer with 250 mM $N_2H_8SO_4$) (1 vol 2 M $N_2H_8SO_4$ to 8 vols digestion buffer).

16. Incubate on ice for 5 min. Remove reagent.

17. Add 1–2 ml/well ammonium sulfate extraction buffer at 0°C for 1–2 min. Remove reagent.

18. Add fixative 3.7% (v/v) formaldehyde in digestion buffer + 0.5% Triton X-100 (v/v) at 0°C for 10 min, gradually warming to room temperature during the incubation period.

19. Wash twice with PBS.

10. Wash twice with PBSA.

Protocol 10. *Continued*

Antibody staining

As indicated in the introduction to this section, antibody selection is very important.

11. Add antibody to PBSA to prevent adhesion to sides of tube, then vortex.

12. Add 0.5–1.0 ml/well of primary antibody in PBSA (dilution tests done previously) or to conserve antibody, invert coverslip on a 20 μl drop of diluted antibody on parafilm.

13. Incubate for 60 min at 37 °C or overnight at 4 °C in a humidified chamber.

14. Wash four times with PBSA.

15. Add secondary antibody in PBSA (dilutions done previously).

16. Incubate for 60 min at 37 °C in a humidified chamber

17. Wash four times with PBSA.

18. Washes: 1 × with PBSA + 0.1% (v/v) Triton + 0.05 μg/ml DAPI (DNA stain) for 5 min (a 1 mg/ml stock solution of DAPI may be aliquoted and stored at –20 °C).

19. Wash once with PBSA + 0.1% Triton (v/v).

10. Wash twice with PBS.

11. Wipe the back of the coverslip with ethanol.

12. Invert the coverslip and mount on a slide with Vectashield.

13. Seal the coverslip to the slide to prevent mixing of mounting medium and immersion oil. Nail varnish works well.

14. Microscopy should be done immediately or as soon as possible as the signal diminishes over time.

15. Store slides in a freezer at –20 °C.

3.2 *In vivo* expression of transcription factors

Transient expression of exogenous DNA permits the identification of functional domains in transcription factors and other proteins that are responsible for DNA binding, nuclear transport, and nuclear matrix targeting. These domains can be expressed autonomously and localized by constructing and analyzing a deletion series of mutant proteins. The expressed proteins may be detected both *in situ* and by biochemical fractionation.

Sequences in transcription factors that are responsible for subcellular trafficking and regulatory activity can be further defined by point mutations in constructs that are exogenously expressed. For example, an AML1B protein with a mutation in the DNA-binding domain that has an intact nuclear matrix-targeting sequence localizes to the nuclear matrix but does not co-localize with RNA polymerase II (29). Molecular dissection of protein domains is critical to understanding the hierarchy of protein–protein interactions that lead to appropriate localization of proteins and fidelity of gene expression.

3.2.1 Overview

- Plate cells on gelatin-coated glass coverslips and grow to 50–75% confluency.

- Transiently transfect cells using a method appropriate for the cell type.

- Allow cells to express protein for 12–48 h.

- Harvest cells as described for *in situ* immunofluorescence or biochemical analysis.

3.2.2 Methods of analysing transient expression of exogenous DNA

Methods for transient transfection vary greatly and the method of choice is dependent on the cell type. Important points to consider are the amount of DNA used in the transfection and the time that the cells are examined following transfection. It is important when assaying protein localization that the exogenous proteins are expressed at low levels. Specific conditions must be determined for each cell type, DNA expression construct, and transfection protocol. It is essential to use high quality DNA. Cesium chloride purification and Qiagen columns (Qiagen) both provide acceptable DNA for transfection. It is critical that no residual ethanol remains in the DNA preparation.

The calcium phosphate method of transfection is based on the formation of a calcium–DNA precipitate in a buffer. When using Hepes buffer, the precipitate is formed in solution and is then gently applied to the cells. When using N,N-bis(2-hydroxyethyl)-2 aminoethanesulfonic acid (BES; Calbiochem) buffer, the precipitate gradually forms in the medium and drops on to the cells. The critical components are the pH of the buffer solution, which must be exact, the concentration of $CaCl_2$, the percentage of CO_2 in the incubator, and the quality and concentration of the DNA (31, 32).

DEAE–dextran transfections combine DNA and DEAE–dextran in medium to form complexes that are taken up by the cells. The efficiency of transfection may be increased by treatment with chloroquine diphosphate and with glycerol shock. Chloroquine may be toxic to many cells and the transfection must be carefully monitored to avoid killing cells. Critical parameters that need to be optimized include DEAE–dextran concentration, DNA concentration, and transfection incubation time (33–35).

Electroporation is best accomplished by following the instructions of the manufacturer of the apparatus. Cells are exposed to a high voltage electric field that results in the temporary formation of pores in the cell membrane that permit DNA to enter the cell. Critical parameters include the duration of current and the maximum voltage of the shock. Voltage and capacitance should be optimized for each cell type. The resistance of the buffer system is also important (36).

New methods for high efficiency transfection in cells that are difficult to transfect are now widely used. These include lipid reagents and polycation molecules. The advantage of these methods is that they are very efficient and reproducible.

Lipid reagents are a combination of a cationic and a neutral lipid that interact with anionic macromolecules (DNA) to form a complex that facilitates efficient uptake into cells. Although numerous lipid reagents are available, Lipofectamine (Gibco-BRL) can be used to transfect numerous cells lines at high efficiency. The method is essentially as described by Gibco-BRL with some modifications that optimize gene expression at low to moderate levels in transfected cells in a single well of a six-well tissue culture plate.

Protocol 11. Lipofectamine transfection for immunolocalization

Reagents

- 0.5% gelatin
- Optimem (Gibco-BRL)

- Lipofectamine (Gibco-BRL)

Method

11. Autoclave glass coverslips for 20 min in 0.5% (w/v) gelatin. Gelatin solution must be made fresh on the day of use. Place one coverslip in each well and allow the gelatin solution to dry completely in a tissue culture laminar flow hood. Drying takes 1–2 h. It is essential that the coverslips be allowed to dry or the cells will grow poorly.

12. Plate cells approximately 24 h before transfecting at 0.08×10^6 cells/well. This number may vary depending on the growth rate of the cells. Cells should be at least 50% confluent at the time of transfection.

13. Dilute 250 ng of expression plasmid and 1.75 µg of empty vector in 100 µl of Optimem (Gibco-BRL) or serum-free medium. DMEM works well and may be used for all cell types during transfection. Higher concentrations of DNA may increase the efficiency of transfection; however, the higher level of expression may not be compatible with protein localization. The pH and buffering conditions of the media are important.

14. Dilute 10 µl of Lipofectamine into 100 µl of Optimem media (Gibco-BRL). Higher concentrations of lipid reagent may increase the efficiency of transfection; however, the expression levels may be excessive for studying protein localization.

15. Combine the diluted DNA and the diluted Lipofectamine and incubate the mixture for 30 min at room temperature to allow DNA–lipid complexes to form.

16. Remove growth medium from the cells and rinse cells with PBS. Serum interferes with the transfection.

17. Dilute the DNA–lipid complexes with 800 μl of Optimem or serum-free medium and transfer to a tissue culture well.

18. Incubate the cells with the diluted DNA–lipid complexes for 2 h. Longer incubation times may increase the efficiency of transfection, but should be kept short to reduce the level of expression in each cell.

19. Remove DNA–lipid complexes and rinse the cells with PBS. Feed the cells with 2 ml of complete medium. At this time the cells should be maintained on normal growth medium.

10. Incubate the cells and harvest as described above for *in situ* immunofluorescence. A time course must be carried out to determine the appropriate time for each cell type and the protein being expressed. Expression of many proteins will be observed within a few hours. 18–24 hours of incubation is often a reasonable time frame. Longer than 48 h may result in artefacts.

Superfect reagent (Qiagen) is a polycation molecule of specific size and shape with branches radiating from a central core and terminating in charged amino groups. Superfect assembles DNA into a compact structure that binds to the surface of cells and is efficiently taken up by the cells and transported to the nucleus. The method is essentially as described by Qiagen with some modifications. The following method is based on a single well of a six-well tissue culture plate and optimized for low to moderate expression of the transfected gene.

Protocol 12. Superfect transfection for immunolocalization

Reagents
- Superfect (Qiagen)
- 0.5% gelatin

Method

1. Autoclave glass coverslips for 20 min in 0.5% (w/v) gelatin. Gelatin solution must be made fresh on the day of use. Place one coverslip in each well and allow the gelatin solution to dry completely in a tissue culture laminar flow hood. Drying takes 1–2 h. It is essential that the coverslips be allowed to dry as the cells will grow poorly otherwise.

2. Plate cells approximately 24 h before transfecting at 0.08×10^6 cells/

Protocol 12. *Continued*

    well. This number may vary depending on the growth rate of the cells. Cells should be at least 50% confluent at the time of transfection.

3. Dilute 250 ng of expression plasmid and 1.75 µg of empty vector in 100 µl of serum-free media. DMEM works well and may be used for all cell types during transfection. Higher concentrations of DNA may increase the efficiency of transfection; however, the expression level in each cell may be too high for studying protein localization. The pH and buffering system of the medium is important.

4. Add 8 µl of Superfect to the diluted DNA. Incubate for 10 min at room temperature to allow DNA–polycation complexes to form.

5. Remove growth medium from the cells and rinse with PBS.

6. Dilute the DNA–polycation complexes with 700 µl of complete medium (includes serum, l-glutamine, and antibiotics) and transfer to a tissue culture well.

7. Incubate the cells with the diluted DNA–polycation complexes for 3 h.

8. Remove DNA–polycation complexes and rinse the cells with PBS. Feed the cells with 2 ml of complete medium. At this time the cells should be maintained on normal growth medium.

9. Incubate the cells for an appropriate time and harvest as described above for *in situ* immunofluorescence. A time course is carried out to determine the appropriate harvest time for each cell type and the protein being expressed. Expression of many proteins will be observed within a few hours. 18–24 hours is often a reasonable time frame. Longer than 48 h may result in artefacts.

3.3 Epitope tags

When designing constructs for exogenous expression in cells it is often useful to include an epitope tag on the protein to be expressed. Epitope tagging adds a short sequence of amino acids to the protein that will assist in identification. Antibodies to numerous epitope tags are commercially available and allow expressed proteins to be easily distinguished from endogenous proteins. Epitope tags also allow proteins to be identified or purified without the time-consuming step of producing specific antibodies. Important points to consider when designing epitope tags are that the additional amino acids will not interfere with protein folding and that the epitope tag will be accessible to antibodies for detection (37–40). Frequently used epitope tags are: FLAG (DYKDDDDK; 41), HA (YPYDVPDYA; 42, 43), Xpress (LYDDDDK; Invitrogen, San Diego, CA), and c-myc (EQKLISEEDL; 44, 45).

    For use in living cells green fluorescent protein (GFP), enhanced green fluorescent protein (EGFP), and blue fluorescent protein (BFP) epitope tags are routinely used. GFP is a 238 amino acid protein isolated from *Aequora victoria*, the marine jelly fish. GFP can be expressed in mammalian cells and its coding sequence can be fused to either the C-terminal or N-terminal portion of a gene as a protein tag (46, 47). GFP fusion proteins are used for studying protein localization and gene expression. GFP can be detected in both fixed and living cells. The fluorescence excitation and emission spectra are similar to those of fluorescein and can be detected with the same equipment. GFP is stable, species independent, and can be observed in real time in living cells (48). Mutant GFP proteins with altered spectral properties have improved fluorescence and higher expression levels than the wild-type GFP in mammalian cells. These variants are referred to as enhanced GFP or EGFP (49). An enhanced blue fluorescent protein (EBFP) has recently been described that has a bright blue light emission (50). Previous BFP variants had very dim fluorescence (51, 52). In addition to alterations in the coding region of the fluorophore, the EGFP and EBFP have silent base changes throughout the coding region that use codons preferentially found in human proteins (53, 54). EBFP and EGFP allow the simultaneous detection of labelled proteins *in situ*.

3.4 Visualization of the components of transcription and assessment of activity

Sites of active transcription may be identified using antibodies to the hyperphosphorylated form of RNA polymerase II (55).

Sites of active transcription may also be identified by labelling nascent RNA with 5-bromouridine 5′-triphosphate (BrUTP) and detection with anti-BrdU (bromodeoxyuridine) antibodies, essentially as described by Wansink *et al.* and Jackson *et al.* (56, 57). Cells are made permeable by treatment with detergents such as Triton X-100 (0.02–0.05%, v/v) or saponin (0.01–0.02%, w/v). Permeabilization should be optimized to approximately 95% of the cells. This can be determined using trypan blue (1%, w/v) staining of cells on coverslips. The permeabilized cells will stain dark blue. If the detergent concentration is too low, cells will not be permeabilized at high efficiency. If the detergent concentration is too high, the cells will detach from the coverslips.

Cells must be permeabilized to allow the existing nucleotide triphosphate (NTP) precursor pools to be removed and allow the modified BrUTP to be used as a substrate by RNA polymerase. Cells are washed following permeabilization to remove endogenous NTPs. Specific concentrations of NTPs and the modified BrUTP are added in the transcription buffer and the endogenous RNA polymerase transcribes RNA. The concentration of precursors and the time of incubation control the rate of elongation of nascent RNA transcripts. It is important to keep the time of incubation short enough so that transcripts are detected at the site of transcription. Cells may be extracted to examine subcellular fractions. Fixation is typically carried out using 4% (v/v) formaldehyde. Formaldehyde fixation maintains cellular morphology and antibody recognition sites. Antibody staining must be done in RNase-free conditions.

Protocol 13.   BrUTP labelling of adherent cells

Equipment and reagents

- transcription buffer: 50mM Tris–HCl (pH 7.4), 10mM MgCl$_2$.6H$_2$O, 0.5mM EGTA, 100mM KCl, 25% glycerol, 0.5mM ATP (Boehringer Mannheim Biochemicals), 0.5mM CTP (Boehringer Mannheim Biochemicals), 0.5 mM GTP (Boehringer Mannheim Biochemicals), 0.2mM BrUTP (Sigma), 25 μM S-adenosylmethionine (Boehringer Mannheim Biochemicals), 1 mM AEBSF (Sigma), 5 units/ml RNasin (Promega)

- glycerol buffer: 20mM Tris–HCl (pH 7.4), 5mM MgCl$_2$, 0.5mM EGTA, 25% glycerol

Method

11. Autoclave glass coverslips for 20 min in 0.5% (w/v) gelatin. Gelatin solution must be made fresh on the day of use. Place one coverslip in each well and allow the gelatin solution to dry completely in a tissue culture laminar flow hood. Drying takes 1–2 h. It is essential that the coverslips be allowed to dry or the cells will grow poorly.

12. Plate cells approximately 24 h before transfecting at 0.08 × 10^6 cells/well. This number may vary depending on the growth rate of the cells. Cells should be at least 50% confluent at the time of transfection.

13. Rinse cells with phosphate-buffered saline (PBS) followed by a 3 min incubation in glycerol buffer.

14. Permeabilize cells with glycerol buffer containing 0.05% (v/v) Triton X-100 and 1 mM AEBSF [4-(2-aminoethyl)-benzenesulfonyl fluoride] (Sigma) for 3 min at room temperature.

15. Add transcription buffer containing NTPs and modified substrate with protease and RNase inhibitors.

16. Incubate for 10–30 min. It is necessary to do a time course to determine the optimal incubation time for each cell type and RNA being studied.

17. Wash cells in PBS containing 0.5% (v/v) Triton X-100 and RNasin (5 units/ml) for 3 min.

18. Wash with PBS containing RNasin (5 units/ml).

19. At this time cells may be fixed or extracted for CSK or NM-IF. It is critical at this time to prevent RNA degradation. Prepare all buffers in

DEPC-treated (diethylpyrocarbonate) water, include RNase inhibitors in all buffers, and use RNase-free BSA.

10. Antibody staining is carried out as described except that RNase-free conditions must be maintained. Anti-BrdU antibodies are available commercially (Boehringer Mannheim).

4. Instrumentation for *in situ* analysis

4.1 Direct immunofluorescence

Epifluorescence microscopy is an invaluable technique that enables us to visualize and identify structures in cells and tissues at the cellular and molecular levels. Immunofluorescence allows us to combine the sensitivity of fluorescence microscopy with the specificity of immunology; antigen/antibody response. Specific fluorescent probes can be generated for many biological molecules with excellent sensitivity. The emitted signal is viewed against a black background, providing high contrast; thus, even weak signals can be realized. Through the use of a multiple staining technique, several probes and their functional interactions can be detected simultaneously.

The most widely used fluorescence microscopes today are equipped with epifluorescence. In this microscope, the illuminator is interposed between the eye pieces and the nose piece, which carries the objectives. This type of illumination has several advantages:

- The objective, first serving as a well-corrected condenser and then as the image-forming light gatherer, is always in correct alignment.

- The area being illuminated is restricted to the area being observed.

- The full numerical aperture of the objective, in Koehler illumination, is utilizable.

- It is possible to combine epifluorescence with other transmitted light techniques such as phase or differential interference contrast (DIC). Often, it is important to combine these techniques to pinpoint specimen areas of interest.

How does it work? The lamp house contains a light source, usually a mercury lamp, which sends light parallel to the table top (perpendicular to the optical axis of the microscope), passes through collector lenses, a variable, centerable aperture diaphragm, and then through a variable centerable field diaphragm. It hits the excitation filter which selects those excitation wavelengths that should reach the specimen and blocks the wavelengths that should not. The selected wavelengths reach the dichromatic beam-splitting mirror. This mirror is a special type of interference filter that reflects shorter wavelength light and efficiently passes longer wavelength light. The dichromatic beam splitter (also sometimes called the dichroic mirror) is tilted at 45°

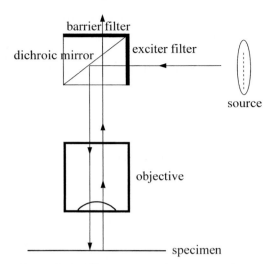

Figure 4. Schematic illustration of fluorescence microscopy optics.

to the incoming excitation light and reflects the excitation light at 90° directly through the objective and on to the specimen. The fluorescent light emitted by the specimen is gathered by the objective, now serving in its usual image-forming function. Since the emitted light consists of longer wavelengths, it is able to pass through the dichroic mirror. Before the emitted light can reach the eye piece, it passes through a barrier filter. This filter blocks any residual excitation light and passes the desired longer emission wavelengths towards the eye pieces. The excitation filter, dichroic mirror, and barrier filter are incorporated into a 'filter cube' in most systems (58) (see *Figure 4*).

4.2 Digital imaging

4.2.1 Z-axis imaging

A cooled CCD camera attached to an epifluorescence microscope can sequentially acquire a set of images in adjacent focal planes. By dividing the depth of the specimen by the thickness of the optical sections you desire, the Z-axis function gathers each sequential plane. This series of images can be sequentially displayed in a movie format to view the three-dimensional archi-tecture of the specimen.

4.2.2 Deconvolution

Computer deconvolution techniques make it possible to see details in a thick specimen without embedding and sectioning, by optically sectioning the specimen and using computer algorithms to remove out-of-focus components from a given focal plane. Most cells are intrinsically three-dimensional. This can be troublesome in resolving fine structure because of the limited depth of field of the light microscope. Computer deconvolution techniques start with a stack of images collected by changing the fine focus of the microscope. Information from adjacent focal planes can be used to correct the central focal plane, thus removing out-of-focus detail.

4.2.3 Green fluorescent protein

In general, most specimens examined by immunofluorescence are fixed, but with the advent of green fluorescent protein (GFP), a 238 amino acid protein which naturally fluoresces green, it is possible to study living, dynamic systems. Cells are transfected with genes encoding the protein and patterns of gene expression can be studied with no side-effects to the cells that contain it.

Technological advances in imaging-contrast methods such as phase and differential interference contrast, CCD cameras, computers with quantitative imaging programs, fluorochrome advances, and microenvironmental control systems, allow the researcher the instrumental capabilities to examine dynamic cultures. To observe living cells, one must use a chamber compatible with the cells and the microscope.

Microenvironmental chambers can be of two types: an open system, generally used for short-term experiments where cells are grown in a culture dish and observed on a heated microscope stage; or a more advanced environmental control system where the pH, CO_2, temperature, and waste products are precisely controlled in an optically designed self-contained perfusable chamber. Several types of environmental control chambers and techniques are well referenced in David Spector's *Light microscopy and cell structure* (59).

4.2.4 Confocal

With the advent of very sensitive electronic imaging sensors, advanced video techniques, and improved capabilities of hardware for digital imaging processing, light microcopy has benefited, as evidenced by the improved resolution of electronically recorded images. These electronically enhanced light microscopes can be used to study specimens that are alive or metabolically active through the use of microenvironmental chambers, and to resolve fine features of fixed or low contrast specimens through high resolution, cooled CCD cameras and quantitative digital imaging programs.

Along with these technological advances came the confocal microscope. In general, a confocal imaging system consists of a laser-scanning microscope with a pinhole aperture in the image plane of the objective. It is an instrument that addresses the three-dimensional structure of biological specimens by optically sectioning them into two-dimensional planes and eliminating any out-of-focus light from the final image. This process increases the contrast and resolution of fine details in the specimen.

References

1. Workman, J.L. and Kingston, R.E. (1998). *Annu. Rev. Biochem.* **67**, 545.
2. Noll, M. and Kornberg, R.D. (1977). *J. Mol. Biol.* **109**, 393.
3. Simpson, R.T. (1991). *Prog. Nucleic Acid. Res. Mol. Biol.* **40**, 143.
4. Bellard, M., Dretzen, G., Giangrande, A., and Ramain, P. (1989). In *Methods in enzymology* (John N. Abelson, Melvin I. Simon, eds.), Vol. 170, p. 317. Academic Press, New York.
5. Montecino, M., Lian, J., Stein, G., and Stein, J. (1996). *Biochemistry* **35**, 5093.
6. Montecino, M., Frenkel, B., van Wijnen, A., Lian, J., Stein, G., and Stein, J. (1999). *Biochemistry* **38**, 1338–1345.
7. Wu, C. (1980). *Nature* **286**, 854.
8. Dingwall, C., Lomonossoff, G.P., and Laskey, R.A. (1981). *Nucleic Acids. Res.* **9**, 2659.
9. Horz, W. and Altenburger, W. (1981). *Nucleic Acids. Res.* **9**, 2643.
10. Mueller, P.R. and Wold, B. (1989). *Science* **246**, 780.
11. Ausubel, F.M., Brent, R., Kingston, R.E., Moore, D.D., Seidman, J.G., Smith, J.A., and Struhl, K.(eds.) (1997). *Current protocols in molecular biology*, p. 15.5.1. John Wiley & Sons, New York.
12. McPherson, C.E., Shim, E.Y., Friedman, D.S., and Zaret, K.S. (1993). *Cell* **75**, 387.
13. Zhang, L. and Gralla, J.D. (1989). *Genes. Dev.* **3**, 1814.
14. Pfeifer, G.P. and Riggs, A.D. (1991). *Genes. Dev.* **5**, 1102.
15. McGhee, J.D. and Felsenfeld, G. (1983). *Cell* **32**, 1205.
16. Meersseman, G., Pennings, S., and Bradbury, E.M. (1992). *EMBO. J.* **11**, 2951.
17. Pennings, S., Meersseman, G., and Bradbury, E.M. (1994). *Proc. Natl. Acad. Sci. USA* **91**, 10275.
18. Pauli, U., Chrysogelos, S., Stein, G., Stein, J., and Nick, H. (1987). *Science* **236**, 1308.
19. Bidwell, J.P., van Wijnen, A.J., Fey, E.G., Dworetzky, S., Penman, S., Stein, J.L., Lian, J.B., and Stein, G.S. (1993). *Proc. Natl. Acad. Sci. USA*, **90**, 3162.
20. Fey, E.G., Capco, D.G., Krochmalnic, G., and Penman, S. (1984). *J. Cell. Biol.* **99**, 203s–208s.
21. Merriman, H.L., van Wijnen, A.J., Hiebert, S., Bidwell, J.P., Fey, E., Lian, J., Stein, J., and Stein, G.S. (1995). *Biochemistry* **34**, 13125.
22. Dignam, J.D., Lebovitz, R.M., and Roeder, R.G. (1983). *Nucl. Acids Res.* **11**, 1475.
23. McCabe, L.R., Banerjee, C., Kundu, R., Harrison, R.J., Dobner, P.R., Stein, J.L., Lian, J.B., and Stein, G.S. (1996). *Endocrinology* **137**, 4398.
24. Zlatanova, J.S. and van Holde, K.E. (1992). *Crit. Rev. Eukaryot. Gene. Exp.* **2**, 211.
25. Stein, G.S., van Wijnen, A.J., Stein, J.L., Lian, J.B., and Montecino, M. (1997). In *Nuclear structure and gene expression* (ed. R.C. Bird, G.S. Stein, J.B. Lian, and J.L. Stein), p. 178. Academic Press, New York.
26. Kingston, R.E., Bunker, C.A., and Imbalzano, A.N. (1996). *Genes. Dev.* **10**, 905.
27. de Jong, L., Grande, M.A., Mattern, K.A., Schul, W., and van Driel, R. (1996). *Crit. Rev. Eukary. Gene. Expr.* **6**, 215.
28. Zeng, C., van Wijnen, A.J., Stein, J.L., Meyers, S., Sun, W., Shopland, L.,

Lawrence, J.B., Penman, S., Lian, J.B., Stein, G.S., and Hiebert, S.W. (1997). *Proc. Natl. Acad. Sci. USA*, **94**, 6746.

29. Zeng, C., McNeil, S., Pockwinse, S., Nickerson, J.A., Shopland, L., Lawrence, J.B., Penman, S., Hiebert, S.W., Lian, J.B., van Wijnen, A.J., Stein, J.L., and Stein, G.S. (1998). *Proc. Natl. Acad. Sci. USA*, **95**, 1585.

30. McNeil, S., Guo, B., Stein, J.L., Lian, J.B., Bushmeyer, S., Seto, E., Atchison, M.L., Penman, S., van Wijnen, A.J., and Stein, G.S. (1998). *J. Cell. Biochem.* **68**, 500.

31. Chen, C. and Okayama, H. (1987). *Mol. Cell. Biol.* **7**, 2745.

32. Chen, C.A. and Okayama, H. (1988). *Biotechniques* **6**, 632.

33. Selden, R.F., Howie, K.B., Rowe, M.E., Goodman, H.M., and Moore, D.D. (1986). *Mol. Cell. Biol.* **6**, 3173.

34. Sussman, D.J. and Milman, G. (1984). *Mol. Cell. Biol.* **4**, 1641.

35. Lopata, M.A., Cleveland, D.W., and Sollner-Webb, B. (1984). *Nucl. Acids Res.* **12**, 5707.

36. Potter, H., Weir, L., and Leder, P. (1984). *Proc. Natl. Acad. Sci. USA* **81**, 7161.

37. von Zastrow, M. and Kobilka, B.K. (1992). *J. Biol. Chem.* **267**, 3530.

38. Reisdorf, P., Maarse, A.C., and Daignan-Fornier, B. (1993). *Curr. Genet.* **23**, 181.

39. Kolodziej, P.A. and Young, R.A. (1991). In *Methods in enzymology* (John N. Abelson, Melvin I. Simon, eds.), Vol. 194, p. 508. Academic Press, New York.

40. Munro, S. and Pelham, H.R. (1984). *EMBO. J.* **3**, 3087.

41. Prickett, K.S., Amberg, D.C., and Hopp, T.P. (1989). *Biotechniques* **7**, 580.

42. Niman, H.L., Houghten, R.A., Walker, L.E., Reisfeld, R.A., Wilson, I.A., Hogle, J.M., and Lerner, R.A. (1983). *Proc. Natl. Acad. Sci. USA* **80**, 4949.

43. Field, J., Nikawa, J., Broek, D., MacDonald, B., Rodgers, L., Wilson, I.A., Lerner, R.A., and Wigler, M. (1988). *Mol. Cell. Biol.* **8**, 2159.

44. Evan, G.I., Lewis, G.K., Ramsay, G., and Bishop, J.M. (1985). *Mol. Cell. Biol.* **5**, 3610.

45. Munro, S. and Pelham, H.R. (1987). *Cell* **48**, 899.

46. Prasher, D.C., Eckenrode, V.K., Ward, W.W., Prendergast, F.G., and Cormier, M.J. (1992). *Gene* **111**, 229.

47. Chalfie, M., Tu, Y., Euskirchen, G., Ward, W.W., and Prasher, D.C. (1994). *Science* **263**, 802.

48. Kain, S.R., Adams, M., Kondepudi, A., Yang, T.T., Ward, W.W., and Kitts, P. (1995). *Biotechniques* **19**, 650.

49. Yang, F., Moss, L.G., and Phillips, G.N., Jr (1996). *Nat. Biotechnol.* **14**, 1246.

50. Yang, T.T., Sinai, P., Green, G., Kitts, P.A., Chen, Y.T., Lybarger, L., Chervenak, R., Patterson, G.H., Piston, D.W., and Kain, S.R. (1998). *J. Biol. Chem.* **273**, 8212.

51. Heim, R. and Tsien, R.Y. (1996). *Curr. Biol.* **6**, 178.

52. Rizzuto, R., Brini, M., De Giorgi, F., Rossi, R., Heim, R., Tsien, R.Y., and Pozzan, T. (1996). *Curr. Biol.* **6**, 183.

53. Haas, J., Park, E.C., and Seed, B. (1996). *Curr. Biol.* **6**, 315.

54. Yang, T.T., Cheng, L., and Kain, S.R. (1996). *Nucl. Acids Res.* **24**, 4592.

55. Mortillaro, M.J., Blencowe, B.J., Wei, X., Nakayasu, H., Du, L., Warren, S.L., Sharp, P.A., and Berezney, R. (1996). *Proc. Natl. Acad. Sci. USA* **93**, 8253.

56. Wansink, D.G., Schul, W., van der Kraan, I., van Steensel, B., van Driel, R., and de Jong, L. (1993). *J. Cell. Biol.* **122**, 283.

57. Jackson, D.A., Hassan, A.B., Errington, R.J., and Cook, P.R. (1993). *EMBO. J.* **12**, 1059.

58. Abramowitz, M. (ed.) (1993). *Fluorescence microscopy: the essentials*. Olympus America Inc., Melville, N.Y.
59. Spector, D.L., Goldman, R.D., and Leinwand, L.A. (ed.) (1998). *CELLS (a laboratory manual): light microscopy and cell structure*, p. 75.1. Cold Spring Harbor Laboratory Press, Plainview.
60. Montecino, M., Pockwinse, S., Lian, J., Stein, G., and Stein, J. (1994). *Biochemistry* **33**, 348.
61. Breen, E.C., van Wijnen, A.J., Lian, J.B., Stein, G.S., and Stein, J.L. (1994). *Proc. Natl. Acad. Sci. USA* **91**, 12902.

10

Antisense oligonucleotides: tools for elucidating gene function

JOHN J. WOLF and W. MICHAEL FLANAGAN

1. Introduction

The goal of antisense technology is to design potent, sequence-specific inhibitors that can be used to elucidate the molecular basis of diseases, validate drug targets, and treat human diseases. Currently, one oligonucleotide (ON) has been approved by the Food and Drug Administration for the treatment of cytomegalovirus-induced retinitis in AIDS patients, and several additional antisense ONs are in various phases of clinical testing for the treatment of inflammatory diseases and cancer (1).

Antisense ONs are short (7–30 nucleotides) synthetic pieces of DNA or modified DNA that are rationally designed to selectively hybridize to a target RNA and interfere with the expression of the encoded protein (*Figure 1*). The formation of the antisense ON–RNA heteroduplex inhibits protein expression by several different mechanisms, including blocking RNA transport, splicing, and translation (2). In most cases, the RNA–ON complex is recognized by RNase H, a nuclear-localized enzyme that degrades the RNA portion of an RNA–ON hybrid (3, 4). Because antisense ONs are complementary to a unique sequence of a gene implicated in the disease state, they hold the promise of being highly specific, efficacious, and less toxic than previously developed human therapeutics.

However, like any evolving technology, the development of antisense ONs has been difficult. Several barriers have had to be overcome, including ON stability, ON affinity for its RNA target, and cellular permeation. Recent technological advances and an understanding of the limitations of antisense-based compounds have resulted in the evolution of a robust biological tool and a new class of drugs.

The goal of this chapter is to help novice antisense researchers design potent antisense ONs, select optimal antisense targets, deliver ONs to different cell types, avoid common ON-mediated experimental artefacts, and use antisense ONs as biological tools for probing gene function.

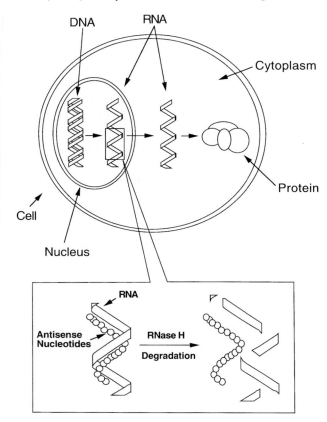

Figure 1. Mechanism of action of antisense oligonucleotides (ONs). Antisense ONs are single-stranded synthetic pieces of DNA or modified DNA that are rationally designed to hybridize to the targeted mRNA in the nucleus or cytoplasm of the cells. Formation of the antisense ON–RNA heteroduplex in the nucleus activates RNase H (lower insert) which cleaves the RNA portion of the ON-RNA hybrid and, thus, inhibits protein expression. Antisense ONs can be used to target any gene of interest as long as a partial sequence of the gene is known. Modified from ref. 2.

2. Chemical modifications of antisense ONs

A successful antisense ON inhibitor must be chemically stable and demonstrate high affinity for its target sequence. The first ONs tested in tissue culture were phosphodiester ONs (5). While phosphodiester ONs demonstrate strong affinity for their RNA target, they were rapidly degraded in culture medium and, thus, showed poor antisense activity (6, 7). This problem was overcome by simply replacing one of the non-bridging oxygen atoms of the phosphodiester linkage with a sulfur atom which created a nuclease-resistant phosphorothioate ON (*Figure 2a*). Phosphorothioate-modified ONs are the most widely

(a)

(b)

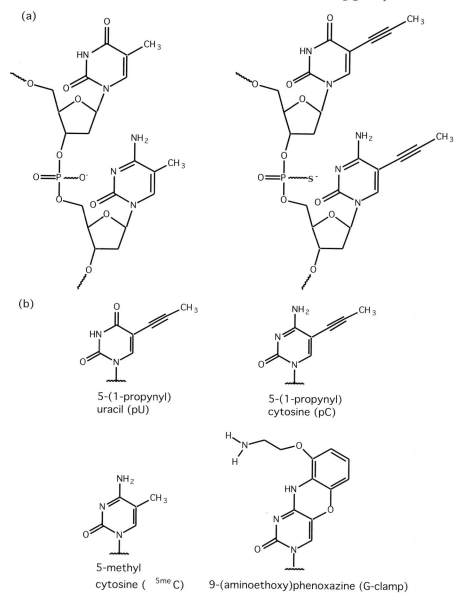

5-(1-propynyl)
uracil (pU)

5-(1-propynyl)
cytosine (pC)

5-methyl
cytosine (^{5me}C)

9-(aminoethoxy)phenoxazine (G-clamp)

Figure 2. Structure and chemical modifications of antisense oligonucletides (ONs). (a) Structure of a phosphodiester dimer containing deoxythymidine and 5-methyl deoxy-cytidine (left) and a phosphorothioate dimer containing 5-(1-propynyl) deoxyuridine and 5-(1-propynyl) deoxycytidine (right). Modification of the phosphodiester linkage with a sulfur atom creates a nuclease-resistant phosphorothioate dimer. (b) Examples of various heterocycle modifications that can be incorporated into antisense ONs to enhance the binding affinity of the ON for its target RNA.

used and commercially available ONs for antisense research. One drawback to phosphorothioate-modified ONs is that they have decreased binding affinity for their RNA target compared with phosphodiester ONs. As a result of the decrease in binding affinity, many researchers screen up to 50 different phosphorothioate ONs to identify a single ON that demonstrates potent antisense activity against its RNA target.

To streamline the identification of active ONs, we have developed ON modifications that enhance the binding affinity of phosphorothioate ONs to their target (reviewed in ref. 8). A systematic evaluation of modifications at the C-5 position of pyrimidines resulted in the identification of C-5 propynyl derivatives of uracil and cytosine (9) (*Figure 2b*). Incorporation of C-5 propynyl modifications on every pyrimidine in a phosphorothioate ON dramatically enhanced RNA–ON heteroduplex stability, as measured by thermal melting temperature and increased antisense ON potency, by greater than 10-fold compared with unmodified phosphorothioate ONs (10). C-5 propyne phosphoramidites can be purchased from Glen Research (www.glenres.com) and incorporated into ONs using standard synthesis conditions.

In addition to C-5 propyne base modifications, we have recently described a novel cytosine analogue, termed G-clamp (11) (*Figure 2b*). Under physiological conditions, antisense ONs hybridize to their RNA target according to Watson–Crick base-pairing rules. The adenosine–thymine interaction is formed by two hydrogen bonds, and the cytosine–guanine interaction is formed by three hydrogen bonds. The G-clamp modification is remarkable because it clamps on to the complementary guanine by forming an additional hydrogen bond. Formation of the additional hydrogen bond resulted in increased binding affinity between the antisense ON and its RNA target. Incorporation of a single G-clamp modification into a previously optimized 20-mer antisense ON targeting c-*raf* increased the potency of the ON by 25-fold (12). Like the C-5 propyne-modified pyrimidines, the G-clamp heterocycle is synthetically compatible with current ON synthesis conditions, although the G-clamp modification is not yet commercially available.

3. Antisense ONs as biological tools for elucidating gene function

3.1 Selection of antisense ONs

One of the allures of antisense technology is the ability to rationally design inhibitors based on the sequence of the disease-causing gene. Thus, it would be expected that choosing an active antisense ON should be trivial. However, the identification of potent antisense ONs to new targets has proven difficult because not all sites on the target RNA are amenable to hybridization with an antisense ON. RNA secondary structure, in which large portions of the mRNA form stable homoduplexes, interferes with RNA–ON heteroduplex formation. For instance, only one of 34 phosphorothioate ONs targeting human c-*raf* decreased c-*raf* RNA levels by 90%, compared with an internal control RNA (13). In most cases, researchers do not have the resources (grant support or time) to screen more than five antisense ONs to any given target. The identification of active ONs can be made more manageable by using ONs that have enhanced binding affinity for their RNA target, such as C-5 propynylpyrimidine or G-clamp-modified ONs (*Figure 2*) that can destablize some RNA secondary structures. In our experience, more than 50% of the C-5 propynyl-modified antisense ONs screened against a target demonstrate antisense activity.

Traditionally, researchers have selected two or three antisense ONs spanning the translational start site of a mRNA with the notion that the translational start site may lack mRNA secondary structure and thus allow easy access for the antisense ON. While active ONs have been found that target the translational start site of mRNAs, any region of the RNA can be targeted including the 5'- and 3'-untranslated regions, introns, and the coding region. Recently, several methods have been published to help predict antisense-accessible regions of the targeted RNA and to identify an active antisense ON (14, 15). However, in most laboratories, the identification of potent antisense ONs is still empirically determined using cell culture-based assays.

3.2 Guidelines for selecting antisense ONs

Potent and specific antisense ONs can be identified from screening as few as six potential antisense ONs using the following criteria for selection. (1) The ON is 15–20 nucleotides in length. An ON of this size is long enough to be stable upon hybridization and is statistically unique in the human genome. (2) The ON sequence is 50–80% pyrimidines, which allows multiple C-5 pyrimidine substitutions and augments ON binding affinity (16, 17). Employing these simple criteria, we have selected potent antisense ONs for over 40 different gene targets without having to screen more than six ONs for any target.

Protocol 1. ON quantitation and preparation

Method

1. Determine the amount of an antisense ON based on the molar extinction coefficient for the ON at A_{260} nm. The molar extinction coefficients (in $M^{-1}cm^{-1}$) for the various bases are shown below.

   thymine (T) = 8.8×10^3

   cytosine (C) = 7.4×10^3

   5-methyl cytosine (^{5Me}C) = 6.0×10^3

Protocol 1. *Continued*

    adenine (A) = 15.2×10^3
    guanine (G) = 11.8×10^3
    propynyl uracil (pU) = 3.2×10^3
    propynyl cytosine (pC) = 5.0×10^3
    G-clamp = 9.0×10^3

2. Calculate the molar extinction coefficient for the ON. Count the number of times each base appears in the ON of interest. Multiply these numbers by the extinction coefficient for that base. The numbers are then added together to obtain the extinction coefficient for that ON.

   For example, the molar extinction coefficient for the ON (5'-pCpUpC AAT GpUG) is calculated below:

   $2pC = (2)(5.0 \times 10^3) = 10 \times 10^3$
   $2pU = (2)(3.2 \times 10^3) = 6.4 \times 10^3$
   $2A = (2)(15.2 \times 10^3) = 30.4 \times 10^3$
   $1T = (1)(8.8 \times 10^3) = 8.8 \times 10^3$
   $2G = (2)(11.8 \times 10^3) = 23.6 \times 10^3$
   $E = 79.2 \times 10^3 \ M^{-1}cm^{-1}$

   Using the Lambert–Beer law the concentration of the absorbing ON can be determined. The Lambert–Beer law is $A = Ecl$, where A is the absorbance, E is the calculated extinction coefficient, c is the concentration, and l is the length of the light path, which is 1 cm.

3. Resuspend or dilute the ON to an approximate final concentration of 500 μM in sterile water based on the above calculations. Measure the absorbance of the ON at 260 nm and determine the true concentration of the ON. Note that the maximum absorbance of the propyne bases is near 290 nm; however, the extinction coefficients given above were determined at 260 nm.

4. Store the resuspended ON at $-20\,°C$. Antisense ONs are stable for many years under these storage conditions.

3.3 Sequence-dependent non-antisense effects

Phosphorothioate ONs are large, polyanionic molecules that can have a wide variety of sequence-dependent, non-antisense effects on the biological processes of cells (18, 19). Phosphorothioate ONs can bind to extracellular receptors and growth factors and dramatically affect cellular proliferation, metabolism, and differentiation (20, 21). Phosphorothioate ONs have also been shown to block virion binding and fusion with cellular membranes (22). ONs containing G-quartets (four guanosines residues that form base-pair interactions) have shown potent anti-proliferative effects following balloon angioplasty in a rat carotid model of restenosis (23). In addition, ONs containing cytosine–phosphate–guanosine (CpG) dimers can activate B cells and natural killer cells, which can cause immune stimulation and anti-tumour effects *in vivo* (24). Since phosphorothioate ONs can have such a wide variety of non-specific effects, the evaluation of antisense activity should never be based solely on biological effects observed *in vitro* or *in vivo*. Instead, these phenotypic changes need to be supported by direct measurement of the targeted RNA and protein (7, 25, 26).

3.4 Guidelines for conducting antisense experiments

Rigorous controls and careful analysis of antisense data are essential to prevent even the most detail-oriented scientist from being mislead by non-antisense-mediated effects by Ons (18). However, with the liberal use of control ONs and analysis of RNA and protein levels, antisense ONs can be important tools for deciphering complex biological systems. Some experimental guidelines for conducting well-controlled antisense experiments are discussed below (26, 27).

1. Direct measurement of the antisense-targeted RNA or protein levels, compared with an internal control RNA or protein that should not be affected by the antisense ON. Identifying internal controls can be difficult, especially when the antisense ONs affect cellular proliferation and/or differentiation. In our experience, cyclophilin RNA provides an important internal control since it is constitutively expressed under many different *in vitro* and *in vivo* conditions.

2. Direct measurement of the effect of the mismatch ON on the expression of targeted RNA or protein levels, compared with its effect on the internal control RNAs or proteins. The effect of the mismatch ON on the targeted protein should also be compared, within the same experiment, with the effect of the antisense ON on the targeted protein.

3. Demonstration that antisense inhibition of the targeted RNA or protein is dose dependent.

4. Demonstration that several (at least two) different antisense ONs inhibit expression of the targeted RNA or protein. Evidence that more than one antisense ON can specifically block expression of the target strongly suggests that the observed biological effects are sequence dependent.

5. The use of several mismatch ONs that retain the identical base composition as the antisense ON. Employing the sense strand sequence as the control ON is not appropriate since the base composition of the ON can have dramatic biological effects, as discussed previously for CpG and

G-quartet sequences. One of the most rigorous controls is the use of mismatch ONs that contain an increasing number of incorporated base mismatches. Correlation of the inhibition of the targeted RNA or protein with the extent of the mispaired bases provides strong evidence that the ON is functioning through an antisense mechanism of action.

3.5 Antisense targets

The optimal antisense target has the following attributes.

1. The sequence of the target is known. This characteristic seems obvious; yet, antisense ONs have been designed based on the human gene sequence and tested in murine cell lines or in mice without knowledge of the precise sequence of the murine homologue. In some cases, the target sequences have been different (28).

2. Reagents for analysing the targeted RNA and protein levels must be readily available. In most cases, nucleic acid probes for measuring RNA levels are easily obtained; yet, mRNA destruction fails to demonstrate that the targeted protein is also destroyed, since protein levels are dependent on the half-life of the protein. Thus, analysis of protein levels is required. For example, levels of the long-lived c-Raf protein were virtually unaltered 24 h after antisense ON transfection. However, when c-Raf protein levels were analysed 48 h after antisense ONs transfection, c-Raf levels were significantly reduced. In contrast, the cyclin-dependent inhibitor p27kip1, which has a half-life of 20 min in tissue culture cells, was completely degraded 12 h after antisense ON transfection.

3. The biological effects of inhibiting the target protein should be easy to examine. Target proteins that act as a rheostat for the biological system are the best targets since small changes in the targeted protein level have dramatic biological effects. For example, a 50–70% antisense-specific inhibition of p27kip1 in 3T3 murine cells allowed these cells to proliferate in the absence of mitogenic factors (29, 30). Similarly, p27kip1 antisense effects were observed in IL-4-treated astrocytoma cells (31) and 1,25-dihydroxyvitamin D_3-treated HL60 promyelocytic leukaemia cells (32). In these cases a modest antisense-mediated decrease in p27kip1 resulted in cellular proliferation that was easily monitored by a variety of methods.

4. ON delivery

One of the major barriers to the development of antisense ONs as robust laboratory reagents and potent human therapeutics is their inability to permeate eukaryotic cell membranes. Several reports have cited antisense effects using ONs added directly to cell medium; however, direct proof that ONs enter cells and affect gene expression by an antisense mechanism of action is lacking (33). Incubation of cells with fluorescently labelled phosphorothioate ONs showed punctate cytoplasmic fluorescence that corresponded to endocytic vesicles. No fluorescently labelled ON was observed in the nucleus, the major site of action for phosphorothioate ONs (10). Thus, for most cells, with the exception of differentiating keratinocytes (34, 35), delivery reagents are required to introduce biologically active antisense ONs into tissue culture cells.

Several methods have been developed to reliably deliver ONs into the cytoplasm and nuclei of cells, including microinjection (6, 36), electroporation (37), cationic lipids (38, 39), streptolysin O (40), scrape loading (41), and small peptides (42). Of these, cationic lipids are the most widely used reagent to deliver antisense ONs to cells *in vitro*.

4.1 ON delivery using cationic lipids

Most commercially available lipid-mediated transfection reagents have three major drawbacks: poor transfection efficiency, cytotoxicity, and they can not be used in serum-containing growth media. To circumvent these problems, GS3815 cytofectin was developed (*Figure 3*) (12). GS3815 cytofectin is a formulation of two moles of a cationic lipid (dimyristylamidoglycyl-*N*w-isopropoxycarbonyl-arginine dihydrochloride, GS3815) with one mole of the zwitterion L-α-dioleoylphosphatidylethanolamine (DOPE; Avanti Lipids, Alabaster, AL). The molecular weight (MW) of GS3815 is 779 g/mole and the MW of DOPE is 744 g/mole. GS3815 cytofectin is sold by Glen Research, Sterling, Virginia (www.glenres.com). GS3815 cytofectin replaces GS2888 cytofectin because it has long-term stability at both 4 °C and –20 °C. The only difference between GS3815 and GS2888 cytofectin is the replacement of the butyloxycarbonyl group of GS2888 with isopropyloxycarbonyl. GS3815 cytofectin efficiently delivers ONs and plasmids to a wide variety of cell lines in the presence of serum-containing growth medium with little or no cytotoxicity.

The following is a protocol for using GS3815 cytofectin to transfect antisense phosphorothioate ONs into adherent cells grown on 100 mm tissue culture dish (Falcon #3003, Becton–Dickinson Labware, Plymouth, England). This protocol is a modification of previous protocols (16, 39). Cells cultured in suspension are handled in an analogous manner (32).

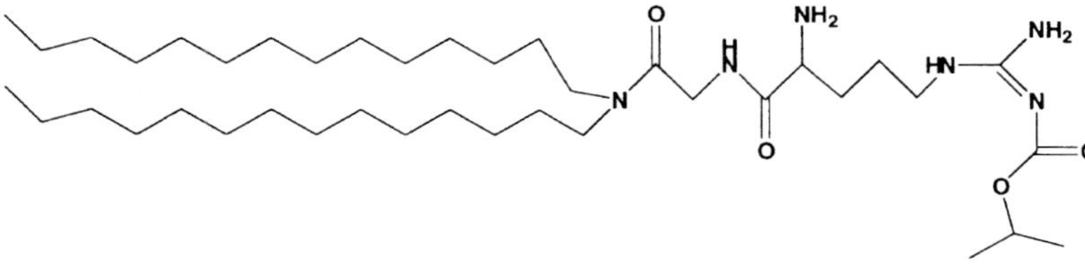

Figure 3. Chemical structure of GS3815 cytofectin.

Protocol 2. Transfection of ONs using GS3815 cytofectin

Equipment and reagents
- Opti-MEM medium (GibcoBRL Life Technologies, Inc., Grand Island, NY)
- GS3815 cytofectin

Method

1. Split the cells the day before the transfection and seed 100 mm plates (Falcon #3003, Becton–Dickinson Labware, Plymouth, England) at a cell density that will result in 60–80% cell confluence on the day of transfection. For most cell lines, this ranges between 1–3 × 10^6 cells per plate.

2. Determine the optimal GS3815 cytofectin concentration for delivering ONs to your cell line of interest. GS3815 cytofectin delivers ONs across a broad range of charge ratios (cytofectin to DNA charge ratios). For most cell lines, the optimal GS3815 cytofectin concentration is 2.5 μg/ml and the optimal antisense ON concentration is between 100 and 1 nM.

   Suggested concentrations for using ONs in cell culture are:

   (a)   unmodified phosphorothioate ONs: 500–100 nM

   (b)   propyne- or G-clamp-modified phosphorothioate ONs: 30–1 nM

3. Store ONs at –20°C. If you are going to be using the ON frequently, then the ON can be stored at room temperature to avoid repeated freeze–thaw cycles.

A. *Preparing the ON/lipid complex*

1. Vortex ONs well and dispense into microcentrifuge tubes (1.5 ml size) at the bench. The antisense ONs should be tested over a range of concentrations (30–1 nM). The final volume of transfection is 4 ml.

2. Transfer the tubes to a tissue culture hood, and add Opti-MEM, a reduced serum medium, to each tube for a final volume of 200 μl. For example, transfecting an antisense ON at a final concentration of 10 nM, the microcentrifuge tube will contain 4 μl of a 10 μM antisense ON stock and a 196 μl Opti-MEM.

3. In a tissue culture hood, dispense GS3815 cytofectin into a 12-well *polystyrene* tissue culture plate. For a 100 mm plate, pipette 10 μl of a 1 mg/ml stock of GS3815 cytofectin. The use of polystyrene plates is essential as the lipid–ON complex binds to polypropylene, which will reduce the transfection efficiency.

4. Add Opti-MEM to each well in the 12-well plate to a final volume of 200 μl (i.e. to 10 μl of a 1 mg/ml GS3815 cytofectin stock add 190 μl Opti-MEM).

5. Combine the 200 μl of the ON–Opti-MEM mixture in the micro-centrifuge tube from step 2 above to the corresponding well in the 12-well polystryrene tissue culture plate containing the lipid–Opti-MEM mixture.

6. Incubate for 10–15 min at room temperature.

7. Add 3.6 ml complete tissue culture growth mediu (10% FBS and antibiotics do not adversely affect transfection efficiency). Mix well. The final transfection volume will be 4 ml. In this example, the final transfection mix (4 ml) contains GS3815 cytofectin at 2.5 μg/ml and the antisense ON at 10 nM final concentration. See *Table 1* for the determination of the final transfection volume.

B. *Transfection of antisense ONs*

1. Remove medium from the cells and replace with the 4 mls of the ON–lipid transfection mixture from step 7 in method A. Do not wash cells with PBS or Tris-buffered saline at any time before or after the addition of the transfection mix as this may affect transfection efficiency.

2. Incubate the transfection mixture with cells for 4–6 h, then add an additional 4 ml of complete medium. Do not remove the lipid–ON transfection mixture at this time unless intolerable toxicity is observed, as indicated by the cells lifting off the plate.

3. Incubate the cytofectin–ON complex for 24–48 h before analysing cells or preparing protein extracts. The time at which extracts are prepared depends on the half-life of the protein of interest. The $t_{1/2}$ of phosphorothioate ONs is ~35 h (43).

C. *GS3815 cytofectin stability and storage conditions*

1. Store GS3815 cytofectin at 4°C. Mix well before use. GS3815 cytofectin is stable for 18 months at 4°C . For longer term storage, keep GS3815 cytofectin at –70°C.

Table 1. Determination of the final transfection volume for commonly used tissue culture plates

Plate size	Transfection vol (ml)	Volume GS3815 cytofectin (1 mg/ml) (in μl)
6-well plate	1	2.5
60 mm plate	2	5
100 mm plate	4	10

4.2 Comments about GS3815 cytofectin

Although GS3815 cytofectin has been used to transfect a broad spectrum of cell lines (*Table 2*), some cell lines have been transfected with limited success. In general, T and B cell lines are impervious to ON transfection using GS3815 cytofectin as well as many of the other commercially available cationic lipids. To test whether your cell line of interest is permeable to antisense ON delivery, a fluorescently labelled ON (FL-ON) is delivered to cells under a variety of transfection conditions and the location of the fluorescent ON in the cell is monitored using a fluorescence microscope (see *Figure 4*). The cells are viewed live 6–24 h following the FL-ON transfection. Cells that are transfected with the ON demonstrate bright nuclear fluorescence due to accumulation of the FL-ON. Successfully transfected cell lines show intense nuclear fluorescence in 80–90% of the cells. As a note of caution, dead and dying cells also demonstrate intense nuclear fluorescence so ensure that the cells are healthy and viable.

4.3 Electroporation of ONs into cells

For cell lines that are impermeable to cationic lipid-mediated ON transfection, electroporation provides an alternative method by which ONs can be introduced into a variety of cells (44, 45). Electroporation is a transfection method in which a high-voltage electrical field is delivered to cells and causes the temporary disruption of the cellular membrane, which allows the entry of

Table 2. Cell lines successfully transfected with GS3815 cytofectin

Cell line	Species	Type
Cos-7	African Green Monkey	Kidney
CV-1	African Green Monkey	Kidney
BalbC-3T3	Mouse	Embryo fribroblast
Rat-2	Rat	Embryo fibroblas
RTLGA	Human	Astrocytoma
Saos-2	Human	Osteogenic sarcoma
HeLa	Human	Cervical carcinoma
NHDF	Human	Normal dermal fibroblast
MCF-7	Human	Breast carcinoma
A549	Human	Lung carcinoma
T24	Human	Bladder carcinoma
HCT116	Human	Colon carcinoma
H460	Human	Lung carcinoma
WiDR	Human	Colon carcinoma
HT-29	Human	Colon carcinoma
CasKi	Human	Cervical carcinoma
SiHa	Human	Cervical carcinoma

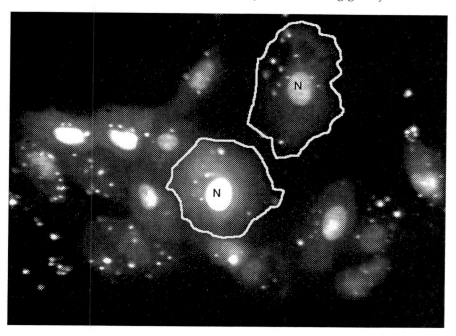

Figure 4. Delivery of fluorescently labelled phosphorothioate oligonucleotides (FL-ONs) to African Green Monkey kidney cells (CV-1) using GS3815 cytofectin (2.5 µg/ml) and 100 nM of the FL-ON. Several of the CV-1 cells have been outlined and the nucleus (N) of the cells denoted. Successfully transfected cell lines demonstrate intense nuclear fluorescence in greater than 90% of the cells as shown in this image. CV-1 cells were viewed live using a Nikon Diaphot inverted fluorescent microscope with 50 watt mercury illumination and a 60 × phase 4 oil emersion len.

ONs. Despite the successful transfection of T and B cells with ONs using electroporation, it has several limitations. First, a large number of cells (1–3×10^7) are required for each transfection sample. The high voltage electrical pulse required to destabilize the cell membrane causes irreparable damage to a large proportion of the cells. Cell death following electroporation usually ranges between 20 and 50%. Secondly, 100- to 1000-fold more ON is required for each sample compared with GS3815 cytofectin-mediated transfections. This is because electroporation does not concentrate the ON during the incubation period, as occurs using GS3815 cytofectin. Despite these limitations, electroporation provides a viable alternative method for introducing ONs into cells that are resistant to GS3815 cytofectin and other available cationic lipid formulations.

Protocol 3. Transfecting antisense ONs by electroporation

Method

1. Dilute confluent cells the day before transfection so that the cells are rapidly proliferating the day of the transfection. For suspension cells such as Jurkat T cells, the cell density should be between 5 and 8 $\times$ 10^5 cells/ml on the day of the transfection.

2. Collect cells for transfection. Adherent cells are trypsinized, resuspended in normal growth media, and centrifuged at 1000 r.p.m. in a Beckman GPR table-top centrifuge for 5 min at room temperature.

3. Remove the medium and resuspend cells in pre-warmed complete growth medium. Centrifuge again as indicated in step 2 to remove any residual trypsin.

4. Remove the growth medium and resuspend cells in complete growth medium at a cell concentration of 3–9 $\times$ 10^7 cells/ml.

5. Pipette cells (300 µl) into a sterile microcentrifuge (1.5 ml).

6. Add ONs to the cells in step 5 so that the final ON concentration will be between 1 and 10 µM. The total volume of the cell suspension and ON should not exceed 350 µl.

7. Incubate the cells and ON together for 10 min at room temperature.

8. Transfer the cell/ON mixture to a 0.4 cm gap cuvette (Bio-Rad Gene Pulser cuvette, cat. no. 165-2088). Place the cuvette into the Bio-Rad electroporation chamber and connect it to the Bio-Rad power supply. The power supply delivers a high voltage pulse of electricity at a defined magnitude and pulse length. The appropriate settings need to be empirically determined. For most cell lines, settings of 300 volts and 960 µF have been successfully used to introduce both ONs and plasmids.

9. Deliver the electrical charge. Remove the cuvette and tap it several times to equilibrate the pH gradient that forms due to electrolysis. Transfer the cuvette to ice. Allow cells to recover for 10 min.

10. Resuspend cells in 10 ml of pre-warmed growth medium and transfer the cells to a single well of a six-well plate.

11. Harvest cells 24–48 h later for analysis.

---

5. Conclusions

For the last two decades, the allure of antisense technology has been the promise of potent, specific, rationally designed inhibitors for any gene of interest. And despite doubts by many researchers whether sequence-specific antisense even existed (18), an antisense renaissance is occurring. Recently, the first phosphorothioate ON has been approved by the FDA for the treatment of CMV-retinitis in AIDS patients and several antisense ONs are used in the clinic for the treatment of inflammatory diseases and cancer. Technological advances in the antisense field have overcome many of the barriers that have limited the use of antisense ONs. Antisense is now a robust technology that can be routinely used in the laboratory to identify and validate new therapeutic targets, and elucidate the roles of proteins in complex biological pathways.

References

1. Flanagan, W.M. (1998). *Cancer Metast. Rev.* **17**, 169.
2. Wagner, R.W. and Flanagan, W.M. (1997). *Mol. Med. Today* **3**, 31.
3. Walder, R.Y. and Walder, J.A. (1988). *Proc. Natl. Acad. Sci. USA* **85**, 5011.
4. Moulds, C., Lewis, J.G., Froehler, B.C., Grant, D., Huang, T., Milligan, J.F., Matteucci, M.D., and Wagner, R.W. (1995). *Biochemistry 34*, 5044.
5. Zamecnik, P.C. and Stephenson, M.L. (1978). *Proc. Natl. Acad. Sci. USA* **75**, 280.
6. Fisher, T.L., Terhorst, T., Cao, X., and Wagner, R.W. (1993). *Nucl. Acids Res.* **21**, 3857.
7. Wagner, R.W. (1994). *Nature* **372**, 333.
8. Matteucci, M.D. and Wagner, R.W. (1996). *Nature* **384**, 20.
9. Froehler, B.C., Wadwani, S., Terhorst, T.J.and Gerrard, S.R. (1992). *Tetrahedron Lett.* **33**, 5307.
10. Wagner, R.W., Matteucci, M.D., Lewis, J.G., Gutierrez, A.J., Moulds, C., and Froehler, B.C. (1993). *Science* **260**, 1510.
11. Lin, K.-Y. and Matteucci, M., D. (1998) *J. Am. Chem. Soc.* **120**, 8531.
12. Flanagan, W.M., Wolf, J.J., Olson, P., Grant, D., Lin, K.-Y., Wagner, R.W., and Matteucci, M.D. *Proc. Natl. Acad. Sci. USA* **96**, 3513 (1999).
13. Monia, B.P., Johnston, J.F., Greiger, T., Muller, M., and Fabbro, D. (1996). *Nature Med.* **2**, 668.
14. Ho, S.P., Bao, Y., Lesher, T., Malhotra, R., Ma, L.Y., Fluharty, S.J., and Sakai, R.R. (1998). *Nature Biotechnol.* **16**, 59.
15. Patzel, V. and Sczakeil, G. (1998). *Nature Biotechnol.* **16**, 64.
16. Flanagan, W.M. (1999). In *The manual of antisense methodology* (ed. S. Endres), in press. Kluwer Academic Publishers, Norwell.
17. Flanagan, W.M. and Wagner, R.W. (1998). In *Applied antisense oligonucleotide technology* (ed. C.A. Stein and A.M. Kreig), p. 175. Wiley-Liss, New York.
18. Stein, C.A. (1995). *Nature Med.* **1**, 1119.
19. Stein, C.A. (1996). *Trends Biotechnol.* **14**, 147.
20. Guvakova, M.A., Yakubov, L.A., Vlodavsky, I., Tonkinson, J.L., and Stein, C.A. (1995). *J. Biol. Chem.* **270**, 2620.
21. Benimetskaya, L., Loike, J.D., Khaled, Z., Loike, G., Silverstein, S.C., Cao, L., El Khoury, J., Cai, T.-Q., and Stein, C.A. (1997). *Nature Med.* **3**, 414.
22. Azad, R.F., Driver, V.B., Tanaka, K., Crooke, R.M., and Anderson, K.P. (1993). *Antimicrob. Agents Chemother.* **37**, 1945.

23. Burgess, T.L., Fisher, E.F.and Ross, S.L. (1995). *Proc. Natl. Acad. Sci. USA* **92**, 4051.

24. Krieg, A.M., Yi, A.K., Matson, S., Waldschmidt, T.J., Bishop, G.A., Teasdale, R., Koretzky, G.A., and Klinman, D.M. (1995). *Nature* **374**, 546.

25. Cowsert, L.M. (1997). *Anti-cancer Drug Design* **12**, 359.

26. Dean, N.M., McKay, R., Miraglia, L., Geiger, T., Muller, M., Fabbro, D., and Bennett, C.F. (1996). *Biochem. Soc.Trans.* **24**, 623.

27. Stein, C.A. and Krieg, A.M. (1994). *Antisense Res. Dev.* **4**, 67.

28. Morishita, R., Gibbons, G.H., Ellison, K.E., Nakajima, M., Zhang, L., Kaneda, Y., Ogihara, T., and Dzau, V.J. (1993). *Proc. Natl. Acad. Sci. USA* **90**, 8474.

29. Coats, S., Flanagan, W.M., Nourse, J., and Roberts, J.M. (1996). *Science* **272**, 877.

30. St. Croix, B., Floerenes, V.A., Rak, J., Flanagan, W.M., and Kerbel, R.S. (1996). *Nature Med.* **2**, 1204.

31. Liu, J., Flanagan, W.M., Drazba, J.A., Estes, M.L., Barnett, G.H., Haqqi, T., Kondo, S., and Barna, B.P. (1997). *J. Immunol.* **159**, 812.

32. Wang, Q.M., Chen, F., Luo, X., Moore, D.C., Flanagan, M., and Studzinski, G.P. (1998). *Leukemia* **12**, 1256.

33. Stein, C.A. (1997). *Antisense Nucl. Acid Drug Dev.* **7**, 207.

34. Nestle, F.O., Mitra, R.S., Bennett, C.F., Chan, H., and Nickoloff, B.J. (1994). *J. Invest. Dermatol.* **103**, 569.

35. Giachetti, C. and Chin, D.J. (1996). *J.Invest. Dermatol.* **107**, 256.

36. Chin, D.J., Green, G.A., Zon, G., Szoka, F.C., and Straubinger, R.M. (1990). *New Biol* **2**, 1091.

37. Bergan, R., Connell, Y., Fahmy, B., and Neckers, L. (1993). *Nucl. Acids Res* **21**, 3567.

38. Bennett, C.F., Chiang, M.Y., Chan, H., Shoemaker, J.E., and Mirabelli, C.K. (1992). *Mol. Pharmacol.* **41**, 1023.

39. Lewis, J.G., Lin, K.Y., Kothavale, A., Flanagan, W.M., Matteucci, M.D., DePrince, R.B., Mook, R.A., Hendren, R.W., and Wagner, R.W. (1996). *Proc. Natl. Acad. Sci. USA* **93**, 3176.

40. Giles, R.V., Spiller, D.G., Grzybowski, J., Clark, R.E., Nicklin, P., and Tidd, D.M. (1998). *Nucl. Acids Res.***26**, 1567.

41. Partridge, M., Vincent, A., Matthews, P., Puma, J., Stein, D., and Summerton, J. (1996). *Antisense Nucl. Acid Drug Dev.* **6**, 169.

42. Pichon, C., Freulon, I., Midoux, P., Mayer, R., Monsigny, M., and Roche, A.-C. (1997). *Antisense Nucl. Acid Drug Dev.***7**, 335.

43. Flanagan, W.M., Kothavale, A., and Wagner, R.W. (1996). *Nucl. Acids Res.* **24**, 2936.

44. Bergan, R., Connell, Y., Fahmy, B., and Neckers, L. (1993). *Nucl. Acids Res* **21**, 3567.

45. Flanagan, W.M. and Wagner, R.W. (1997). *Mol. Cell. Biochem.* **172**, 213.

11

Human leukaemia cells as a model differentiation system

DOROTHY C. MOORE and GEORGE P. STUDZINSKI

1. Introduction

Cell differentiation usually proceeds at the same time as cessation of cell proliferation, and *in vitro* systems for cell differentiation provide excellent models to study orderly down-regulation of growth-related cellular activities. Many such models are available, and these are illustrated in this volume with reference to systems frequently used to study differentiation of human malignant cells. Differentiation of colon cancer cells is described in Chapter 12, of human prostate cancer in Chapter 13, and of human leukaemia in this chapter.

Cultured human leukaemia cells offer a great system to initiate studies of cell differentiation. These cells can be grown in suspension, since this is the natural way in which haematopoietic cells exist, and suspension cultures are both convenient from the tissue culture labour point of view, and quite inexpensive compared with growing cells on monolayers. Also, several leukaemia cell lines can be induced to differentiate into different mature phenotypes, permitting studies directed to mechanisms of switching from one cell type to another. Another advantage is that there are innumerable monoclonal antibodies that have been developed for clinical purposes which allow recognition of differentiation-related expression of surface markers.

2. Cell types

An outline of human haematopoietic cell differentiation is presented in *Figure 1*. All of these steps, or nearly all, can be reproduced in culture systems. Lymphoid cell differentiation falls more properly under the purview of immunology and is considered in specialized texts. Myeloid cell types widely used for differentiation studies include HL60 cells (promyelocytic leukaemia capable of differentiation to granulocytic or monocytic phenotypes), U937 promonocytic cells, K562 cells (considered by some to represent the prototype of a myeloid stem cell with the capacity to differentiate to erythroid, megakaryocytic, and granulocytic/monocytic phenotypes) (1, 2), HEL cells, KU 812 cells, MEG 01 cells, and a number of other human

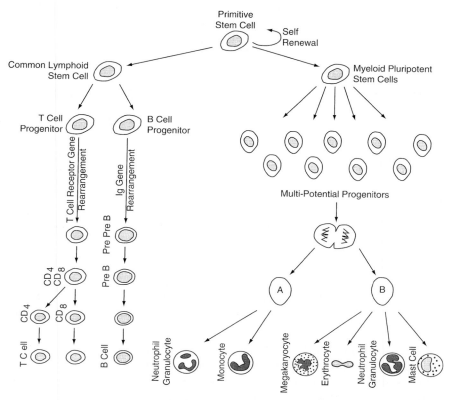

Figure 1. An outline of developmental stages of haematopoietic cells. A primitive stem cell, capable of slow self-renewal, gives rise to both lymphoid and myeloid lineages. The pluripotent myeloid stem cells are shown to give rise to two daughter progenitor cells; progenitor cell A is bipotential, and differentiates to either neutrophil granulocytes or monocytes, while progenitor cell B is multipotential and differentiates to megakaryo-cytes, erythrocytes, and other cells shown. Established leukaemia cell line HL60 corresponds to cell A, and K562 line to cell B.

leukaemia cell lines, which can be induced to more mature forms and allow study of mechanisms of cell growth and differentiation. The HL60 and K562 leukaemia cell lines are two of the human cell types commonly used by investigators, and thus will serve as models to demonstrate the various modalities used to assess leukaemic cell differentiation.

3. Assessment of monocytic differentiation

Phenotypic expression of differentiation can be demonstrated by several methods, documented to provide the evidence of myelomonocytic differenti-ation in previous reports (e.g. 3–7)

Protocol 1. Cellular morphology in air-dried smears stained with May–Grunwald–Giemsa stain

Equipment and reagents
- May–Grunwald–Giemsa stain (Harleco, Gibbstown, NJ)
- IEC PR-7000 centrifuge

Method

1. Use 2×10^6 cells.

2. Harvest cells by centrifuging at 2000 r.p.m. (750 $\times$ g) for 10 min at room temperature.

3. Make a smear and let air-dry.

4. Stain with May–Grunwald–Giemsa solution at room temperature.

Interpretation

The cardinal cytological parameter of granulocytic maturation is the presence of a nuclear constriction or segmentation. Differentiation towards the monocytic phenotype is recognized by an oval or indented nucleus and a decrease in cell size from the undifferentiated form (6).

Protocol 2. Nitroblue tetrazolium reduction (NBT)

Equipment and reagents
- IEC PR-7000 centrifuge
- PBS: 66 mM phosphate, 100 mM NaCl, pH 7.4
- 0.2 % NBT stock solution at 4°C, in the dark (light sensitive)
- 0.01mg/ml phorbol-12-myristate-13-acetate (PMA) also known as 12-*O*-tetradecanoyl phorbol-13-acetate (TPA) stock solution at −20°C
- ethanol

Method

1. Use 2×10^6 cells.

2. Harvest cells by centrifuging at 2000 r.p.m. (750 $\times$ g) for 10 min.

3. Add 0.1 ml of water containing 0.2 % NBT and 0.1 mg/ml of PMA freshly dissolved in the presence of ethanol.

4. Vortex the solution.

5. Incubate at 37°C for 30 min.

6. Prepare a smear by resuspending the cells in a drop of 0.5% albumin and spreading on a glass slide.

227

Protocol 2. *Continued*

7. Count at least 300 cells, depending on experimental reproducibility in the system.

Interpretation

Cells showing blue deposits of formazan are scored as positive for myelomonocytic differentiation.

Protocol 3. Phagocytosis of opsonized sheep erythrocytes

Equipment and reagents

- IEC PR-7000 centrifuge
- rabbit anti-sheep haemolysin (Difco, Detroit, MI)
- Tris-buffered (pH 7.2) 0.83% ammonium chloride

- sheep RBC (red blood cells) (Hazelton Dutchland)
- May–Grunwald–Giemsa stain (Harleco, Gibbstown, NJ)
- PBS: 66 mM phosphate, 100 mM NaCl, pH 7.4

Method

1. Wash sheep RBC three times by centrifuging at 750 g with PBS.
2. Place on ice for 30 min with rabbit anti-sheep haemolysin diluted 1:500 with PBS.
3. Mix HL60 cells and opsonized sheep RBC in a 1:10 ratio.
4. Incubate at 37 °C for 90 to 20 min.
5. After the incubation period, the adherent but not phagocytosed RBC are lysed by resuspending the cells in Tris-buffered 0.83% ammonium chloride for 2–5 min at room temperature.
6. Make air-dried smears and stain with May–Grunwald–Giemsa stain.
7. Count 300 cells.

Interpretation

The proportion of cells containing ingested RBC are enumerated and interpreted as positive. This is most marked for the macrophage phenotype, but is also seen in monocytic and occasionally granulocytic differentiation.

Protocol 4. The non-specific esterase stain (NSE)

Equipment and reagents

- IEC PR-7000 centrifuge
- Sorrenson's buffer (67 mM at pH 7.0) stock solution

- 37 % buffered formalin (pH 6.6) at 4 °C
- acetone at 4 °C
- water

- 4% NaNO$_2$ solution
- hexazotized pararosaniline (stored at 4°C)
- ethylene glycol monoethyl ether (stored at 25°C)

- non-specific esterase substrate: α-naphthyl butyrate (Sigma Chemical Co., St. Louis, MO)

Method

1. Use 1 × 10^6 cells.

2. Adherent cells are detached by gentle pipeting.

3. Harvest cells by centrifugation at 750 × g for 10 min.

4. Fix dried smears in an ice-cold mixture of 37% buffered formalin, acetone, and water (25:45:30) at 4°C for 30 sec.

5. Wash the slide in running distilled water five times.

6. Stain for NSE activity by the method of Yam *et al.* (8), using α-naphthyl butyrate in place of α-naphthyl acetate, without using a counterstain. Keep in the staining solution for 45 min at 37°C.

7. Wash again in running distilled water five times and air-dry.

8. Examine at least 300 cells for a positive NSE reaction.

Interpretation

Macrophages and monocytes show a diffusely scattered reddish-brown precipitate throughout the cytoplasm (*Figure 2*). Neutrophils and their precursors give a negative reaction, no colour. Plasma cells and many lymphocytes demonstrate one or two focal precipitations. Since the NSE reaction can be combined with autoradiography (see *Figure 3*), and is well established as an excellent marker of monocytic differentiation, it is used commonly as an investigative tool.

Protocol 5. Monocyte surface markers

Equipment and reagents

- flow cytometer (Coulter Electronics, Hialeah, FL)
- PBS

- monoclonal antibodies CD14 (MY4-RD1) and CD11b (Mol-FITC) (Coulter Electronics, Hialeah, FL)

Method

1. Use aliquots of 3 × 10^6 cells with a high viable count.

2. Harvest cells and keep on ice for 10 min.

3. Centrifuge cells and wash twice with PBS (750 × g for 10 min at 4°C).

Protocol 5. *Continued*

4. Incubate with 0.5 µl each of CD11b and CD14 for 45–60 min at room temperature in the dark.

5. Wash the cells twice with PBS (750 × g for 10 min 4°C).

6. Resuspend cells in 0.5 ml of PBS.

7. Analyse cells on a flow cytometer using the manufacturer's guidelines.

8. Set the threshold using isotypic mouse IgG.

4. Assessment of erythroid differentiation

The phenotypic expression of erythroid differentiation can be monitored by several procedures (1, 9–12).

Protocol 6. Cellular morphology in air-dried Wright–Giemsa stain

Equipment and reagents
- IEC PR-7000 centrifuge
- cytospin (Shandon, Pittsburgh, PA)
- PBS
- Wright–Giemsa stain (Beckman, Brea, CA)

Method

1. Use 3 × 10⁶ cells.

2. Harvest cells by centrifuging at 750 × g for 10 min at room temperature.

3. Use cytospin to prepare a smear.

4. Stain with Wright–Giemsa solution at room temperature.

Interpretation

Several morphological changes are noted (see *Figure 4*); an increase in the apparent cell size to approximately twice that of the untreated cells, vacuolization of the cytoplasm with some cells showing a prominent perinuclear vacuole of cytoplasmic inclusion, lighter colour of the cytoplasm with a pink tinge, accentuation of cytoplasmic blebbing, and a decreased nuclear-to-cytoplasmic ratio. The nuclei appear very granular but with prominent nucleoli.

Cells with these changes are interpreted to be of erythroid lineage.

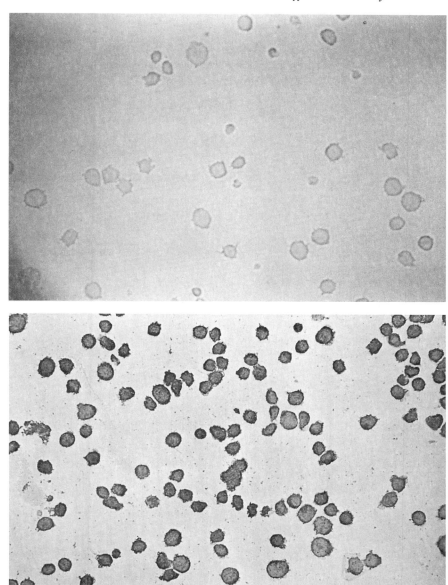

Figure 2. Examples of monocytic differentiation of HL60 cells induced by 1,25-dihydroxyvitamin D_3 (1,25D_3). (a) Proliferating HL60 cells, untreated, stained by non-specific esterase (NSE) procedure, characterizing monocytic phenotype. The cells are negative for the differentiation marker. (b) The appearance of cells after 96 h exposure to 2.4 $\times$ 10^{-8} M 1,25D_3 . Note the positive stain (red, see also *Figure 3*) and changed cell shape indicating monocytic differentiation.

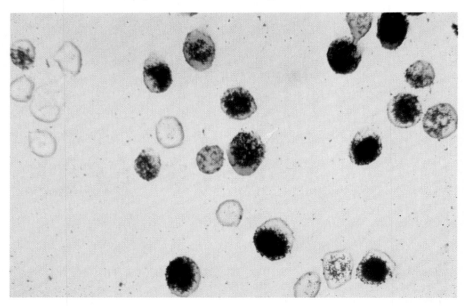

Figure 3. HL60 cells at high magnification following exposure to 1,25D$_3$ as described above, then pulse-labelled with [^{3}H]thymidine, stained by NSE, followed by visualization of DNA replication by autoradiography for [^{3}H]thymidine incorporation. Note DNA replication (nuclear grains) in differentiating cells (cytoplasmic red staining for NSE).

Protocol 7. Benzidine dihydrochloride test for haemoglobin (Hb) (13)

Equipment and reagents
- Neubauer haemocytometer (Thomas, Swedesboro, NJ)
- benzidine dihydrochloride
- 3% acetic acid
- distilled water
- 1% hydrogen peroxide

Method

1. Prepare a flask of 3 × 10^5 cells/ml in fresh medium to initiate experiments.

2. Remove 0.5 ml of cell suspension at scheduled times.

3. Prepare benzidine dihydrochloride in 3% acetic acid in distilled H$_2$O, and keep refrigerated.

4. Add hydrogen peroxide (1%) just prior to use.

5. Mix the cell suspension with the benzidine solution in a 1:1 ratio.

6. Within 4–5 min, evaluate cells using a Neubauer haemocytometer.

7. Count 300 cells.

Interpretation

Blue colouring in the cytoplasm is considered to demonstrate positivity for Hb.

Protocol 8. Spectrophotometric measurement of the Hb content of cell culture (12, 14)

Equipment and reagents

- IEC PR-7000 centrifuge
- Beckman DU-64 spectrophotometer
- PBS
- 0.5% Nonidet P-40

- lysing buffer: 0.81% NaCl, 0.03% magnesium acetate, 0.12% Tris (pH 7.4)
- pure human Hb, Sigma (St. Louis, MO)

Method

1. Wash 5×10^6 cells twice with ice-cold phosphate-buffered saline (PBS).

2. Centrifuge the cells at $750 \times g$ for 10 min.

3. Resuspend in 1 ml of lysing buffer and 0.5% Nonidet P-40.

4. Incubate on ice for 15 min.

5. Remove the nuclei by centrifugation at 750 *g* for 20 min.

6. Obtain an absorption curve for Hb from the supernatant fluid in a spectrophotometer from a reading of the change in OD of the curve at 414 nm.

7. To standardize the procedure, dissolve 5,10, and 20 μg of pure human Hb in water and then add to the lysate of 5×10^6 untreated cells to obtain a calibration curve.

8. Add 10 μg of exogenous Hb to each group after the initial reading, and obtain a second curve, which is used to calculate the unknown amount of Hb present in the test group.

9. Calculate the cell Hb content on the basis of the total cells and viable cells in the sample.

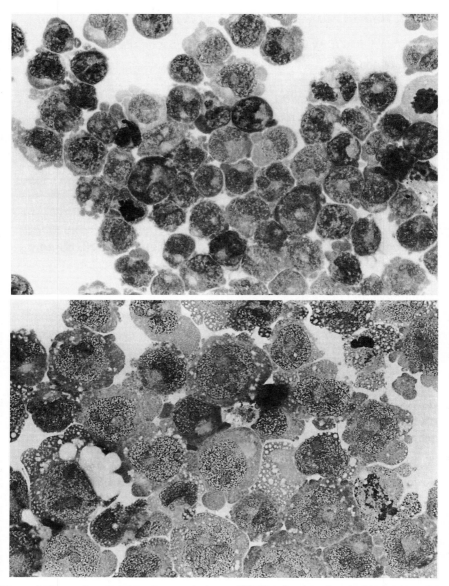

Figure 4. Examples of change in morphology of K562 cells after chemical induction of the erythroid lineage by arabinocytosine (Ara-C). (a) Untreated K562 cells (control) stained by the Wright–Giemsa procedure. The cells are small with large round nuclei, a small rim of blue-grey cytoplasm, and a high nuclear-to-cytoplasmic ratio. Occasional nucleoli are evident and the nuclear chromatin has a granular, open pattern. The percentage of haemoglobin (Hb)-containing (benzidine-positive) cells was 1%. (b) K562 cells treated with 1 $\times$ 10^{-6} M Ara-C for 72 h. Morphological changes noted are: increase in cell size, vacuolization of the cytoplasm with some cells showing a prominent perinuclear vacuole or cytoplasmic inclusion, lighter colour of the cytoplasm with a pink tinge, accentuation of cytoplasmic blebbing, and a decreased nuclear-to-cytoplasmic ratio. The nuclei have a very granular appearance with prominent nucleoli. The percentage of Hb-containing cells was 60%.

Protocol 9. Erythroid surface markers

Equipment and reagents

- flow cytometer (Coulter Electronics, Hialeah, FL)
- PBS
- monoclonal antibody glycophorin A (Coulter Electronics, Hialeah, FL)

Method

1. Use aliquots of 3×10^6 cells with a high viable count.
2. Harvest the cells and keep on ice for 10 min.
3. Centrifuge the cells and wash twice with PBS at (750 $\times$ g for 10 min at 4°C).
4. Incubate with 0.5 μl of glycophorin A for 45–60 min at room temperature in the dark.
5. Wash the cells twice with PBS (750 $\times$ g for 10 min at 4°C).
6. Resuspend in 0.5 ml of PBS.
7. Perform the analysis using a flow cytometer.
8. Set the threshold using isotypic mouse IgG.

Protocol 10. Globin mRNA production

Equipment and reagent

- IEC PR-7000 centrifuge
- −70°C freezer
- Beckman DU-64 spectrophotometer
- PBS
- guanidium isothiocyanate procedure (15)
- CsCl gradient
- phenol
- chloroform
- 200 mM NaCl
- ethanol (ice-cold)
- Tris–EDTA buffer (pH 7.4)
- 6% formaldehyde/1% agarose gel
- nitrocellulose membrane
- 50% formamide
- 5 $\times$ Denhardt's solution
- 50 mM Tris (pH 7.5)
- 0.8 M NaCl
- 10 % dextran sulfate
- 75 μg/ml salmon sperm DNA
- 0.5 % sodium dodecyl sulfate (SDS)
- ^{32}P-labelled probe for globin genes
- 2 $\times$ sodium chloride–sodium citrate
- 0.1 % SDS
- X-ray film

A. RNA isolation

1. Isolate total cellular RNA by a guanidium isothiocyanate procedure with a CsCl gradient centrifugation.
2. Extract RNA with a 1:1 phenol/chloroform mixture.
3. Precipitate RNA in 200 mM NaCl and ice-cold ethanol.
4. Collect by centrifugation at 20°C.

Protocol 10. *Continued*

5. Air-dry and dissolve in Tris–EDTA buffer.

6. Quantitate RNA by absorbance at 260 nm, and assess the purity by the ratio of the absorbance at 260 and 280 (this ratio should be close to 2.0).

7. Store in a –70 °C freezer.

B. *Northern analysis*

1. Size fractionate RNA by loading 15 μg/well of denatured RNA on 6% formaldehyde/1% agarose gels.

2. Transfer to a nitrocellulose membrane.

3. Pre-hybridize the membrane for 4 h at 42 °C in 50% formaldehyde, 5 × Denhardt's solution, 50 mM Tris, 0.8 mM NaCl, 10% dextran sulfate, 75 μg/ml salmon sperm DNA, and 0.5% SDS.

4. Hybridize for 24 h at 42 °C using 2×10^6 cpm/ml of ^{32}P-labelled probe.

5. Prepare the probe by the random priming labelling technique, according to the instructions provided by Boehringer Mannheim Co. (Indianapolis, IN).

6. Wash the membrane twice in sodium chloride–sodium citrate, and 0.1% SDS three times, for 30 min each, at room temperature.

7. Place the membrane in a plastic bag, and expose to an X-ray film.

8. Develop after 2–3 days to see the appropriate bands.

5. Assessment of megakaryocytic differentiation

Many leukaemic cell lines, such as the K562 and HEL, can be induced to differentiate to the megakaryocytic cell phenotype using TPA or other agents (2, 16–18).

Protocol 11. Cellular morphology by May–Grunwald–Giemsa stain, by phase contrast microscopy, and by electron microscopy

Equipment and reagents

- cytocentrifuge
- phase contrast microscope
- electron microscope (model 400T, Philips Electronic Instruments, Inc., Mahwah, NJ)
- May–Grunwald–Giemsa stain (Harleco, Gibbstown, NJ)
- trypsin–EDTA
- 0.1 M cacodylate buffer (pH 7.4)
- 2% cacodylate-buffered OsO_4
- 4% 0.1 M cacodylate-buffered gluteraldehyde
- propylene oxide
- Epon/propylene oxide
- Epon
- gelatin capsules
- toluidine blue
- uranyl acetate
- lead citrate

A. *May–Grunwald–Giemsa stain*

1. Use 2×10^6 cells.
2. Harvest the HEL cells by centrifugation.
3. Use the cytospin to make a smear
4. Make a smear and let air-dry.
5. Stain with May–Grunwald–Giemsa solution at room temperature.

Interpretation

Within 3–5 days after TPA stimulation, HEL cells show a marked increase in cell size, cytoplasmic content, and nuclear complexity. The nuclei of these cells become large and lobulated; the cytoplasm undergoes maturational changes, becoming less basophilic, and contains large regions of eosinophilia (17).

B. *Phase contrast microscopy*

1. Harvest adherent K562 cells by washing three times and enumerating the cells still adherent to the flask.
2. Remove the remaining adherent cells with trypsin–EDTA.
3. Examine with a phase microscope.

Interpretation

Morphological changes indicating megakaryocytic differentiation of K562 cells are that cells become more adherent, highly elongated, and partially spread (16).

C. *Electron microscopy*

11. Fix cells in 4% 0.1 M cacodylate-buffered gluteraldehyde for 1 h at room temperature.
12. Wash twice in 0.1 M cacodylate buffer.
13. Post-fix in 2% cacodylate-buffered OsO_4 for 1 h at room temperature.
14. Wash twice in the 2% buffer
15. Dehydrate the cells in graded alcohol to propylene oxide.
16. Infiltrate the cells in graded concentrations of Epon/propylene oxide for 3 h.
17. Pellet the cells in microfuge tubes.
18. Embed in pure Epon, utilizing gelatin capsules.
19. Allow the samples to harden overnight at 65 °C.
10. Cut plastic sections (1.0 μm) and stain with toluidine blue to identify areas of interest.

Protocol 11. *Continued*

11. Stain ultra thin sections with uranyl acetate and lead citrate, and examine using an electron microscope.

Interpretation

Ultrastructural changes associated with megakaryocytic differentiation consist of lobulation of the nucleus, a marked increase in the cytoplasm, and an increase in mitochondria and free ribosome numbers. Additionally, a few cells show membranous vesicles, which appear similar to regions of the nascent demarcation membrane system. This system is felt to eventually delimit the developing platelets (2, 17).

Protocol 12. Histochemical staining

Equipment and reagents

- cytospin
- Dako's APAAP staining kit (Capinteria,CA)
- 4% paraformaldehyde (PFA)
- moAbs GP IIb/IIIa or GP IIIa (Dako Co., Glostrop, Denmark)
- PBS

Method

11. Air-dry the specimen and fix with PFA at 4°C overnight.

12. Perform enzyme immunostain with moAbs against GP IIb/IIIa or GP IIIa using Dako's APAAP staining kit according to the manufacturer's recommendation.

13. Incubate cytospin slides with moABs against GP IIb/IIIa or GP IIIa for 30 min.

14. Wash twice with PBS for 5 min.

15. Overlie the linking antibody on the specimen for 30 min.

16. Wash with PBS twice for 5 min.

17. Add APAAP complex for another 30 min.

18. Wash twice with PBS.

19. Add substrates and incubate the specimen for an additional 15 min.

10. Use normal rabbit serum as a negative control for these experiments.

11. Count at least 200 cells and repeat three times.

Background staining with normal mouse serum or irrelevant isotypic IgG should be less than 2%.

Protocol 13. Megakaryocytic surface markers by flow cytometry

Equipment and reagents

- flow cytometer (Coulter Electronics, Hialeah, FL)
- PBS

- Monoclonal antibody CD41 (glycoprotein IIb/IIIa) (Coulter Electronics, Hialeah, FL)

Method

1. Use aliquots of 3×10^6 cells with a high viable count.
2. Harvest the cells and keep on ice for 10 min.
3. Centrifuge the cells and wash twice with PBS at $750 \times g$ for 10 min at 4°C.
4. Incubate with 0.5 μl of glycophorin A for 45–60 min at room temperature in the dark.
5. Wash the cells twice with PBS ($750 \times g$ for 10 min at 4°C).
6. Resuspend in 0.5 ml of PBS.
7. Perform the analysis using a flow cytometer.
8. Set the threshold using isotypic mouse IgG.

---

6. Summary

These detailed procedures for several leukaemia differentiation systems exemplify the approaches to, and the performance of studies of leukaemia cell differentiation. Further examples can be found in refs 11, 12, 16, and 17.

Acknowledgments

The laboratory studies in the authors' laboratories have been supported by grants R29DK45270 (to D.C.M.) and R01-CA 44722 (to G.P.S.) from the National Cancer Institute. We are also grateful to Dr Milan Uskokonic, Hoffmann-La Roche, Nutley, NJ, who graciously provided vitamin D analogues as differentiation-inducing agents.

References

1. Moore, D.C., Carter, D.L., Bhandal, A.K., and Studzinski, G. P. (1991). *Blood*, **77**, 1452.
2. Nakamura, M., Kirito, K., Yamanoi, J., Wainai, T., Nojiri, H., and Saito, M. (1991). *Cancer Res.*, **51**, 1940.

3. Wang, Q.M., Chen, F., Luo, X., Moore, D.C., Flanagan, M., and Studzinski, G.P. (1998). *Leukemia*, **12**, 1256.

4. Studzinski, G.P., Rathod, B., Wang, Q.M., Rao, J., and Zhang, F. (1997). *Exp. Cell Res.*, **232**, 376.

5. Wang, Q.M., Jones, J.B., and Studzinski, G.P. (1996). *Cancer Res.*, **56**, 264.

6. Studzinski, G. P., Bhandal, A.K., and Brelvi, Z.S. (1985). *Proc. Soc. Exp. Biol. Med.*, **179**, 288.

7. Studzinski, G.P., Bhandal, A.K., and Brelvi, Z.S. (1985). *Cancer Res.*, **45**, 3898.

8. Yam, L.T., Li, C.Y., and Crosby, W.H. (1971). *Am. J. Clin. Pathol.*, **55**, 283.

9. Moore, D.C., Carter, D.L., and Studzinski, G.P. (1992). *J. Cell. Physiol.*, **151**, 539.

10. Hata, Y., Yamaji, Y., Shiotani, T., Fujita, J., Kamano, H., Ikeda, K., Takahara, J., and Irino, S. (1991). *Biochem. Pharmacol.*, **42**, 2307.

11. Dean, A., Ley, T.J., Humphries, R.K., Fordis, M., and Schechter, A.N. (1983). *Proc. Natl. Acad. Sci. USA*, **80**, 5515.

12. Cioe, L., McNab, A., Hubbell, H.R., Meo, P., Curtis, P., and Rovera, G. (1981). *Cancer Res.*, **41**, 237.

13. Fibach, E., Treves, A., and Rachmilewitz, E.R. (1983). *Cancer Res.*, **43**, 4136.

14. Rutherford, T.R. and Weatherall, D.J. (1979). *Cell*, **16**, 415.

15. Maniatis, T., Fritsch, E.F., and Sambrook, J. (ed.) (1982). *Molecular cloning: a laboratory manual*. Cold Spring Harbor Laboratory Press, Cold Spring Harbor.

16. Burger, S.R., Zutter, M.M., Sturgill-Koszycki, S., and Santoro, S.A. (1992). *Exp. Cell Res.*, **202**, 28.

17. Long, M.W., Heffner, C.H., Williams, J.L., Peters, C., and Prochownik, E.V. (1990). *J. Clin. Invest.*, **85**, 1072.

18. Murate, T., Hotta, T., Tsushita, K., Suzuki, M., Yoshida, T., Saga, S., Saito, H., and Yoshida, S. (1991). *Blood*, **77**, 3168.

Induction of differentiation of human intestinal cells *in vitro*

HEIDE S. CROSS and ENIKÖ KALLAY

1. Introduction

The intestinal epithelium is, in principle, a valuable tissue for the study of cell differentiation, as it is spatially organized around its proliferative units, the crypt (small intestine) or the basal section of the crypt (large intestine). The progeny of clonal undifferentiated proliferating cells express eventually at least four very different phenotypes (1), e.g. enterocytes, mucin-producing goblet cells, endocrine cells, and Paneth cells. The determinants of this phenotypic commitment are largely unknown. During the orderly upward migration of cells from crypt to villus along the crypt length in the colon, the temporal sequence of differentiative events can be well defined. However, this situation unfortunately cannot be recreated in intestinal culture systems, probably because of the absence of supportive tissues, and the inability to recapitulate this complex dynamic process with continuous cell proliferation and cell loss within a few days. Thus, *in vitro* studies of differentiation markers in normal intestinal cells are difficult to perform.

An additional problem is the frequent absence of polarity in culture. Like all simple epithelia, the intestinal mucosa is theoretically a very good model for the study of cell polarity, since fully differentiated intestinal cells are strictly polarized. This is, from a physiological point of view, essential: only under such circumstances can nutritional uptake from the lumen, as mediated by brush border enzymes and transport systems in the apical brush border, function properly and a compromise of the epithelial barrier be avoided.

In order to obtain *in vitro* model systems, three strategies were used in the past:

- isolation and culture of normal intestinal cells (2)
- *in vitro* culture of intestinal explants
- culture of tumour cells, mainly from the colon

However, as Quaroni *et al.* (2) demonstrated with normal rat intestinal cells, cells can fail to differentiate and express brush border enzymes even though they form epithelial monolayers in culture. Explants have been frequently used for *in vitro* experimentation and have provided evidence for differentiation – related expression of intestinal transport (e.g. calcium and phosphate) (3) and of brush border enzymes (4). However, normal entero-cytes and explants in culture have a very limited life-span, at most five days. The most successful strategy appears to be to culture the neoplastic cells. Fogh and Tempe (5) have isolated several human colon cancer cell lines which appear to have unlimited life-spans and are, at least with certain limitations, useful models for the study of differentiation of intestinal cells. In several of these cell lines, e.g. HT 29 and Caco-2, differentiation can be manipulated in culture. Some of these experimental approaches for the Caco-2 cell line will be described in this chapter. However, one has to keep in mind that immortal cells are genetically altered. Thus, these cell lines are probably far removed from normal human physiology.

The most difficult, but probably the closest to the *in vivo* situation is the isolation of primary cultures from the normal, pre-cancerous, and cancerous human intestine. These cells can be used for a limited time span only, but at least those derived from tumours can be frozen in liquid nitrogen after successful culturing, and can be thawed for specific experimentation (see refs 6 and 7, for example). In this chapter several of our experimental approaches to study differentiation of colon cells will be described in detail.

2. Differentiation agents

2.1 Vitamin D compounds

$1\alpha,25$-Dihydroxyvitamin D_3 ($1,25$-D_3), the active metabolite of vitamin D_3 produced by hydroxylation steps in liver and kidney, is a key calcium- and phosphate-regulating hormone with effects on bone, intestine, and kidney. In recent years $1,25$-D_3 has been shown to possess cell-regulating properties in a number of tissues and cells not directly involved in intestinal absorption and bone mineralization: it has potent pro-differentiating activities, e.g. myeloid leukaemia cells are induced into non-proliferating monocytes/macrophages (8). These vitamin D receptor (VDR)-mediated actions are due to specific gene transcription activation. The VDR has been detected in a variety of normal and cancerous cells from different organs (see ref. 9). Expression levels of this receptor may be related to the extent of vitamin D responses.

It is suggested that vitamin D regulates induction of terminal differentiation and passage of cells through the cell cycle. It is not known however, whether the pro-differentiating action of the steroid causes the inhibition of cell growth, or whether the two activities are distinct. In any case, both actions can be exploited for clinical applications, for instance in the treatment of malignancies or of psoriasis.

The active metabolite $1,25$-D_3, as well as several side chain-altered analogues can decrease proliferation and increase differentiation of colon cancer cell lines (10) and of primary intestinal cultures obtained from human colon surgical material (7).

2.2 High levels of calcium (1.8–2.4 mM) in the culture medium

In normal epithelial cells, low extracellular calcium (below 0.25 mM) generally promotes cell proliferation, whereas high calcium elicits terminal differentiation. However, it was suggested that intestinal pre-neoplastic cells lose the ability to respond to the differentiation signal of high calcium by the time they have advanced to a benign tumour state (adenoma). This theory does not hold up any longer since in the colon adenocarcinoma-derived cell line Caco-2 high calcium is indeed anti-mitotic and pro-differentiating (11). In a recent paper (12) it was determined that the growth inhibitory effect of high calcium concentration can be demonstrated *in vitro*, and that the effect, mediated via the apical colonocyte membrane, is the reduction of c-*myc* expression. This sheds a new light on the suggestion of Garland *et al.* (13) that serum levels of the precursor 25-D_3, as well as calcium levels in the gut lumen, may partially determine the risk in humans of colon cancer.

3. Model systems: establishment and use

3.1 Cell lines (Caco-2)

The human adenocarcinoma-derived Caco-2 cell line, a good model for *in vitro* studies, is available from the repository of the ATCC collection. Grown under standard culture conditions, it spontaneously exhibits signs of structural and functional differentiation. The monolayer is covered with typical brush border microvilli. The asymmetry of the cells and the presence of tight junctions suggest that the monolayer is polarized. Functionally, the differentiation is characterized by high levels of the brush border-associated enzymes, including alkaline phosphatase and sucrase. This line appears to be polyclonal since during prolonged growth, cells at different levels of differentiation can give rise to clonal progeny. The group of Quaroni and co-workers (14) have isolated clones 2/15 and 2/3 (the former grows slowly and reaches a high level of differentiation after confluence, the latter grows faster and its differentiation level is low). In the author's laboratory, another clone (AQ) was isolated by dilution plating. This grows fast, reaches medium levels of differentiation, and has a uniquely altered expression of the vitamin D receptor.

3.1.1 Growth conditions

All clones are routinely cultured in vented tissue culture flasks (Falcon plasticware, Becton–Dickinson) at 37°C in a humidified atmosphere of 95% air and 5% CO_2 in Dulbecco's modified Eagle's medium (DMEM) containing

4 mM glutamine, 10% FCS (heat-inactivated at 56°C for 30 min), 20 mM Hepes, 50 U/ml penicillin, and 50 µg/ml streptomycin. Cultures are fed every 48 h and subcultured serially when approximately 80% confluent. Cells are used between passages 10 and 30.

3.2. Primary cultures

All types of tissue are processed within 1 to 2 h after surgery.

Protocol 1. Isolation of colon crypts from normal mucosa and adenoma biopsies

Reagents

- culture medium: DMEM with 10 mM glucose, 4 mM glutamine, 1% FCS, 2% Luria broth (LB), 0.2 U/ml insulin, 10 µg/ml human transferrin, 10 ng/ml sodium selenite, 200 U/ml penicillin, 200 µg/ml streptomycin, 50 U/ml nystatin, and 50 µg/ml gentamycin

- wash medium: DMEM, 200 U/ml penicillin, 200 µg/ml streptomycin, 50 U/ml nystatin and 50 µg/ml gentamycin
- Luria broth: 5 g/l NaCl, 10 g/l bactotryptone, and 5 g/l yeast extract

Method

1. Wash specimens extensively with wash medium.

2. Scrape mucosa from submucosa and chop into pieces of 1–2 mm in diameter using sterile scalpels.

3. Transfer the cut pieces to a 30 ml Sterilin tube. Wash several times by centrifuging at 100 g for 5 min at room temperature (to remove mucosal debris: single cells and mucus remain in the supernatant).

4. Add wash medium containing 0.4 U/ml collagenase P (Boehringer Mannheim) and 1.2 U/ml dispase I (Boehringer Mannheim) and digest for 5–15 min at 37°C.[a]

5. Add wash medium to 10 ml and pipette up and down several times to separate crypt-like structures mechanically from interstitial tissue. Filter through a 100 µm nylon filter.

6. Wash again as above several times.

7. Resuspend the crypts with culture medium and incubate cultures at 37°C, 5% CO_2, and 95% humidity.

[a] It is essential to use very small tissue pieces and an extremely short time of enzyme exposure to ensure survival of cells in culture for up to 3 weeks (adenomas) and up to 5 days (normal mucosa).

The establishment of primary cultures from tumour tissue differs slightly from the method used for normal tissue.

Protocol 2. Establishment of primary cultures from colon adenocarcinomas

Reagents

- wash medium (see *Protocol 1*)
- medium I: DMEM with 10% FCS, 4 mM glutamine, 10mM Hepes, 100 U/ml penicillin, 100 μg/ml streptomycin, 50 μg/ml gentamycin, 50 U/ml nystatin, 1 μg/ml hydrocortisone, 0.2 U/ml insulin, 2 μg/ml transferrin, 5 nM sodium selenite

- medium II: like medium I without gentamycin and nystatin
- medium III: DMEM with 10% FCS, 4 mM glutamine, 10 mM Hepes, 100 U/ml penicillin, 100 μg/ml streptomycin

Method

1. Wash specimens extensively with wash medium.
2. Remove fatty and necrotic parts and cut tissue into 1 mm slices.
3. Transfer cut pieces to a 30 ml Sterilin tube. Wash several times (see *Protocol 1*, step 3).
4. Dissociate mechanically by gentle rotation, shaking, and agitation with a Pasteur pipette.
5. Plate resulting cell clumps and single cells in medium I.
6. Digest remaining tissue enzymatically with 0.3 mg/ml collagenase and 0.8 U/ml dispase in wash medium for 1–2 h at 37 °C.
7. Add wash medium and dissociate mechanically (see *Protocol 1*, steps 5 and 6).
8. Seed isolated cells and cell clumps on to round glass coverslips in medium I.
9. Use medium II after 1 week of uncontaminated growth.
10. Use medium III when epithelial growth is well established.

Preferential growth of epithelial cells without fibroblast overgrowth is initiated on glass coverslips by using human foreskin 3T3 fibroblasts lethally irradiated with 60 Gray (6000 rads) of gamma radiation as a feeder layer.

For initial passaging, primary cultures are subcultured only when tumour cell areas become confluent. This, in some cases, is achieved only after several months. For initial passaging, cells are scraped off from coverslips, in subsequent passaging, dispase (2 U/ml) is used to detach the cells in sheets. Generally, after three passages enough cells for proliferation assays are available.

3.3 Freshly isolated cells

With a new method recently developed in our laboratory ('cold cocktail method'), single cell suspensions from colon mucosa with viability up to 45 h after isolation can be obtained and used for investigating changes in metabolic activity.

Protocol 3.   Cold cocktail method

Reagents

- disruption buffer: DMEM with 10 mM Hepes, 200 U/ml penicillin, 200 μg/ml streptomycin
- cold cocktail: DMEM containing 1000 U/ml collagenase (Sigma), 0.5% trypsin (Gibco), 0.02% ethylendiaminetetraacetic acid disodium salt (EDTA)
- dissociation cocktail: DMEM containing 1000 U/ml collagenase (Sigma), 0.5% trypsin (Gibco), 0.02% 5 μg/ml DNase
- culture medium: DMEM with 10 mM Hepes, 200 U/ml penicillin, 200 μg/ml streptomycin, and 10% FCS[a]

Method

1. Mince tissue pieces with a scalpel and wash several times in disruption buffer by centrifuging at 200 *g* for 10 min at 4°C.
2. Resuspend the pellet in cold cocktail and incubate at 4°C for a minimum of 2 h or overnight.
3. Wash again and centrifuge as above.
4. Resuspend the pellet in warm (37°C) dissociation cocktail and shake strongly for 10–60 min at 37°C in a water bath to obtain single cells.
5. Cells are seeded in tissue culture dishes in culture medium.

[a] FCS is generally beneficial; however, if experimentation demands serum-free conditions, 0.1% BSA (fatty acid free) can be used.

To test for viability, perform a trypan blue exclusion test. Check the metabolic activity of cells by using the resazurin metabolic activity test. Viable cells reduce the amount of the oxidized dark blue form of resazurin into its red fluorescent intermediate. The amount of dye conversion in solution is measured fluorometrically or spectrophotometrically.

Protocol 4.   Metabolic activity test

Equipment and reagents

- CytoFluor fluorescent plate reader (Millipore)
- *in vitro* toxicology assay kit containing resazurin (Sigma)

Method

1. Add 10% v/v resazurin dye solution to the culture medium of cells (optimal 10^6 cells/cm^2).
2. Put cultures in an incubator for 2–4 h. Keep incubation times consistent.

---

3. Measure samples spectrophotometrically by monitoring decrease in absorbance at 600 nm (measure absorbance of plates at 690 nm, subtract from 600 nm reading). Alternatively, measure fluoro-metrically[a] at 590 nm (excitation 560 nm).

[a] Fluorometric measurement is more sensitive.

3.4 Frozen human tissue

Clean human tissue from surgery of necrotic and fatty parts, cut into 3 mm × 5 mm pieces, and freeze directly in liquid nitrogen (freeze before transfer into cryotubes!). Alternatively, tissue can be fixed by standard methods before freezing which provides better preservation of tissue structure.

4. Markers of differentiation and markers of cessation of proliferation

It is difficult to describe intestinal cell markers of differentiation without describing also markers of proliferation. The following aspects will therefore be discussed:

- morphology
- enzymatic activities: sucrase–isomaltase (SI), alkaline phosphatase (AP)
- evaluation of proliferation: proliferating cell nuclear antigen (PCNA) expression, thymidine incorporation into DNA, cell cycle distribution
- cell cycle-associated markers: cyclin D1 protein expression, c-Myc expression
- markers occurring during colon cancer progression: VDR expression, epidermal growth factor receptor (EGFR) expression
- E-cadherin expression

4.1 Morphology and related markers

The apical brush border of intestinal absorptive cells is designed to expand the membrane surface area available for hydrolysis and transport. This brush border provides a good model system for the study of membrane–cytoskeletal interactions. The isolated cytoskeleton of brush borders can be divided into two distinct areas, the microvilli and the terminal web, which is a cytoskeleton-rich region in the apical cytoplasm beneath the microvillus. The core of the microvilli is composed of microfilaments made of actin isoforms. The terminal web cytoskeleton consists of three distinct filamentous domains oriented perpendicularly to the microvilli core bundles. Villin is a globular actin-binding protein which is found in large amounts in cells having a well-organized brush border. It is a monomeric calcium-binding (three sites per molecule) protein in solution. When calcium is below micromolar concentrations, villin acts as a bundling factor. Above 10 mM, villin severs actin microfilaments into fragments.

Differentiation of intestinal cells is a continuous and gradual process throughout their life-span. Morphogenesis of the brush border occurs during migration of cells from the crypt base to the top. During this process microvilli are gradually and synchronously elongated. Villin is expressed in immature cells at the crypt base as well as in mature cells at the crypt top. However, there is a gradient along the crypt axis: the amount of villin in differentiated cells is 10-fold higher than in immature cells. Interestingly, the level of villin mRNA is low at the crypt base, reaches a maximum at the villus bottom, and decreases towards the villus tip. In the colon, where no villi exist, villin is still present in the crypt cells, but in smaller amounts. Enterocytic differentiation appears to be accompanied by the targeting of villin into the developing brush border structure. Since colon enterocytes, and also colon cancer cells, display a brush border to a greater or lesser extent, localization of villin may also be a marker for differentiation, which could be influenced by differentiating substances like vitamin D.

Calcium, which previously has been described as an anti-proliferative substance in the colon, may very well influence the morphology and differentiation of cells via alteration of cytoskeletal structures and components like villin (see also Section 4.6).

Protocol 5. Brush border membrane isolation

Equipment and reagents

- Optima TLX ultracentrifuge (Beckman)
- Waring blender
- phosphate-buffered saline (PBS)
- 1 M $MgCl_2$

- mannitol buffer: 50 mM mannitol, 2 mM Tris–HCl pH 7.1, 1 mM PMSF, 5 μg/ml leupeptin, 5 μg/ml antipain, 10 μg/ml aprotinin

Method[a]

1. Wash cells in culture with PBS and scrape into mannitol buffer.

2. Homogenize in a Waring blender.

3. Add 1 M $MgCl_2$ to a final concentration of 10 mM.

4. Incubate at 4°C for 30 min to agglutinate and separate basolateral and intracellular membranes from brush border membranes.

5. Centrifuge for 15 min at 3000 *g*.

6. Centrifuge supernatant for 30 min at 27 000 *g*.

7. Resuspend the pellet in mannitol buffer, rehomogenize, and centrifuge at 27 000 *g*.

> **8.** Resuspend the resulting pellet with a 25 gauge syringe, determine the protein concentration (*Protocol 6*), and use for further assays.
>
> [a] All steps are carried out at 4°C.

It is important to determine the total amount of protein present in samples needed for further analysis. One of the alternatives is the use of the BCA protein assay (Pierce) based on the well-known reaction of proteins with Cu^{2+} in an alkaline medium (yielding Cu^{+}) with a highly sensitive detection reagent for Cu^{+}, namely bicinchonic acid (BCA).

Protocol 6. BCA protein assay

Equipment and reagents

- ELISA reader
- 96-well plate
- BCA protein assay kit (Pierce)

- working reagent: mix 50 parts of reagent A (sodium carbonate, sodium bicarbonate, bicinchonic acid, sodium tastrate in 0.2 N sodium hydroxide) with 1 part of reagent B (4% cupric sulfate)

Method

1. Prepare a set of protein standards of known concentration by diluting the BSA standard solution in the same diluent as your unknown samples.
2. Pipette 10 μl of standard or unknown sample into the wells of the 96-well plate.
3. Add 200 μl working reagent into each well.
4. Mix samples for 30 sec on a microtitre plate shaker.
5. Incubate at 37°C for 30 min.
6. Read the absorbance at or near 562 nm with an ELISA reader.

4.2 Enzymatic activities

4.2.1 Sucrase–isomaltase

Differentiated absorptive cells of the small intestine (enterocytes) and of the colon (colonocytes) are morphologically very similar. However, the respective brush border membranes contain only a small number of common glycoproteins. Only some of the digestive hydrolases, such as SI and dipeptidyl peptidase IV (DPP IV) are transiently expressed in the human fetal colon and in some differentiated colonic tumours (15). Even this is surprising, since expression of the two hydrolases can normally be correlated with development of villi, and these are certainly absent from the colon. Therefore, expression of the two hydrolases may rather be a sign of oncofetal development and not of differentiated colon cells (14, 16).

For enzymatic assays of sucrase–isomaltase the method of Messier and Dahlqvist is used (17).

Protocol 7.   One-step ultra micro method for disaccharidase activity assay

Equipment and reagents

- U-2000 spectrophotometer (Hitachi)
- substrate solutions: 62 mM disaccharide (sucrose and isomaltose)
- standard glucose solution: 1.1 mM (mix 200 mg glucose, 2.7 g benzoic acid, and water to 1000 ml)
- 50% v/v sulfuric acid in water

- phosphate–glucose oxidase (PGO) reagent: add 2 mg glucose oxidase (Boehringer Mannheim) to 9.8 ml 0.5 M sodium phosphate buffer, pH 6, 0.1 ml 1% *o*-dianisidine (in 95% ethanol), and 0.1 ml 0.1% peroxidase

Method

1. Mix 100 μl PGO reagent with 20 μl brush border membrane proteins (see *Protocol 5*) and incubate at 37 °C for 2 min.
2. Add 100 μl substrate solution and incubate at 37 °C for 75 min.
3. Stop the reaction by adding 100 μl 50% H_2SO_4.
4. Incubate a second tube containing an identical mixture for only 15 min after addition of the disaccharide, stop with H_2SO_4.
5. Measure the absorbance at 530 nm against a blank solution containing 100 μl PGO reagent, 120 μl water, 100 μl 50% H_2SO_4.
6. Standard curve: mix 100 μl PGO reagent with 10, 15, or 20 μl standard glucose solution, with water to a total volume of 220 μl. Incubate for 15 min and stop with 100 μl H_2SO_4. Measure the absorbance at 530 nm against a glucose-free blank.
7. Calculate the activity of the disaccharidases, expressed as μmol disaccharide hydrolysed/ml/per min from the formula: $(G_{75}-G_{15})d \times 50/n \times 60 \times 180$. G_{75} and G_{15} are the amounts of glucose, in μg, found after 75 and 15 min, respectively, *d* is the dilution factor of the brush border membrane solution, and *n* is the number of glucose molecules liberated per molecule of substrate.

4.2.2 Alkaline phosphatase

In contrast to the hydrolases mentioned above, alkaline phosphatase activity appears to be a good marker for colonocyte differentiation (18). Colon cell lines such as Caco-2, which differentiate spontaneously in culture after reaching confluency, also display a gradual increase of the marker during this time. A maximum is reached by 15 days after confluency. This spontaneous elevation of activity can be elicited earlier by addition of a differentiating agent to the culture medium, such as 1,25-D_3 or high calcium (11).

Protocol 8. Alkaline phosphatase activity

Equipment and reagents
- CytoFluor fluorescent plate reader (Millipore)
- 96-well plate
- standard *p*-nitrophenol (pNP) (Sigma) in 0.5% Triton X-100: 5, 10, 20, 40, 60 nmol/50 µl per well
- PBS
- Triton X-100 (0.5% in water)
- *p*-nitrophenylphosphate (pNPP) solution pH 9.5: 20 mM pNPP (Sigma), 100 mM diethanolamine (Sigma), 150 mM NaCl, 2 mM MgCl$_2$

Method

1. Wash cells twice with PBS.

2. Remove PBS and freeze plates at −20 °C.

3. During subsequent thawing add 0.5% Triton X-100[a] and shake for 5 min at room temperature.

4. To a 96-well plate add 50 µl 0.5% Triton X-100 as blank, 50 µl from each of the pNP standards, and 50 µl of the unknown samples.

5. Add 150 µl pNPP solution to each well.

6. Incubate for up to 30 min at room temperature.[b]

7. Stop reaction by adding 50 µl NaOH (0.5 M).

8. Measure the absorption at 405/490 nm in a microplate reader.

[a] Add 200 µl when cells are subconfluent, and 500 µl when cells are confluent.
[b] Check the intensity of the staining.

4.3 Evaluation of proliferation

Although these procedures ar described in Chapter 1, slightly different individual steps have been found to be optimal in studies of colon cells.

As stated previously, measurement of proliferation has to be included when studying differentiation of cells. In general, cessation of growth also suggests an advance in differentiation. However, this does not necessarily apply in colon cancer: though cells show some differentiation markers, they still can be highly proliferative. Proliferative activity is frequently a useful indicator of malignant potential: increased proliferative activity ultimately leads to the development of multiple clonal subpopulations, and in turn, this increases the chance that some will acquire the ability to invade, migrate, and metastasize.

A major problem in attempting to assess proliferation is chosing the appropriate method, since some methods will only provide data on the proliferative status of cells and not on the proliferative rate. Three methods are demonstrated here: PCNA measurement, thymidine incorporation into

DNA, and cell cycle distribution. It is advisable to use at least two of these for correct assessment.

4.3.1 Proliferating cell nuclear antigen

PCNA is a 36 kDa nuclear protein which progressively accumulates from G1 to M phase of the cell cycle and disappears at the end of mitosis. It is a useful semi-quantitative marker in immunohistochemical analysis of cell proliferation. With haematoxylin staining as a background, the monoclonal antibody shows positive nuclei in the lower 30% of a normal colonic crypt, whereas during malignant disease the proliferative compartment extends also to the upper part of the crypt.

Protocol 9. Immunohistochemistry[a]

Reagents

- PBS
- fixative solution:[b] 4% formaldehyde in PBS
- peroxidase inhibitor solution (Dako)
- 0.5% BSA in PBS
- blocking solution (protein block serum-free, Dako)
- antibodies
- AEC substrate kit (Dako)

Method

1. Wash cells with PBS and fix with fixative solution for 5 min at room temperature. Wash with PBS again.
2. Add peroxidase inhibitor solution, incubate for 5 min at room temperature, and wash with PBS.
3. Block cells with blocking solution for 1 h at room temperature.
4. Add antibody in blocking solution (working dilution must be determined individually). Incubate overnight at 4°C.
5. Wash cells three times for 20 min each with 0.5% BSA.
6. Add (secondary) horseradish peroxidase (HRPO)-coupled antibody diluted in 0.5% BSA and incubate for 1 h at room temperature.
7. Wash cells three times for 20 min each with PBS.
8. AEC staining: mix 1 drop substrate with 2 ml buffer, add to the cells over 2–10 min depending on the intensity of the staining.
9. Wash twice with water.
10. Mount with aqueous mounting fluids.

[a] Immunohistochemical studies of cells and tissues (paraffin or frozen) can be performed also with specific kits [e.g. Histostain™-Plus kit (ZYMED)].
[b] Other comonly used fixatives are: acetone, ethanol, methanol, Bouin's.
Note. Another variant of this procedure is described in Chapter 1, *Protocol 8.*

4.3.2 Thymidine incorporation into DNA

Protocol 10. [³H]Thymidine incorporation

Equipment and reagents
- liquid scintillation counter, e.g. Wallac 1410 (Pharmacia)
- [³H]thymidine
- liquid scintillation cocktail (Beckman)
- PBS
- 5% trichloroacetic acid (TCA)
- 0.1 M NaOH

Method

1. Add [³H]thymidine (4 μCi/ml/well) to cells grown in 24-well plates, and incubate for 5 h at 37°C, 5% CO_2, and 95% humidity.

2. Wash twice with warm (37°C) PBS for 2 min.

3. Precipitate with ice-cold TCA (5%), twice for 10 min.

4. Wash with ice-cold water twice for 5 min. Drain excess water and remove by aspiration.

5. Solubilize cells in 500–1000 μl 0.1 M NaOH. (You can store cell lysate at –20°C.)

6. Neutralize 0.5 ml cell lysate with 50 μl glacial acetic acid, add 10 ml liquid scintillation cocktail for aqueous samples, and measure the counts in a liquid scintillation counter.

Note. Slightly different details are provided in Chapter 1, *Protocol 5.*

[³H]Thymidine incorporation is expressed in counts per minute (cpm) per mg of protein. Determine the protein concentration from the cell lysate (see *Protocol 6*).

4.3.3 Cell cycle analysis
Measurement of DNA content of a cell gives a static view of the cell cycle. The method is based on the fact that cells in G2 and M phases of the cell cycle have double the DNA content of those in G0 and G1 phases. Cells in S phase will have a DNA content between these extremes.

4.4 Cell cycle-associated markers
Progression of cells through the cell cycle is governed by the sequential formation and degradation of proteins like c-Myc (for entrance into the G1 phase from the resting phase) and of a series of cyclins that complex with and

activate several cyclin-dependent kinases. These cyclin and cyclin-dependent kinase complexes play a critical role in cell proliferation and differentiation. There are at least 11 distinct cyclin genes in the human genome that regulate different parts of the cell cycle. For example, immunoneutralization of cyclin D1 results in cell cycle arrest in the G1 phase. Cyclin D1 overexpression has been associated with neoplasia, it decreases cell size, and makes cells less dependent on exogenous growth factors. Though in other types of neoplasia rearrangement and increased expression of the cyclin D1 gene has been observed, it is rarely found in colon cancer cells. However, there is definitive evidence that cyclin D1 is more highly expressed in pre-malignant and malignant colon than in normal colon, and, interestingly, the immunohistochemical stain is not only found in the nucleus, but also sometimes in the cytoplasm without appearing in the nucleus. This suggests an additional, currently undefined, physiological role for the protein (19).

Protocol 11. Cell cycle analysis

Equipment and reagents

- flow cytometer FACSCalibur (Becton Dickinson)
- STC solution, pH 7.6: 3.4 mM trisodium citrate, 0.1% v/v Nonidet P-40, 1.5 mM spermine–tetrahydrochloride (STC), 0.5 mM Tris
- trypsin solution: 0.03 mg/ml trypsin in STC solution
- trypsin inhibitor solution, pH 7.6: 100 ml STC, 50 mg trypsin inhibitor, 60 mg RNase
- STC* solution, pH 7.6: add 200 mg STC to 200 ml STC solution
- propidium iodide (PI) solution: dissolve 4.2 mg PI in 10 ml STC* solution

Method

1. Harvest cells by mild trypsinization, wash them twice in PBS, and resuspend in 2.0 ml trypsin solution and incubate for 15 min at room temperature.

2. Add 2.0 ml trypsin inhibitor solution, and incubate again for 15 min at room temperature.

3. Add 20 ml of STC* buffer and collect nuclei by centrifuging at 1000 *g* for 20 min.

4. Resuspend in 300 μl PI solution.

5. Analyse stained nuclei on a flow cytometer using linear amplification. Cell cycle distribution can be analysed using CELLFIT software.

The expression of cell cycle-associated marker proteins can be determined by immunoblotting (*Protocol 14*) of total cell lysates (*Protocol 12*) or nuclear and cytoplasmic cell fractions (see *Protocol 13*) with specific antibodies.

Protocol 12. Total cell lysate

Reagents
- lysis solution: 1% SDS, 1 mM sodium vanadate, 10 mM Tris, pH 7.4
- PBS

Method

1. Rinse adherent cells grown on tissue culture dishes with PBS.

2. Drain excess PBS and add 1 ml of boiling lysis solution per 10 cm diameter Petri dish.

3. Scrape cells from the dish, transfer to a microcentrifuge tube, and boil for an additional 5 min. The viscosity of the sample can be reduced by brief sonication or by several passages through a 26 gauge needle.

4. Centrifuge the samples for 5 min to remove insoluble material. At this point you have the opportunity to measure the protein concentration in a small aliquot of sample using the BCA method. (*Protocol 6*).

Protocol 13. Nuclear/cytoplasmic protein preparation

Equipment and reagents
- Hettich universal centrifuge 30 RF
- PBS
- Triton X-100 (0.3% in nuclear buffer)
- 1 M NaCl
- nuclear buffer, pH 6.4: 150 mM NaCl, 1 mM KH_2PO_4, 5 mM $MgCl_2$, 1mM EGTA, 0.2 mM DTT, 10% glycerol, 1mM PMSF, 20 μg/ml aprotinin, and 1 μg/ml leupeptin

Method[a]

1. Wash adherent cells with PBS. Add nuclear buffer and scrape with a rubber policeman.

2. Centrifuge cells for 10 min at 150 *g*. Supernatant is the cytoplasmic fraction.

3. Resuspend the pellet in nuclear buffer and centrifuge again for 10 min.

4. Solubilize the pellet in nuclear buffer with 0.3% Triton X-100 and incubate for 30 min with gentle rocking. Spin down for 10 min at 200 *g*.

5. Resuspend the pellet in nuclear buffer and centrifuge again. Repeat this step twice.

6. Resuspend the pellet in nuclear buffer and add 1 M NaCl to an end concentration of 0.35 M.

Protocol 13. *Continued*

7. Shake for 30 min, centrifuge at 300 *g* for 20 min.

8. Centrifuge the supernatant for 10 min at 20μ.000 *g*. Add glycerin up to 20% to the supernatant (this is the nuclear fraction) and store at –70°C.

ᵃAll steps are done on ice.

Measure the protein concentration of total cell lysates, nuclear, and cytoplasmic cell fractions with the BCA method (*Protocol 6*).

Protocol 14. Western blotting

Equipment and reagents

- electrophoresis apparatus
- electroblotting apparatus
- nitrocellulose
- Kodak film
- 2 × sample buffer: 250 mM Tris pH 6.8, 4% SDS, 10% glycerol, 0.006% bromphenol blue, 2% mercaptoethanol
- polyacrylamide gel
- blocking buffer (SuperBlock from Pierce)

- running buffer: mix 3.0 g Tris, 11.4 g glycin, 1.0 g SDS with water to 1000 ml
- transfer buffer: 25 mM Tris, 190 mM glycine, 20% MeOH
- PBST (0.1% Tween 20 in PBS)
- antibodies
- ECL chemiluminescence substrate (Amersham)

Method

1. Add an equal volume of 2 × sample buffer to protein samples and boil for 3–5 min.

2. Apply equal amounts of protein to each well of a polyacrylamide gel. Separate proteins by electrophoresis (SDS–PAGE) in running buffer.

3. Transfer proteins from gel to nitrocellulose at 20 mA for 4 h and 10 mA for 14 h in transfer buffer.

4. Remove the blot from the transfer apparatus, place into blocking buffer, and incubate for 2 h at room temperature.

5. Decant the blocking buffer from the blot, add the primary antibody diluted in the blocking buffer, and incubate overnight at 4°C.

6. Decant the primary antibody solution and wash the blot three times for 10 min in PBST.

7. Dilute the HRPO-conjugated secondary antibody in blocking buffer, and incubate for 2 h at room temperature.

8. Wash again, three times for 10 min, in PBST.

9. Place the blot in a tray containing ECL substrate for 1 min and detect specific protein bands by autoradiography.

4.5 Markers occurring during colon cancer progression

4.5.1 Vitamin D receptor

The vitamin D receptor is a 45–50 kDa protein (monomer) generally thought to be located in the nucleus. It can homodimerize or heterodimerize with the Retinoid X receptor and this dimer can bind to vitamin D responsive elements with high affinity. Within each target tissue the level of VDR is not fixed, but rather is dynamically regulated by multiple factors. The level of VDR expression in a cell may be important because it can determine the amplitude of the response evoked by 1,25-D_3 treatment. A number of factors regulate VDR abundance and they include homologous regulation by 1,25-D_3, as well as heterologous regulation by other hormones and growth factors. Various studies have shown that upregulation of the VDR may be due to an increase in transcription, while other studies show that it may be due to stabilization and prolongation of the receptor half-life by ligand binding. It appears that, at least in Caco-2 cells, the latter may be the case (20). Interestingly, in Caco-2/15 cells the VDR protein level increases with differentiation after confluency, whereas in Caco-2/AQ it declines. The difference may be due to secretion of autocrine growth factors.

4.5.2 Epidermal growth factor receptor

Epidermal growth factor (EGF) and transforming growth factor-α (TGF-α) have been implicated in growth regulation of cells by binding to a common membrane receptor (EGFR). Growth stimulation by these substances has been shown in multiple colon cancer cell lines. In human colon cancer it has been documented that overexpression of EGFR may indicate an advanced stage of the disease, and may predict metastatic potential (21). It is also well known that EGFR in normal colonocytes is mainly present at the basolateral and not the apical aspect of cells (22), whereas in enterocytes from fetal or suckling rats the distribution is mainly on the apical side with stimulation by EGF-like substances via the gut lumen. Our recent investigations with human colon cancer-derived primary cultures showed that in cancer cells a similar distribution pattern to that in fetal colonocytes is found (6). The question remains whether treatment of colon cancer cells with differentiating substances like 1,25-D_3 would result in a relocation of EGFR.

Protocol 15. Determination of EGFR on cell surfaces: binding assay

Equipment and reagents

- gamma counter (Beckman)
- binding medium: 0.1% BSA in DMEM
- 0.5 ng/ml [^{125}I]EGF
- 50 ng/ml EGF (unlabelled)

Protocol 15. *Continued*

Method

1. Wash cell monolayers grown on filter units (Transwell polycarbon membranes) and incubate in binding medium with 0.5 ng/ml [^{123}I]EGF for 3 h at 4°C on either the apical or basal filter side, with only binding medium without radioactivity at the opposite chamber.

2. Determine non-specific binding in the presence of 100-fold excess of unlabelled EGF.

3. Rinse monolayers in ice-cold binding medium, trypsinize, and count in a gamma counter.

4. Measure the protein concentration. (*Protocol 6*). Express bound radio-activity, or the amount of receptor, as fmol/mg protein.

4.6 E-cadherin expression

E-cadherin is a120 kDa protein that plays a major role in the establishment of calcium-dependent cell–cell adhesion in epithelial cells. It promotes adhesion by formation of intercellular junctions. It has also multiple other functions, such as establishment of cell polarity and cell sorting, and it plays an important role in signal transduction. The correct function of E-cadherin requires its association with the cytoskeleton. E-cadherin interacts at both sides of the cell membrane. On the extracellular side, different domains have been implicated in Ca^{2+} binding and cell–cell interactions. On the cytoplasmic side, it has been shown that E-cadherin binds several proteins like α-catenin, α-catenin, plako-globin, and p120, involved in linking E-cadherin to the cytoskeleton (23). Many of these cell functions are typical of differentiated intestinal cells. During the process of malignant transformation, E-cadherin function is lost, and epithelial cells become apolar and highly invasive. Similarly, poorly differentiated metastatic human colon carcinomas frequently contain reduced or undetectable levels of E-cadherin (24). Thus, there appears to be a strong connection between high E-cadherin levels, differentiation, and high levels of calcium. Since, in Caco-2 monolayers, high levels of calcium in the medium appear to suppress mitotic activity of cells (13) this is also of interest for E-cadherin expression. To investigate the E-cadherin protein expression, cytoskeleton-associated and -non-associated (Triton-insoluble and -soluble) cellular fractions are prepared and analysed by Western blotting (*Protocol 14*).

The following protocol on semi-quantitative PCR (polymerase chain reaction) is used in general for evaluation of markers in liquid nitrogen-frozen human tissue. This method, instead of Northern blotting, is frequently necessary, owing to the small amounts of tissue needed for the evaluation. However, when using whole tissue, it is essential to consider that these consist of many types of cells. Therefore, when evaluating a marker of only epithelial cells, expression of cytokeratins for example, has to be taken as a reference.

Protocol 16. Extraction of Triton-soluble and -insoluble cellular fractions

Equipment and reagents

- CSK buffer: 50 mM NaCl, 10 mM Pipes, pH 6.8, 3 mM MgCl$_2$, 0.5% Triton X-100, 300 mM sucrose, 1 mM orthovanadate, 20 μM phenylarsine oxide, 1 mM PMSF, 5 μg/ml leupeptin, 5 μg/ml antipain, 10 μg/ml aprotinin
- Hettich universal centrifuge 30 RF
- PBS
- SDS buffer: 20 mM Tris, pH 7.5, 5 mM EDTA, 2.5 mM EGTA, 1% SDS

Method

1. Wash adherent cells with PBS.
2. Solubilize the sample in CSK buffer for 20 min at 4°C with gentle rocking.
3. Scrape cells with a rubber policeman, centrifuge at 18000 *g* for 10 min at 4°C. Supernatant constitutes the Triton-soluble fraction.
4. Triturate the pellet in SDS buffer and boil at 100°C for 10 min.
5. Centrifuge at 18000 *g* for 10 min at 4°C. Supernatant constitutes the Triton-insoluble fraction.
6. Determine the protein concentration (see *Protocol 6*).

Protocol 17. Semi-quantitative PCR

Equipment and reagents

- GeneAmp PCR system 9600 (Perkin–Elmer)
- TRIzol reagent (Gibco)
- oligo (dT)
- 5 × first strand buffer: 250 mM Tris–Cl, pH 8.3, 375 mM KCl, 15 mM MgCl$_2$
- 0.1 M DTT
- 10 mM dNTP mix (10 mM each)
- SUPERSCRIPT II reverse transcriptase (Gibco)
- 10 × PCR buffer: 200 mM Tris–Cl, pH 8.4, 500 mM KCl
- 50 mM MgCl$_2$
- amplification primer (forwards and reverse)
- *Taq* DNA polymerase

A. *First strand cDNA synthesis*

1. Extract total RNA with the TRIzol reagent following the manufacturer's instructions.
2. Add 1 μl oligo (dT) (500 μg/ml), 1–5 μg total RNA, and sterile water to 12 μl.
3. Heat the mixture to 70°C for 10 min and quickly chill on ice.

259

Protocol 17. *Continued*

4. Centrifuge briefly, add 4 μl 5 × first strand buffer, 2 μl 0.1 M DTT, 1 × μl 10 mM dNTP mix, and incubate for 2 min at 42 °C.

5. Add 1 μl (200 units) SUPERSCRIPT II reverse transcriptase. Incubate for 50 min at 42 °C.

6. Inactivate the reaction by heating at 70 °C for 15 min.

B. *Semi-quantitative PCR[a]*

1. Mix 2.5 μl 10 × PCR buffer, 0.75 μl MgCl$_2$,[b] 0.5 μl dNTP mix, amplification primer pairs,[b] 0.5 μl Taq polymerase, cDNA[b] (from first strand reaction), and water to 25 μl.

2. Denature the mixture for 5 min at 94 °C.

3. Perform 20–40 cycles of PCR. Annealing and extension conditions are primer and template dependent and must be determined empirically (e.g. denaturing at 94 °C for 15 sec, annealing at 58 °C for 30 sec, extension at 72 °C for 1 min).

4. Perform a final extension step at 72 °C for 10 min, then cool at 4 °C.

[a] For semi-quantitative PCR one can use the Taq PCR Core Kit from Qiagen.
[b] The concentration of MgCl$_2$, primer, and cDNA must be determined individually.

An endogenous sequence, known to be present at constant levels throughout a series of samples to be compared, can be used as an internal standard in semi-quantitative PCR reactions. Endogenous mRNA standards, such as β-actin, GAPDH, or cytokeratin 8 (a common marker for intestinal epithelial cells), can be used as an internal control for the RT-PCR reaction, to determine relative levels of specific mRNAs. To avoid the differences between tubes, RT-PCR is performed in one tube and an aliquot is used for the amplification of specific mRNAs (E-cadherin, β-catenin, VDR) in the same tube with the internal standards (β-actin, GAPDH, or cytokeratin 8). The amount of input RNA and the number of amplification cycles has to be optimized to ensure that PCR products are in the exponential phase of amplification. Aliquots of the PCR products are resolved by electrophoresis on a 2% agarose gel. For quantitation, the photographs of the ethidium bromide-stained gels are analysed by densitometry. The OD area of the PCR products of interest (E-cadherin, β-catenin, VDR) will be divided by the OD area for the β-actin (or GAPDH or cytokeratin 8) band of the corresponding sample.

References

1. Leblond, O. P. and Cheng, H. (1976). In *Stem cells of renewing cell populations*, p. 7. Academic Press, New York.

2. Quaroni, A., Wands, J., Trelstad, R. L., and Isselbacher,-K. J. (1979). *J. Cell Biol.*, **80**, p. 248.
3. Cross, H. S., Pölzleitner, D., and Peterlik, M.. (1986). *Acta Endocrinol.*, **113**, p. 96.
4. Menard, D. and Malo, C. (1979). *Dev. Biol.*, **69**, p. 661.
5. Fogh, J. (ed.) and Trempe, G. (1975). In *Human tumor cells in vitro*, p.115. Plenum Press, New York.
6. Tong, W. M., Ellinger, A., Sheinin, Y., and Cross, H.S. (1998). *Br. J. Cancer.*, **77**, 1792.
7. Tong, W. M., Bises, G., Sheinin, Y., Ellinger, A., Genser, D., Pötzi, R., Wrba, F., Wenzl, E., Roka, R., Neuhold, N., Peterlik, M., and Cross, H. S. (1998). *Int. J. Cancer.*, **75**, 467.
8. Abe, E. Miyaura, C. Sakagami, H., Takeda, M., Konno, K., Yamazaki, T., Yoshiki, S., and Suda, T. (1981). *Proc. Natl. Acad. Sci. USA*, **78**, 4990.
9. Cross, H. S., Bajna, E., Bises, G., Genser, D., Kallay, E., Potzi, R., Wenzl, E., Wrba, F., Roka, R., and Peterlik, M. (1996). *Anticancer Res.*, **16**, 2333.
10. Bischof, M. G., Redlich, K., Schiller, C., Chirayath, M. V., Uskokovic, M., Peterlik, M., and Cross, H. S. (1995). *J. Pharmacol. Exp. Ther.*, **275**, 1254.
11. Cross, H. S., Pavelka, M., Slavik, J., and Peterlik, M. (1992). *J. Natl. Cancer Inst.*, **84**, 1355.
12. Kallay, E., Kifor, O., Chattopadhyay, N., Brown, E. M., Bischof, M. G., Peterlik, M., and Cross, H. S. (1997). *Biochem. Biophys. Res. Commun.*, **232**, 80.
13. Garland, C., Shekelle, R. B., Barrett-Connor, E., Criqui, M. H., Rossof, A. H., and Paul, O. (1985). *Lancet*, **1**, 307.
14. Cross, H. S. and Quaroni, A. (1991). *Am. J. Physiol.*, **261**, C1173.
15. Zweibaum, A., Hauri, H. P., Sterchi, E., Chantret, I., Haffen, K., Bamat, J., and Sordat, B. (1984). *Int. J. Cancer*, **34**, 591.
16. Gorvel, J. P., Ferrero, A., Chambraud, L., Rigal, A., Bonicel, J., and Maroux, S. (1991). *Gastroenterology*, **101**, 618.
17. Messier, M. and Dahlqvist, A. (1966). *Anal. Biochem.*, **14**, 376.
18. Schwartz, B., Lamprecht, S. A., Polak-Charcon, S., Niv, Y., and Kim, Y. S. (1995). *Oncol. Res.*, **7**, 277.
19. Arber, N., Hibshoosh, H., Moss, S. F., Sutter, T., Zhang, Y., Begg, M. Wang, S. Weinstein, I. B., and Holt, P. R. (1996). *Gastroenterology*, **110**, 669.
20. Hulla, W., Kallay, E., Krugluger, W., Peterlik, M., and Cross, H. S. (1995). *Int. J. Cancer*, **62**, 711.
21. Radinsky, R., Risin, S., Fan, D., Dong, Z., Bielenberg, D., Bucana, C. D., and Fidler, U. (1995). *Clin. Cancer Res.*, **1**, 19.
22. Playford, R. J., Hanby, A. M., Goodlad, R. A., Gschmeissner, S., Patel, K., Peiffer, L. P., and McGarrity, T. (1995). *Gastroenterology*, **108**, A747.
23. Kemler, R. (1993). *Trends Genet.*, **9**, 317.
24. Dorudi, S., Sheffield, J., Poulsom, R., Northover, J. M. A., and Hart, I. R. (1993). *Am. J Pathol.*, **142**, 981.

13

Proliferation and differentiation of human prostate cells

THERESE THALHAMMER and HEIDE S. CROSS

1. Introduction

More than 200000 new cases of prostate cancer are diagnosed anually in the USA. Thus, in incidence and prevalence, prostate cancer is the most common malignancy in American men, and black men in the USA have the highest incidence of prostate cancer in the world. Some epidemiological observations in various populations suggest a vital role for steroid hormones in prostate carcinogenesis. Although androgens are clearly the most influential in this respect, others such as oestrogens, retinoids, and vitamin D compounds might also be important. With advancing age, while there is a decline in serum testosterone levels, a concomitant increase of benign prostatic hyperplasia is observed. However, the incidence of microscopic, latent prostate cancer also shows a very close correlation with increasing age and thus decreasing testosterone. This, and also data from animal experimentation, indicates an additional fundamental change in men with age, which results in the transformation of prostate epithelial cells (1, 2)

Though there are large variations in the incidence of clinical prostate cancer between various regions of the world, there appears to exist a similar incidence of latent microscopic cancer in all male populations. This clearly points to a multistep process of carcinogenesis and also suggests that environmental and/or life-style factors such as nutrition may contribute to clinical manifestations of the disease (3).

The major problem in treatment is presented by patients with hormone-independent tumours. Although prostate epithelial cells are commonly transformed with increasing age to microscopic cancer and then to clinical cancer which may, for a certain period, be treatable by chemical and surgical castration as long as the tumour is inside the capsule, ultimately, hormone-independent cancer cells arise, which leads to uncontrollable metastasis. However, as long as hormone-dependent cells exist, androgens may not directly mediate growth of prostatic epithelial cells, but rather modify indirectly the behaviour of cells. It is becoming clear that, in order to prevent uncontrollable growth, the function of various growth-stimulatory and -inhibitory factors and of their receptors, whose inappropriate expression or loss disrupts normal regulation of cell proliferation and differentiation, needs to be studied in normal, hyperplastic, and cancerous prostate cells. In this chapter, the regulation and interaction of steroid hormones with growth factors with respect to proliferation and differentiation of prostatic cells will be discussed to delineate their relevance during prostate tumour progression.

1.1 Components of prostate tissue

The epithelium of normal prostate tissue is comprised of two major populations: the basal cells and the luminal cells, and each seems to have a unique role in prostate biology. Basal cells are flat and elongated and lie underneath the luminal cells along the basal lamina. Luminal cells are secretory and possess the androgen receptor. These two epithelial cell types can also be differentiated by their distinctive cytokeratin profile. Basal cells express cytokeratins 5 and 14, whereas luminal cells express cytokeratins 8 and 18. Luminal cells also secrete certain differentiated products such as PSA (prostate-specific antigen) with androgen-promoting PSA gene transcription. Stem cells are believed to exist primarily in the basal cell population. Interestingly, following castration, cells with basal properties persist and luminal cells disappear. Antibodies against proliferation-associated antigens predominantly localize to basal cells.

The epidermal growth factor (EGF) as well as transforming growth factor alpha (TGF-α) are potent mitogens for prostatic epithelial cells *in vitro* and are supposed to be synthesized by stromal cells. The receptor for both ligands, the EGFR, is found on basal cells. However, other growth factors, such as insulin-like growth factor, IGF I and IGF II, also appear to be potent mitogens for prostatic cells.

Lately the concept has emerged that androgens do not directly modulate growth of prostatic cells, but rather act via stimulation of stromal cells to secrete growth factors, which then act on epithelial prostatic cells. If, during tumour progression, cells become androgen independent, it is because androgen-sensitive luminal cells disappear and autocrine growth factors may govern proliferation (4, 5).

2. *In vitro* models

In order to study proliferation and differentiation of prostate cells, good *in vitro* models need to be established. Well-known prostate cell lines are LNCaP, PC3, and DU-145, which, according to *Table 1*, may be hormone dependent or independent for growth. Additionally, primary cultures may be established which present the advantage of having further differences in cell type available. However, the method is not easy and maintenance is difficult and relatively short-lived.

2.1 Human prostate cancer cell lines: (LNCaP, PC3, and DU-145)

The following three prostate cancer cell lines, characterized in *Table 1*, are widely used in research:

- The hormone-dependent LNCaP (ATCC CRL 1740) cell line was derived from a supraclavicular lymph node of a 50-year old patient with hormone-refractory prostate carcinoma.

- The hormone-independent PC3 (ATCC CRL 1435) was derived from a bone marrow metastasis of a 62-year old caucasian with prostate cancer.

- The hormone-indepedent DU-145 (ATCC HTB 81) cells were derived from a brain metastasis of a 69-year old caucasian prostate cancer patient.

2.1.1 Growth conditions

The RPMI 1640 culture medium supplemented with 10% FCS, 100 U/ml penicillin, and 100 µg/ml streptomycin is routinely used for the culture of human prostate cancer cell lines LNCaP, PC3, and DU-145.

Table 1. Characterization of prostate cancer cell lines (6, 7)

	LNCaP	PC3	DU-145
Androgen receptor	+	±	±
Androgen-dependent growth	+	−	±
Prostate-specific antigen	+	±	−
Cytokeratin expression			
Cytokeratin 8	+	+	+
Cytokeration 18	+	+	+
Secretion of growth factors			
EGF	++	+	++
TGF-α	++	+	+++
TGF-β	−	++	++
bFGF	−	+	++
IGF-1	++	+++	++
Expression of growth factor receptors			
EGFR	++	+	+++
TGF-β-R	−	++	++
FGF-R	+	+++	+++
IGFR-Typ 1	+	++	+++
Effects of endogenous growth factors on cellular proliferation			
EGF	+	−	±
TGF-β	−	+	+
bFGF	+	−	+
aFGF	−	−	−
Tumorigenicity in nude mice	**+**	**+**	**±**

+: positive; −: negative; ±both positive and negative data have been reported.

Cells are maintained in bicarbonate-buffered RPMI medium under cell culture conditions ($37\,°C$, humidified atmosphere of 95% air and 5% CO_2). The medium is changed twice a week. Cells are subcultured serially by trypsinization when approximately 80% confluent. Viability is measured by trypan blue exclusion. Cells are used between passages 10 and 30.

2.2 Primary culture from human prostate

Epithelial and stromal cells are isolated from human prostate specimen (benign prostate hyperplasia, prostate carcinoma) as described by Kashani *et al.* (8).

Protocol 1.   Isolation of primary cells from prostate tissue

Reagents and equipment

- RPMI 1640 medium (supplemented with 10% FCS, 100 U/ml penicillin, 100 μg/ml streptomycin)
- collagenase (Sigma)
- DNase (Sigma)

- anti-cytokeratin antibodies (Becton–Dickinson, Mountain View, CA)
- anti-vimentin and anti-desmin antibodies (Immunotech, Marseille, FR)

Method

1. Mince tissue samples with scissors into 1–2 mm^3 samples, put into Petri dishes, and wash extensively with RPMI 1640 medium supplemented as described above.

2. Perform digestion overnight at $37\,°C$ using RPMI supplemented with collagenase and DNase under continuous stirring. Collect small, undigested clumps by sedimentation, wash, and culture for 48 h.

3. Enrich adherent cell clumps by sequential trypsinization using trypsin–EDTA at room temperature. Only cells exhibiting strong adhesion properties are referred to as freshly enriched epithelial cells.

4. Prove the epithelial origin by staining the cells with an anti-cyto-keration antibody directed against cytokeratin types 8 and 18.

5. Generate stromal cells by extended culturing of cells with low adhesion properties over several passages to avoid contamination with epithelial cells.

6. After the third passage, harvest the cells using trypsin–EDTA treatment. Stromal origin is indicated by positive staining for vimentin and desmin.

---

3. Viability, metabolic activity, and proliferation

The viability of cells is usually tested by trypan blue exclusion. The metabolic activity can be determined by measuring the reduction of a tetrazolium salt

(MTT) to a coloured formazan product by mitochondrial enzymes present only in living, metabolically active cells, or analogous methods with commercially available kits. Thymidine incorporation into newly synthesized DNA, combined with measurement of the cell cycle distribution, renders a good assessment of the proliferation rate (9).

Protocol 2. Metabolic activity test

Equipments and reagents
- microplate reader (OD at 515 nm)
- 96-well plates
- DMSO (dimethyl sulfoxide)

- MTT [3-(4,5-dimethylthiazol-2-yl)2,5-diphenyl tetrazolium bromide] (Sigma, Munich, Germany)

Method

1. Harvest cells by trypsinization, resuspend in fresh medium, and plate at a density of 10 000 cells/0.2 ml medium in 96-well microtitre plates.

2. After culturing the cells for an appropriate time (4–8 days) add 30 μl of MTT (5 mg/ml phosphate-buffered saline).

3. After 4 h of incubation with MTT at 37 °C, carefully remove the supernatant.

4. Add 100 μl DMSO to each well.

5. Read the absorbance at 515 nm on a microplate reader.

Protocol 3. Thymidine incorporation

Equipment and reagents
- 24-well plates
- [³H]thymidine
- PBS
- 5% trichloroacetic acid (TCA)

- 0.1 M NaOH
- liquid scintillation cocktail (EMULSIFIER Safe™, Meriden, CT)
- liquid scintillation counter

Method

1. Grow cells in 24-well plates.

2. Add [³H]thymidine (2 μCi/ml) to cells in each well and incubate for 5 h at tissue culture conditions.

3. Wash twice with warm (37 °C) PBS.

4. Precipitate cells with ice-cold TCA for 10 min (repeat twice).

5. Wash twice with ice-cold water for 5 min.

6. Solubilize cells in 0.5–1 ml of 0.1 M NaOH.

Protocol 3. *Continued*

7. Neutralize 0.5 ml cell lysate with 50 μl glacial acetic acid and add liquid scintillation cocktail.

8. Measure the radioactivity of the samples in a liquid scintillation counter.

Protein assay. [Equipment and reagents: photometer (ELISA reader), protein kit, bovine serum albumin]. Depending on the samples, use either:

1. a modified assay according to the method of Lowry *et al.* (10), compatible with detergents but incompatible with sulfhydryl reagents (Biorad, Hercules, CA),

   or

2. a method described by Bradford (11), compatible with sulfhydryl reagents, but incompatible with detergents (Biorad, Hercules, CA)

Both assays can be automated for rapid determinations in 96-well microtitre plates. Bovine serum albumin is used as a standard.

4. Measurement of prostate-specific antigen (PSA)

Human prostatic epithelial cells constitutively secrete prostate-specific antigen (PSA), a kallikrein-like serine protease, which is a normal component of the seminal plasma. PSA is currently used as a specific diagnostic marker for the early detection of prostate cancer. PSA degrades the extracellular matrix glycoproteins fibronectin and laminin and, thus, may facilitate invasion by prostate cancer cells (12).

Protocol 4. PSA measurement

Equipment and reagents

- 24 well plates
- PSA ELISA (Immunocorp, Montreal, Canada)
- ELISA reader (450 nm)

Method

1. LNCaP cells (2×10^4 cells/well) are grown in 24-well microtitre plates in RPMI medium supplemented as described in Section 2.1.1.

2. Collect supernants after 72 h of incubation with hormones.

3. Carry out the PSA ELISA assay according to the manufacturers' descriptions. Measure absorbance at 450 nm.

---

4. PSA values are quantified as nanograms of PSA per ml. PSA production in the individual cell line can either be expressed as ng/mg protein/h or ng/cell/h.

Protocol 5. Hormonal stimulation of LNCaP cells

Equipment and reagents
- RPMI 1640 medium (Gibco, BRL)
- FCS
- hormones (e.g. 5α-dehydrotestosterone, 1 μM stock solution in ethanol)
- charcoal (activated, hydrochloride acid washed, cell culture tested) (Sigma, Munich, Germany)

Method

1. Maintain LNCaP cells for 3 days in 10% 'charcoal stripped' FCS containing medium. Under these conditions the proliferation is markedly reduced.

2. Plate 2 × 10⁴ pre-treated cells in multiwell plates in medium containing the hormone and culture for an appropriate time (see above).

---

5. Signal transduction

Phosphorylation and dephosphorylation of proteins mediate the signal transduction events that control a multitude of cellular processes. Therefore, the quantitation of kinase activities is essential for the study of cellular proliferation and differentiation.

5.1 Protein kinase C (PKC)

Protein kinase C (PKC), a family of homologous serine/threonine kinases with at least 11 isoenzymes, is a mediator of signal transduction that is activated following ligand stimulation of transmembrane receptors by hormones, neurotransmitters, and growth factors (13–16).

The three subfamilies comprise: the conventional PKC proteins (cPKC), activated by Ca^{2+}, phospholipid, diacylglycerol (DAG), and phorbol esters; the novel (nPKC); and the atypical subfamilies (aPKC), both of which are independent of Ca^{2+}. Furthermore, aPKC are independent of DAG and phorbol esters.

The cPKC isoforms α, β, γ are activated *in vivo* by Ca^{2+} and DAG. DAG is either generated as a result of hydrolysis of phosphatidylinositol diphosphate by phospholipase C, which also generates inositol triphosphate (IP3), which in turn releases intracellular Ca^{2+}, or via hydrolysis of phosphatidylcholine (PC). DAG activates PKC by decreasing the K_m for Ca^{2+}. PKC can also be activated by phorbol ester (PMA), for which it acts as the intracellular receptor. cPKCs appear to provide a control cell point and are involved in a wide range of cellular responses in cell proliferation and differentiation.

The regulatory region of the PKC contains a 'pseudo-substrate' region, which is highly conserved for the α, β, γ PKC isoforms, and is apparently involved in maintaining the enzyme in the inactive state. Peptides containing the pseudo-substrate sequence are inhibitors of the PKC, whereas antibodies against the pseudo-substrate region can activate the PKC.

Protocol 6. Total PKC enzymatic activity

Equipment and reagents

- water bath (30°C)
- liquid scintillation counter (Beckman)
- centrifuge
- DEAE column (0.5 g DE52 cellulose (Sigma) suspended in buffer B)
- PKC assay system (Gibco BRL)
- 1% phosphoric acid
- scintillation cocktail
- 20–25 μCi/ml [γ-^{32}P]ATP

- buffer A: 0.5 mM EDTA, 0.5 mM EGTA, 0.5% Triton X-100, 25 μg/l aprotinin, 25 μg/l leupeptin, 20 mM Tris–HCL, pH 7.5
- buffer B: 0.5 mM EDTA, 0.5 mM EGTA, 20 mM Tris–HCl, pH 7.5
- buffer C: 0.5 mM EDTA, 0.5 mM EGTA, 10 mM NaCl, 10 mM β-mercaptoethanol, 20 mM Tris–HCl, pH 7.5
- substrate solution: 20 mM MgCl$_2$, 50 μM acetyl-myelin basic protein 4–14 (Ac-MBP)

Method

1. Harvest subconfluent cells by scraping and centrifuge at 500 *g* for 10 min. The pellet is washed twice with PBS.

2. Homogenize cells in buffer A.

3. Partially purify PKC by ion-exchange chromatography on a DEAE column. Add the homogenate to the DEAE column , wash twice with buffer B (5 ml total volume). Elute the PKC-containing fraction with 2 ml of buffer B.

4. The enzymatic reaction is carried out in 0.5 ml Eppendorf vials (0.1–0.5 μg protein) according to manufacturers' directions. In brief: first add the phosphatidylserine and PMA dissolved in buffer C, then incubate at 30°C and add the substrate. After mixing and incubation at 30°C for 5 min spot an aliquot of the mixture (10–50 μl) on to the phosphocellulose filter supplied with the kit.

5. Wash the filter twice with 1% phosphoric acid and twice with distilled water.

6. Transfer the phosphocellulose filters into scintillation vials, add 10 ml scintillation cocktail, and count radioactivity in a liquid scintillation counter.

To ensure that only the activity of the protein kinase C is measured, control experiments with an inhibitory peptide specific for the activating region of protein kinase C (supplied with the kit) should be done in parallel with each assay. Before starting the enzymatic reaction, pre-incubate the partly purified PKC with the peptide (20 min at 25 °C).

Alternatively, the activity of PKC [and protein kinase A (PKA)] can be measured with a *non-radioactive method*. In an enzyme-linked immuno-sorbent assay (ELISA) supplied by, for example, Calbiochem (La Jolla, CA) or Upstate Biotechnology (Lake Placid, NY), PKC and PKA activity can be measured simultaneously. A synthetic pseudo-substrate for PKC/PKA coated on microwell plates is phosphorylated through PKC/PKA on serine. A biotinylated monoclonal antibody that recognizes the phosphorylated form of the peptide is used for the detection together with streptavidin conjugated to peroxidase. The peroxidase substrate is added to the reaction mixture on the microtitre plate and the colour intensity of the conjugate is measured.

This assay is useful for screening of inhibitors and activators of PKA/PKC or detection of pharmacological effects of compounds on PKA/PKC in prostate cancer cell lines. It is recommended that partially purified enzymes (see above) are used.

The equipment and reagents needed are as follows:

- 96-well plate
- microplate (ELISA reader, opitical density setting 490 nm)
- multichannel pipette
- water bath (25 °C)
- adenosine 3′,5′-cyclic monophosphate sodium (for PKA assays)
- adenosine 5′-triphosphate disodium salt

All other reagents are supplied with the kit.

5.1.1 Identification of PKC isoforms

Identification of PKC isoforms is done by standard methods: at the mRNA level by RT-PCR (see below) and at the protein level by Western blotting using isoenzyme-specific antibodies. Calbiochem, San Diego, CA; Santa Cruz, CA; Transduction Laboratories, Lexington, KT; and Upstate Biotechnology, Lake Placid, NY, offer PKC isoenzyme-specific antibodies with minimal cross-reactivity, inhibitory peptides, and control samples expressing particular isoenzymes (cell lysates from Jurkat cells and rat brain homogenate).

The molecular weight of individual isoforms ranges between 72 and 115 kDa.

To discriminate between membrane-associated and cytosolic PKC isoforms by Western blot, analysis can be done on isolated cytoplasmic and particulate fractions of prostate cancer cell lines.

Protocol 7. Preparation of a protein lysate for Western blot analysis of PKC isoforms

Equipment and reagents

- lysis buffer: 1mM EDTA, 2 mM EGTA, 1% Triton X-100 (Sigma), 0.3% β-mercaptoethanol, 1mM dithiothreitol (Sigma), 1 mM phenymethylsulfonyl fluoride, 1 μM aprotinin, 1 μM leupeptin, 50 mM Tris–HCl, pH 7.5

- centrifuge (Hettich)
- cell scraper (Costar)
- syringe with 26 G neddle (Becton–Dickinson)

Method[a]

1. Wash cells twice with ice-cold PBS, and harvest by scraping.

2. Centrifuge at 500 g for 5 min.

3. Resuspend the pellet in lysis buffer and homogenize with a syringe (26 G needle).

4. Wait for 30–45 min.

5. To remove particles insoluble in the lysis buffer centrifuge for 5 min at 2000 *g*. Collect the supernatant.

6. Determine the protein concentration in the supernatant.

7. Seperate 20–30 μg protein per lane on 8% PAGE for Western blot analysis.

[a] All steps are done at 4°C.

Protocol 8. Isolation of particulate and cytoplasmic fractions from prostate cancer cell lines

Equipment and reagents

- universal centrifuge (Hettich 30 RF)
- buffer A: 10 mM *N*-2-ethylpiperizine-*N'*-2-ethanesulfonic acid (Hepes), 0.25 mM saccharose, 1 μM leupeptin, 1 μM calpain, and 1 μM phenylmethylsulfonyl fluoride (PMSF), pH 7.5

- buffer B: 10 mM *N*-2-ethylpiperizine-*N'*-2-ethanesulfonic acid, 0.25 mM saccharose, pH 7.0
- table-top centrifuge (Hettich 1394)
- ultracentrifuge (Beckman)

Method[a]

1. Collect approx. 10×10^6 exponentially growing cells by centrifugation (1000 g for 5 min), aspirate the remaining medium, and wash cells with PBS

2. Homogenize in 4 × buffer A by repeated freezing and thawing.

3. After removing cell debris at low speed (1000 *g* for 5 min), separate the particulate from the cytosolic fraction at 100000 *g* for 60 min. Resuspend the pellet in buffer B. Store the resuspended pellet and cell supernatant at −80°C until use.

a All steps are done at 4°C.

5.2 Protein tyrosine kinases

Analogous to the procedure described above for PKC/PKA, protein tyrosine kinase activity can be measured in cell lysates using synthetic substrates. While protein tyrosine kinases are known to exhibit strict substrate specificity *in vivo*, a variety of synthetic substrates can be used *in vitro*. Random tyrosine-containing co-polymers with high tyrosine content, which are phosphorylated by various protein tyrosine kinases to a different extent, are commercially available (e.g. Oncogene Research Products, Cambridge, MA). Detection is made by a peroxidase-coupled anti-phosphotyrosine antibody in an ELISA assay.

Sensitive quantitative measurements can also be done by measuring the incorporation of ^{32}P-labelled ATP in the presence of Mg^{2+} into a natural or synthetic polypeptide phosphotyrosine substrate.

5.3 Measurement of $[Ca^{2+}]_i$ in single cells

An increase in intracellular calcium ion concentration controls a wide range of cellular functions, including cellular adhesion, motility, gene expression, and proliferation (17–19).

Protocol 9. Calcium imaging

Equipment and reagents

- apparatus for fluorescence video analysis. A perfusion chamber in which the coverslips are mounted is equipped with a heating system to keep the temperature at 37°C. The recording system includes an inverted microscope with an objective lens (Fluor, × 40) and a xenon lamp providing a dual excitation light of 340 and 380 nm. The excitation light is selected by a spinning chopper mirror and directed to the cell by a dichroic mirror. The emitted light (510 nm) is monitored at a resolution of 3/sec and the signal is stored in a computer

- Fura-2AM (Molecular Probes, Eugene, OR)
- EGTA
- ionomycin (Sigma, Munich)
- medium L: 136.89 mM NaCl, 5.36 mM KCl, 1.26 mM CaCl₂, 0.98 mM MgCl₂.6 H₂0, 5 mM D(+)-galactose (Merck, Darmstadt, G), 5 mM Na-pyruvate, 0.81 mM MgSO₄.7 H₂0, KH₂PO₄, Na₂HPO₄.2H₂0, 10 ml vitamins (Gibco BRL, Paisley, SC), 20 ml essential amino acids (Gibco BRL)

Method

1. Grow cells on sterile coverslips in 4 ml medium L for 2 days.

2. Before starting the experiment, load cells with Fura-2/AM for 1 h at

---

Protocol 9. *Continued*

37°C. Using 5 μM Fura-2/AM (1h) the fluorescence of the loaded cells is five times that of the autofluorescence of unloaded cells.

3. Wash once with medium L.

4. Transfer the coverslip with the Fura-2-loaded cells to the perfusion chamber (0.3 ml volume) with medium L at 37°C. Continuously perfuse the cells at 10–12 ml/min. Excitation at 380 and 340 nm is automatically exchanged at a rate of 100 Hz, and the emitted light is detected through a 510 nm filter with a photomultiplier.

5. Start the experiment. The ratio of fluorescence intensity due to excitation at 340 nm (F340) and that at 380 nm (F380) is calculated as a function of time.

6. After 5 sec add the test substance [for control experiments: addition of histamine (100 μM) results in a transient increase in the $[Ca^{2+}]_i$ in PC3 cells within 5 sec].

7. Calibration. For maximal $[Ca^{2+}]_i$ add ionomycin (5 μM) to the medium. For minimal $[Ca^{2+}]_i$ deplete Ca^{2+} by adding 5 mM EGTA to the medium.

8. The ratio is converted to $[Ca^{2+}]_i$ as described by Grynkiewicz *et al.* (20) using the equation:

$$[Ca^{2+}]_i = (R-R_{min})/(R_{max}-R)K_dS$$

where R = Fura-2 fluorescence ratio (F340/F380), R_{max} and R_{min} are determined by addition of ionomycin followed by addition of EGTA, K_d is the dissociation coefficient of the Fura-2/Ca^{2+} complex (assumed to be 224 nM), S (constant) is the ratio of the fluorescence at 380 nm in Ca^{2+}-free solution to the fluorescence at 380 nm in Ca^{2+}-containing solution.

5.4 Growth factors in the prostate

Members of the EGF receptor family and its ligands are overexpressed, or expressed as an autocrine, loop in many tumours. Besides EGF, transforming growth factor-α (TGF-α), betacellulin, heregulin, heparin-binding EGF, and amphiregulin belong to the EGF-family and bind to the EGFR. EGF growth factors respond to hormonal stimuli, e.g. TGF-α has oestrogen-responsive elements and the heparin-binding domain containing amphiregulin expression is inducible by oestrogen (21).

Another heparin-binding growth factor found in prostate cancer cell lines (22) is the basic fibroblast growth factor (bFGF), which stimulates growth and migration of a number of cell types, including fibroblasts and endothelial cells (23)

Expression of amphiregulin and bFGF studied at the mRNA level by semi-quantitative RT-PCR, or by protein evaluation in Western blot analysis is described in *Protocols 10* and *11*, respectively.

Protocol 10. Semi-quantitative RT-PCR

The general protocol on semi-quantitative PCR is used for the evaluation of growth factor expression in human prostate cancer cell lines, primary cultures of human epithelial and stromal prostate cells, and liquid nitrogen-frozen human tissues. Endogeneous sequences known to be present at a constant level can be used as internal standards (e.g. β2-microglobulin, β-actin).

A. *RNA isolation*
Equipments and reagents

- Hettich microcentrifuge
- RNAzol B (Molecular Research Center, Cincinnati, OH)
- chloroform
- DEPC (diethylpyrocarbonate; Sigma, Munich)
- DEPC–H_2O: 1 ml DEPC is diluted up to 1 litre with H_2O; after 24 h at room temperature, the solution is autoclaved at 180°C.
- isopropanol
- ethanol 75%

Method

1. Lyse cells and tissues with RNAzol

2. Extract RNA by mixing lysed cells with chloroform (0.2 ml/ml RNAzol) for 5 min on ice and centrifuge at 10000 g for 15 min. The organic phase contains DNA and proteins, the aqueous phase contains RNA.

3. Transfer the aqueous phase to a fresh tube, precipitate RNA with isopropanol (0.6 ml/ml RNAzol) for at least 45 min on ice.

4. Collect the pellet and wash it with 75% ethanol/ RNAzol at 10000 g

5. Resupend the pellet in 10–25 μl DEPC–H_2O.

6. Pellet can be stored (at –70°C) in 70% ethanol.

B. *First strand cDNA synthesis*

Reagents

- 'RNase out' ribonuclease inhibitor (Gibco BRL, Paisley, SC)
- multicore (10) buffer (Promega)
- RQ1 RNase-free DNase (Promega)
- oligo (dT) (Gibco-BRL)
- ribonuclease inhibitor (Gibco-BRL)
- 5 × first strand buffer: 250 mM Tris–HCl, pH 8.33, 375 mM KCl, 15 mM $MgCl_2$
- 0.1 M DTT (dithiotreitol)
- 10 mM d NTP mix
- SUPERSCRIPT II™ reverse transcriptase (Gibco-BRL)
- 10 × PCR buffer: 200mM Tris–HCl, pH 8.4, 500 mM KCl
- 50 mM $MgCl_2$
- amplification primers
- Taq DNA polymerase
- water-saturated phenol (100 g sodium acetate is dissolved in 40 ml DEPC–H_2O at 60–65°C. The upper phase must be aspirated)

275

Protocol 10. *Continued*

- TAE buffer, pH 7.2 (50): 222 g Tris, 129.9 g sodium acetate, and 19 g EDTA are dissolved in 1 litre Aqua bidest
- ethidium bromide (1 mg/ml)
- 2% agarose gels (0.6 g agarose are dissolved in 30 ml TAE buffer and heated for 1 min)

Method

1. Dilute 10 μl RNA (1:5 with DEPC-treated H_2O)

2. Add 1 μl RNase Out ribonuclease inhibitor (40 U/μl)

3. To remove contaminating genomic DNA before the cDNA synthesis, add 2.5 μl multicore (10) buffer and 1 μl RQ1 RNase-free DNase (1 U/μl) and incubate for 15 min at 37 °C.

4. Extract once with an equal volume of phenol/chloroform (1:1) and centrifuge at 10 000 *g* for 10 min at 4 °C.

5. Collect the aqueous phase containing the RNA (about 100 μl) and transfer it to a fresh tube.

6. Precipitate the RNA by adding 2 vols of ice-cold 100% ethanol and 2 M sodium acetate (pH 5.2) to a final concentration of 0.2 M (12 h at 4 °C).

7. To 1–5 μg RNA add 1 μl Oligo (dT 12–18) (0.5mg/ml) and sterile water. Incubate at 70 °C for 10 min.

8. Add 1 μl ribonuclease inhibitor (40 U/μl), 4 μl 5 × first strand buffer, 2 μl DTT, 1 μl (10 μM) NTP mix, and incubate for 2 min at 42 °C in a water bath.

9. Add 1 μl SUPERSCRIPT II™ RT (200 U/μl) and incubate for 50 min at 42 °C.

10. Inactivate the reaction by heating at 70 °C for 15 min.

11. The sample containing the cDNA is diluted 1:10 in ddH_2O and stored at −20 °C until use.

C. *Semi-quantitative PCR*

1. To 25 μl cDNA add 25 μl 'master mix' [2 × concentrated, containing 10 μl PCR buffer, 0.5 μl forward primer (25 μM), 0.5 μl reverse primer (0.25 μM), 1 μl NTP mix, 4 μl Taq polymerase (0.25 U/μl), and 14 μl sterile water].

2. Denature the mixture for 5 min at 94 °C.

3. Perform 20–40 cycles of PCR. Annealing and extension conditions are primer and template dependent and must be determined empirically (see below).

4. Perform a final extension step at 72 °C for 10 min, then cool at 4 °C.

The amount of input RNA and the number of amplification cycles is optimized to ensure that PCR products are in the exponential phase of amplification.

PCR conditions and primers for β2-microglobulin (internal mRNA standard), bFGF, and amphiregulin are as follows.

β2-Microglobulin (28 cycles)

1. 'Hot start' 4 min, 94 °C.

2. 30 sec, 94 °C (denaturation of the DNA double strands).

3. 30 sec, 61 °C (annealing of the primers)

4. 1 min, 72 °C (Taq polymerase is synthesizing a new strand)

5. 7 min, 72 °C (final extension, then cool at 4 °C).

    5′-ACCCCCACTGAAAAAGATGA

    3′-CAACCATGCCTTACTTTATC

    (Product size 549 bp)

bFGF (30 cycles, conditions as described above)

    5′-GTCCGGGAGAAGAGCGACCCT

    3′-AGGTCCTGTTTTGGATCCAAG

    (Product size 270 bp)

Amphiregulin (30 cycles, steps 1 and 2 as above)

3. 30 sec at 54 °C

4. 50 sec at 72 °C

    5′-CTCGGGAGCCGACTATGA

    3′-GGACTTTTCCCCACACCG

    (Product size 327 bp)

Aliquots of the PCR products (200–1000 bp) are resolved by electrophoresis on 2% agarose gels. For documentation and quantification PCR products fluorescent bands on the gel are photographed and analysed by densitometry.

5.5 Western blot analysis of amphiregulin and bFGF

Western blot analysis of growth factors is performed in a total cell lysate or in cell culture supernatants. For preparation of a total cell lysate (see *Protocol 7*, page 272). Cell culture supernatants are first treated with 10% trichloroacetic acid to precipitate the protein used in Western blot analysis.

Protocol 11. Western blot analysis for the detection of bFGF and amphiregulin in prostate cancer cells

Equipment and reagents
- electrophoresis apparatus (vertical slab gel unit, power supply)
- electroblotting chamber
- nitrocellulose or polyvinylfluoride (PVDF) membrane
- electroblotting buffer: 25 mM Tris, pH 8.3, 192 mM glycine, 15% methanol; bring the volume to 3 litre
- blocking buffer (depending on the antibody use either 3–5% BSA or 1–5% non-fat dry milk in PBS)
- SDS–polyacrylamide gel
- PBST (PBS with 0.5–1% Triton X-100)
- primary antibodies against bFGF (Calbiochem/Oncogene, Cambridge, MA), amphiregulin (Neomarkers, Freemont, CA)
- peroxidase-labelled secondary antibody
- ECL chemiluminescence substrate (Amersham)
- chemiluminescene or X-ray film (Amersham)

Method

1. Separate proteins on an SDS–polyacrylamide gel: typically a 0.7 mm 7.5–15% gel depending on the size of the protein of interest.

2. After running the gel, remove and immerse it in the electroblotting buffer.

3. Pre-wet membranes in electroblotting buffer and put on the gel

4. Assemble the apparatus and place it into the electroblotting apparatus with the membrane facing the anode.

5. Electroblot for 2–4 h using a constant voltage of 25–35 V.

6. Remove the blot from the apparatus and wash it twice with PBST.

7. Place the blot into the blocking buffer and gently shake it for 2–12 h.

8. Incubate the blot in the primary antibody (diluted in PBST) overnight at 4°C.

9. Wash three times with PBST.

10. Dilute secondary antibody with PBST and incubate for 1 h.

11. Wash three times with PBST.

12. Gently shake the blot in ECL substrate for 1 min and detect immunoreactive bands on the X-ray film.

13. Immunoreactive bands can be quantified by densitometry.

6. Summary

Thus, it is evident that studies of mammalian cell differentiation employ a wide range of techniques of proven value in investigations of cell growth, but the unique focus is on the orderly cessation of activities associated with cell proliferation, thus allowing expression of genes required for specialized functions

References

1. Wilding, G. (1995) *Cancer Surv.* **23**, 43.
2. Kumar, M.V. and Tindall, D.J. (1997) In *Principles of genitourinary oncology.* Lippincott-Raven Publishers, Philadelphia.
3. Miller, A.B., Berrino, F., Hill, M., Pietinen, P., Riboli, E., and Wahrendorf, J. (1994) *Eur. J. Cancer* **30A**, 207.
4. Steiner, M.S. (1993) *Urology* **42**, 99.
5. Peehl, D.M. (1996) *Prostate* **6**(S), 74.
6. Lalani, el-N., Laniado, M.E., and Abel, P.D. (1997) *Cancer Metastasis Rev.* **16**, 29.
7. Webber, M.M., Bello, D., and Quader, S. (1997) *Prostate* **30**, 58.
8. Kashani, M., Steiner, G., Haitel, A., Schaufler, K., Thalhammer, T., Amann, G., Kramer, G., Marberger, M., and Schöller, A. (1998) *Prostate* **37**, 98.
9. Romijn, J.C., Verkoelen, C.F., and Schroeder, F.H. (1988) *Prostate* **12**, 99.
10. Lowry *et al.* (1951) *J. Biol. Chem.* **193**, 265.
11. Bradford, (1976) *Anal. Chem.* **72**, 248.
12. Gao, X., Porter, A.T., Grignon, D.J., Pontes, J.E., and Honn, K.V. (1997) *Prostate* **31**, 264.
13. Nishizuka, Y. (1988) *Nature* **334**, 661.
14. Yasuda, I., Kishimoto, A., Tanaka, S., Tominaga, M., Sakurai, A., and Nishizuka, Y. (1990) *Biochem. Biophys. Res. Commun.* **166**, 1220.
15. Blobe, G.C., Obeid, L.M., and Hannun, Y.A. (1994) *Cancer Metastasis Rev.* **13**, 411.
16. Dekker, L.V, Palmer, R.H., and Parker, P.J. (1995) *Curr. Opin. Struct. Biol.* **5**, 396.
17. Clapham, D.E. (1995) *Cell* **80**, 259.
18. Dolmetsch, R.E., Lewis, R.S., Goodnow, C.C., and Healy, J.I. (1997) *Nature* **386**, 885.
19. Wasilenko, W.J., Cooper, J., Palad, A.J., Somers, K.D., Blackmore, P.F., Rhim, J.S., Wright, G.L., and Schellhammer, P.F. (1997) *Prostate* **30**, 167.
20. Grynkiewicz, G., Poenie, M., and Tsien, R.Y. (1985) *J. Biol. Chem.* **260**, 3440.
21. Martinez-Lacaci, I., Saceda, M., Plowman, G.D., Johnson, G.R., Normanno, N., Salomon, D.S., and Dickson, R.B. (1995) *Endocrinology* **136**, 3983.
22. Han, I.S., Sylvester, S.R., Kim, K.H., Schelling, M.E., Venkateswaran, S., Blanckaert, V.D., McGuinness, M.P., and Grisworld, M.D. (1993) *Molec. Endocrinol.* **7**, 889.
23. Strawn, L.M. and Shawver, L.K. (1998) *Exp. Opin. Invest. Drugs* **7**, 553.

Senescence and immortalization of human cells

KAREN HUBBARD and HARVEY L. OZER

1. Introduction

Normal vertebrate diploid cells in culture have a finite life-span (1). The maximum number of population doublings is dependent on cell type and species. It has also been shown to vary as a function of age of the donor organism, resulting in its utilization as a model for the proliferative aspects of cellular ageing. Although cells from all species eventually cease to proliferate and become 'senescent', the capability to achieve infinite or continuous cell growth (i.e. became 'immortal') by mechanisms that overcome cellular senescence is strikingly dependent on the animal species. Rodent cells readily become 'spontaneously' immortal such that there are multiple established fibroblastoid cell lines; for example, mouse 3T3 cells. In marked contrast, normal chicken and human fibroblasts (HF) rarely, if ever, become immortal. The observation that many human tumours are immortal cell lines when introduced into tissue culture suggests that overcoming senescence is a stage in carcinogenesis. One experimental manipulation that has been successful in facilitating the immortalization of HF has been through the introduction of genes from DNA tumour viruses into otherwise normal cells. In particular, the gene encoding the large T antigen (or *A* gene) from SV40 has been especially useful in this regard. Microinjection of large T antigen will reactivate DNA synthesis in senescent cells, although mitosis is not induced. Stable introduction of the *A* gene (as by viral infection or DNA-mediated gene transfer) prior to senescence will extend the limited life-span for multiple generations; however, such cells typically undergo cell death, termed 'crisis', as distinct from senescence. These SV40-transformed cells can also be considered pre-immortal cells since they have an increased likelihood of becoming immortal, albeit with a varied and, most often, low frequency. Indeed, many continuous cell lines have been developed *in vitro*. It should be emphasized that SV40-immortalized cells are still dependent on large T antigen function. We (2) and others (3) have developed a two-stage model. In the first stage, large T antigen functions as a mitogenic agent and transiently overcomes the basis for cellular senescence. A cellular mutation is required for the second stage. A gene has been localized to the long arm of chromosome 6, whose inactivation is consistent with being responsible for this step (4). The remainder of this chapter describes each of these stages of cellular proliferation; senescence, crisis, and immortalization. In addition, we will summarize how SV40 and, especially its *tsA* mutants encoding a heat-labile large T antigen, have been used to develop cell lines for multiple lines of experimentation.

2. Cellular replicative senescence

Primary cells introduced into culture often proliferate when given appropriate medium and other culture conditions. Such media have been developed for multiple cell types and include examples of fully defined media, as well as those supplemented with animal sera, most often bovine sera. This period may include an initial interval of adaptation but is often followed by a stage of repeated and sometimes extensive cell proliferation. In the case of human fibroblasts derived from fetal tissue, it may exceed 50 population doublings (PD) (and an approximately equal number of cell generations). This is followed by a plateau period during which the population appears to slow down due to an increase in generation time and a decreased likelihood of cell division, resulting finally in a cessation of cell proliferation, or senescence.

Loss of cellular proliferation in senescence is a stochastic process at the individual cell level. The mechanism is the subject of some controversy. Earlier models suggesting an accumulation of mutations in nuclear DNA or other errors have become less favoured because of the absence of strong supporting data, although there has been evidence of alteration in mitochondrial DNA as a contributing factor. On the other hand, senescence does share features consistent with replicative inhibition associated with terminal differentiation. Furthermore, an important function of cellular senescence is to prevent carcinogenesis. Senescent cells that have undergone an irreversible growth arrest would be expected to be precluded from exhibiting deregulated cell growth, even if such cells harbour mutation(s) in oncogenes and/or growth suppressor genes. This tumour suppressive mechanism bears a price. Senescent cells may be resistant to some (although not all) apoptotic signals. Hence the retention of senescent cells in tissues as they age may have serious consequences for tissue integrity in older individuals.

In the case of human fibroblasts, replicative senescence would appear to be at least an important experimental model for cellular ageing, regardless of the underlying mechanism. Cellular proliferative capacity exhibits a range of values as determined by the size and number of single cell-derived colonies at low cell density, consistent with behaviour of the population at higher cell density. Similarly, assessment of DNA synthesis by [³H]thymidine autoradiography indicates a heterogeneous population of cells with regard to generation time and number of replicatively active cells. This heterogeneity decreases with the proliferative history of culture. Both the percentage of cells undergoing DNA synthesis (*Protocol 1*, part B) and colony formation can therefore be used to assess a particular point in time in the life-span of a population of cells in comparison to a reference cell line and the same cell line early in its life-span. It has also been well documented that somatic cells undergo a shortening of the repetitive sequences at the telomeric ends of chromosomes. This is due to the special requirements for replication of DNA termini and several factors, including a specific replicase called telomerase, are involved. There is no detectable telomerase activity in HF (and many other cell types), such that HF loses approximately 40 bp per cell generation. Senescent cells have considerably shortened telomeric sequences. Hence the length of telomeric sequences by Southern blot analysis after digestion with appropriate restriction enzymes may be useful in determining the replicative history and, therefore, the future capacity of the cell population.

It should be noted that models based on loss of telomeric sequences have been proposed to explain senescence as well (5). The role of telomeres and telomerase during ageing will be discussed in greater detail later. In any case, when fetal human fibroblasts are studied remarkable similarity is observed, with a total life-span of 60–70 PD not only for a given cell line within a laboratory but also between cell lines (WI-38, IMR90, HS74BM, TIG) in different laboratories, when subculture ratios of 1:3 to 1:5 in conventional growth medium containing 10–15% fetal bovine serum are employed. None the less, it is necessary and appropriate for a particular laboratory to determine life-span by serial passage. Senescence is defined by the absence of one PD in 2–4 weeks at terminal passage. Cells assume a typical morphology characterized by marked cellular enlargement and a flattened appearance. Less than 5% of the cells undergo DNA synthesis in a 48 h period, shown by the incorporation of [³H]thymidine into nuclear DNA as determined by autoradiography. Cells have a G1 content of DNA, although it should be recognized that tetraploidy and aneuploidy commonly occur late in the fibroblast life-span. In addition to these methods for life-span determination, the senescence-associated marker, acid form of β-galactosidase (SA-βgal) has recently been found to be particularly useful as a rapid indicator for replicative senescence (6). As populations of cells become older, individual cells display an increase in staining for SA-βgal (*Protocol 1*, part D). The determination of life-span and the methodology for assessing whether a culture has become senescent are described in *Protocol 1*.

Recently, it has been shown that senescent cells are defective in expression or function of several important growth-regulatory molecules associated with progression of cells through G1 into S. It was first reported that the induction of c-fos and the activity of its transcription factor form AP-1 (composed of c-fos and jun) were decreased in senescent fibroblasts. However, microinjection of purified c-fos or its cDNA in a suitable expression vector did not in itself induce DNA synthesis. It was also found that the growth suppressor pRb-1 is aberrantly regulated. In G0 or early G1 the pRb-1 protein is hypophosphorylated and becomes phosphorylated as cells progress through G1 into S. In this manner the association between pRb-1 and the transcription factor E2F-1 becomes altered. E2F-1 stimulates transcription of several genes involved in DNA synthesis and although multiple other proteins may be involved, it is clear that the phosphorylation of pRb-1 and release of E2F-1 is a key step in the regulation of E2F-1. Most interestingly it has been shown that pRb-1 is underphosphorylated in senescent cells, even after such cells are stimulated to express other factors by treatment with serum-derived growth factors (7). Cellular cyclin-dependent kinases (cdks) have been identified which are responsible for the phosphorylation of pRb-1. Furthermore, inhibitors of such cdk have also been identified and two such genes (p21 and p16) show increased expression as HF approach and enter senescence (8). p21, variously termed CIP, WAF, and sdi-1, has been shown to be over-expressed in senescent and quiescent cells. Furthermore, it inhibits cell proliferation when expressed at high levels in growing cells, such as when its gene expression is induced by the growth suppressor p53. Null mutations of p21 in human diploid fibroblasts result in a temporary bypass of senescence (9). Recently, it has been demonstrated by mouse knock-out studies that members of the INK4a family of protein kinase inhibitors (p16 and p19^{ARF}) have major roles in cell growth and senescence (10). p16 and p19^{ARF} protein accumulate in senescent fibroblasts and induce growth arrest when overexpressed. Both proteins act upstream to p53 and pRb-1 in their functions. Inactivation of pRb-1, p53, and p16 are commonly observed in human tumours. Taken together, these results and others support the model that cellular senescence is dependent on normal growth regulatory factors.

Protocol 1. Determination of life-span and preparation of senescent cell populations

Equipment and reagents

- growth medium
- FBS (fetal bovine serum)
- Coulter counter
- haemocytometer
- [³H]thymidine
- PBS
- 5% TCA
- Kodak emulsion
- Kodak D-19 developer
- Kodak acid fixer
- 2% formaldehyde

- lysis buffer: 0.2% SDS, 20 mM Tris–HCl, pH 7.5, and 2 mM EDTA
- 0.2% glutaraldehyde
- staining solution: 1 mg/ml X-gal in dimethylformamide, 0.2 M citric acid/Na phosphate buffer, pH 6.0, 5 mM potassium ferrocyanide, 5 mM potassium ferricyanide, 150 mM sodium chloride, and 2 mM magnesium chloride
- citric acid/Na phosphate buffer (0.2 M): 36.85 ml 0.1 M citric acid solution, 63.15 ml 0.2 M sodium phosphate (dibasic) solution

The solution used to dissolve the X-gal can be stored at −20 °C. X-gal is not stable in aqueous solution and should be freshly made the day of the assay. X-gal is light sensitive; thus solutions containing X-gal should be covered with foil and kept in the dark.

A. *Cell cultures*

Human fibroblast and other cell lines of limited life-span at determined PD can be obtained from the National Institute of Aging Collection in the Cell Culture Repository at the Coriell Institute, Camden, NJ, or the American Type Culture Collection (ATCC), Rockville, MD.

Method

1. For estimation of life-span, subculture human fibroblasts at 1:4 ratios in growth medium containing 10% fetal bovine serum (FBS).

2. Determine cell numbers either electronically using a Coulter counter or manually with a haemocytometer.

3. Calculate the number of population doublings achieved using the following formula: $N_C/N_S = 2X$ (where C = number cells harvested at confluency; S = initial number of cells seeded; X = population doublings). Cells are considered senescent when the labelling index of nuclei and the rate of DNA synthesis are less than 5% of that found for early passage cells, or greater than 90% of the cells are positive for SA-βgal.

Once the population doubling for senescent cells has been determined, this information can be used for larger scale culture preparations. Frozen cultures at intermediate passage levels represent useful starting cultures.

B. *Labelling index of nuclei*

1. At each subcultivation to be tested, seed cells at $1–5 \times 10^4$ cells/35 mm tissue culture dish.

2. After 24 h, add [^{3}H]thymidine at a final concentration of 0.1–1 µCi/ml.

3. Wash cells at time intervals of 1 to 3 days and process for auto-radiography (all operations are on ice).

4. Wash cells twice with cold phosphate-buffered saline (PBS) followed by a 15 min incubation with 5 ml cold methanol.

5. Remove methanol and incubate cells with 5 ml cold 5% trichloro-acetic acid (TCA) for 30 min (this allows for extraction of un-incorporated radioactivity).

6. Wash cells twice with distilled H_2O and allow to air-dry.

Protocol 1. *Continued*

7. For autoradiography, add 0.5 ml pre-warmed Kodak emulsion (50 °C) to each dish followed by gentle rocking to ensure entire coverage of the surface (all operations are done in the dark).

8. Pour excess emulsion off the dish. Cover dishes with foil and leave in the dark for 1 week.

9. Add 1:1 dilution of Kodak D-19 developer (pre-warmed to 59 °C) to each dish and incubate for 4 min at room temperature.

10. Wash dishes with distilled H_2O.

11. Add Kodak acid fixer and incubate dishes for 5 min. Pour off the fixer and add distilled water to each dish.

12. Allow dishes to air-dry and count the number of labelled nuclei using an inverted microscope. (Staining of cells facilitates detection of unlabelled cells.)

C. *Rate of DNA synthesis*

1. Seed 1×10^5 cells per 100 mm dish 24 h prior to [^{3}H]thymidine incubation, as described in Chapter 1, *Protocol 5*.

2. Wash cells at time intervals of 1 to 3 days with cold PBS.

3. Harvest cells by trypsinization and dilute with PBS. Take an aliquot for cell counting; centrifuge the remainder and resuspend in 1 ml cold PBS by vortexing.

4. Add 1 ml of 2 × lysis buffer, vortex the tube, and incubate at room temperature for 10 min.

5. Add 2 ml of 20% TCA and store the sample on ice for 1 h.

6. Filter precipitated DNA through pre-wet GF/A filters (Whatman) and wash with 5% TCA (three times) followed by cold 95% ethanol.

7. Count filters in a scintillation cocktail. (Sensitivity can be improved by solubilization of samples prior to the addition of scintillant.)

D. *Senescence associated (SA)-β-galactosidase assay [modification of Dimri et al., (6)]*

1. At each subcultivation to be tested, seed cells at 1×10^4/35 mm tissue culture dish.[a]

2. Wash cells at time intervals of 2–5 days with PBS (twice) and process for *in situ* β-galactosidase assay.

3. Fix cells for 3–5 min at room temperature with 2% formaldehyde + 0.2% glutaraldehyde in PBS.

4. Wash cells twice with PBS.

5. Add staining solution (1–2 ml per 35 mm dish).

6. Incubate at 37 °C in a humidified chamber. Do not use an incubator with a 5–10% CO_2 environment such as a CO_2 incubator used to propagate cell cultures. Blue colour is detectable in some cells within 2 h, but staining is maximal in 12–16 h.

[a] It is important that non-confluent cultures be used since non-senescent cells which are confluent or maintained at confluence have been observed to express SA-βgal in the laboratories of the authors.

Studies with SV40 large T antigen, which will be discussed in more detail in subsequent sections, further support this interpretation. Large T antigen has multiple effects on cellular gene expression in different experimental systems. It has direct, as well as indirect, effects on viral and cellular DNA synthesis. This protein can induce cellular DNA synthesis in both quiescent and senescent cells. It is capable of forming complexes with pRb-1 and p53. When large T antigen complexes with pRb-1, E2F-1 is released from pRb-1. Mutants of large T antigen, which cannot bind pRb-1, do not induce cell DNA synthesis (11). Furthermore, the subunit composition of cyclin–cdk complexes is rearranged in SV40-transformed cells (12). These changes may either be direct effects of large T antigen or mediated through inhibition of p53-activated transcription of the cdk inhibitor p21 (13).

3. SV40 transformation and crisis

Introduction of an SV40 genome into multiple cell types extends the proliferative life-span of such cells and increases the likelihood that such cells will become immortal. Although this has been demonstrated in rodent cells as well as human cells, this approach has been most useful in studies with human cells. It is relatively simple to obtain cultures or cloned cell lines expressing an SV40 genome since most human cells are susceptible to infection with SV40 virus particles or can be transfected by purified DNA. The SV40 genome is 5.2 kb. Its sequence is known and multiple mutations in all of its functions are available. Its genome can be divided into three regions: a control region (C), the region expressed early in infection (E), and the region expressed late in infection (L). The control region contains the sequence required for DNA replication (origin) flanked by the promoter and the enhancer for the E region as well as the promoter for the L region. The E region is required for transformation and DNA synthesis. It encodes two polypeptides: large T antigen (100 kDa) and small t antigen (21 kDa). The L region encodes the viral structural proteins and is not involved in SV40-mediated transformation. The course of SV40 infection is species dependent. It replicates efficiently in several monkey cell lines, especially CV-1, BSC-1, or VERO cells. Infection of rodent cells is considered non-permissive, as it does not produce progeny virus. Replication of its DNA is undetectable in mouse or rat established cell lines; it is markedly reduced in Chinese hamster cell lines. SV40 infection of human cells is semi-permissive (14). Low but readily detectable levels of progeny virus are produced (typically 10^6 infectious units per culture of 10^6 cells). However, viral DNA synthesis is quite efficient in many cells with as many as 10000 copies per cell reported. Infection of human fibroblasts by SV40 virus results in a mixed cell population. Some cells are productively infected, producing high levels of DNA and virus, and die within a few days due to the cytopathic effect of the virus. Most cultures establish a persistent infection with stable expression of a low level of virus production. The viral genome is integrated into many cells, resulting in their transformation. However, these cells are capable of excision and subsequent replication of the viral genome. Finally, some cells are not infected even at high multiplicity of infection but can be subsequently infected by progeny virus. Including antibody to neutralize SV40 virus particles in the culture medium can minimize virus spread (15).

Non-productive SV40 infection of permissive and human cells can be obtained by the introduction of defective viral genomes which cannot produce virus particles due to deletion or disruption of L region coding sequences. This commonly occurs as part of the cloning process in generating a recombinant DNA in a plasmid vector. Such constructs can still replicate DNA efficiently in human cells but are not expected to produce extracellular virus. (If the L region is only interrupted, recombinational mechanisms may result in an intact viral genome.) We have emphasized in this laboratory the use of cloned origin-defective mutants of SV40, especially those in which the *Bgl*1 site has been deleted due to removal of a few base pairs. These ori-minus constructs were prepared by Y. Gluzman. They express both SV40 T antigens at normal levels but do not replicate DNA or produce progeny virus. They integrate stably into human fibroblasts at 1–2 copies per cell. Integrants have been selected based on the transformed phenotype or using a co-selectable marker for resistance to the neomycin analogue G418. Constructs in which the SV40 origin has been deleted altogether provide a similar benefit. Such examples utilize the long terminal repeat (LTR) regulatory sequences from Rous sarcoma virus (RSV) or murine leukaemia virus (MLV), the gluco-corticoid-responsive LTR from mouse mammary tumour virus (MMTV), or the cellular metallothionein promoter. It should be noted that the advantage would be negated by the inclusion of a drug-resistant sequence containing an SV40 origin (such as SVneo, for example) in the DNA construct.

It was recognized early that SV40-transformed human cells were not typically immortal. In the original studies, mixed cultures of transformed and non-transformed cells were passaged by conventional subcultivation until all the cells were positive for large T antigen by immunofluorescence. However, after multiple successful passages, the culture underwent a series of morpho-logical changes with marked cell rounding and cell death, termed crisis: 'a period of balanced cell growth and cell death followed by a decrease in the total number of surviving cells' (16). Our laboratory has isolated a series of independent SV40-transformed human fibroblasts (SV/HF) generated with cloned ori-minus SV40 DNA (2, 4, 17). All transformants were 100% large T antigen-positive upon isolation (as a transformed focus or colony in agarose). All showed an extended life-span, 80–90 PD versus the parental HF, which senesced at 64 PD. However, all underwent crisis despite the absence of viral DNA synthesis, ruling out virus production or viral-mediated cell death being responsible for the crisis. Although the mechanisms of extended life-span and crisis are still not understood in detail, several aspects are evident. First, extension of life-span requires expression of large T antigen but does not require small t antigen, since we have found that constructs in which a small deletion affecting small t antigen but not large T antigen still show extended life-span. Such transformants do not show classical transformed foci; the transfectants were isolated by including the gene for resistance to G418 using a non-SV40 promoter (i.e. RSVneo) in the vector. Secondly, persistent growth requires continuous large T antigen expression. HF transformed by origin-defective SV40 encoding a heat-labile, temperature-sensitive (ts) large T antigen (SVtsA58) display extended life-span when cultured at 35 °C, the permissive temperature for large T function. When cultures of such transformed cells (SVtsA/HF) are shifted to 39 °C, large T function is lost and the cells undergo growth arrest, exhibiting morphological changes reminiscent of senescence. Thirdly, large T functions involved in the inactivation of pRb-1 and p53 are likely to be involved. Large T antigen forms complexes with both proteins at 35 °C; however, no complexes are observed at 39 °C with either protein. Immunoreactive large T antigen is still present although at reduced levels. Fourthly, it has also been demonstrated that the large T–p53 complex is important since HF infected by a mutant SV40 which is unable to bind human p53 does not have the extension in life-span (18). In view of the role of pRb-1- and p53-induced gene products in senescence, it is not surprising that such functions are important to SV40-mediated extension of life-span as well.

A major unanswered question remains as to why the initial ability of large T antigen to extend the life-span ceases to be sufficient. Although no answers are currently available, two observations are relevant. SV/HF show an increased requirement for growth factors, suggesting a decreased responsiveness to them. Although the mechanism is unknown, it is known that reducing the concentration of growth factors can induce an injury response pathway and trigger apoptosis. Indeed, several features of crisis are similar to apoptosis and unpublished data obtained in this laboratory show internucleosomal DNA fragmentation, typical of apoptosis, in SV/HF at crisis. Also, shortening of cell telomeric DNA sequences continues during this period of extended life-span. Therefore, telomeres may have reached a critical length that results in major changes in chromosome stability or chromatin structure. Alternatively, the persistent inactivation of p53 (through complex formation with large T antigen) and the loss of normal checkpoint controls may have accumulated sufficient DNA damage that cells are no longer viable. A limited number of studies have examined gene expression in cells during or approaching crisis. Several mRNAs characteristic of senescent cells are elevated; however, the significance of these changes is unclear.

4. SV40-immortalized cell lines

A rare SV40-transformed human cell becomes immortal. Since all SV/HF cells express large T antigen and there is no evidence for major changes in SV40 function in the immortal compared with the pre-immortal SV/HF (4), attention has focused on non-SV40 functions. Several lines of evidence support this approach. First, immortalization is recessive to limited life-span in cell hybrids. Hybrids between senescent cells and normal cells have reduced a life-span consistent with the presence of an inhibitor of cell proliferation in senescent cells. Hybrids between immortal and normal cells indicate that immortal cells are not resistant to such inhibition since the life-span is limited, consistent with that of the normal cell. Rather, the data support the model that immortal cells have inactivated their own senescence gene. Secondly, hybrids between immortal cells are often not immortal. Pereira-Smith and Smith (19) have shown that hybrids between independent immortal cell lines behave in a fashion consistent with complementation of loss of function. Although all but one of the SV/HF that they tested did not complement each other, multiple cell lines did complement SV/HF and each other. The data fit a minimum of four complementation groups, with SV/HF and some human tumour cell lines in group A. HeLa and several other tumour lines were in group B. It should be noted that HeLa cells are from a patient with cervical carcinoma infected by human papillomavirus (HPV). HPV is related to SV40; both have proteins which bind to pRb-1 and p53 and inactivate their functions, although by different mechanisms. The fact that SV/HF and HeLa are immortal for different reasons further emphasizes that although these are growth suppressors, which are pivotal to extended life-span, their inactivation is insufficient for immortalization.

As a prelude to identifying the gene involved in immortalization, several laboratories have employed genetic approaches that have been useful in identifying genes critical in carcinogenesis. We (4, 17) and others (20) have found that loss of sequences on the long arm of chromosome 6 is associated with immortalization of matched sets of pre-immortal and immortal SV/HF. The model is that the grossly intact chromosome has a mutated copy of the gene responsible for senescence (*SEN6*) and the other copy is deleted on the rearranged chromosome 6. Hence, there is no functional copy of *SEN6*. Consistent with this model, introduction of a normal chromosome 6 by microcell-mediated cell transfer experiments (MMCT) results in restoration of senescence and the inhibition of cellular proliferation (21). Although immortal tumour cell lines typically have multiple chromosome rearrangements, which complicate the first approach, MMCT can be used. Such experiments have been successfully employed by others to demonstrate a putative senescence gene for complementation group B cells on chromosome 4 (22), for group C cells on chromosome 1 (23), and for group D cells on chromosome 7 (24). Furthermore, data thus far also show that introduction of these chromosomes does not suppress the growth of other tested complementation groups, ruling out the idea that they are general growth suppressors. It should be noted that these studies do not preclude the presence of multiple growth suppressors on each chromosome being tested. These data explain, at least in part, the observation that immortalization has a virtually undetectable frequency in normal fibroblasts. Since multiple growth suppressors have to be inactivated in human fibroblasts by SV4O, multiple mutational events would presumably be required for HF when SV40 is not involved. Indeed, the immortal human fibroblasts used in the studies on chromosome 7 had undergone over 20 treatments with mutagen. Fibroblasts from patients with Li–Fraumeni syndrome have a mutant p53 and have been reported to 'spontaneously' become immortal (25). Such cell lines are highly aneuploid, consistent with loss of p53 checkpoint function and accumulation of genomic rearrangements.

This interpretation also explains why even SV40-transformed HF do not readily become immortal. Indeed in studies in this laboratory and others in which pSV3neo-transformed HF were studied, the frequency of immortalization for individual transformants differed considerably from one to another. About 40% of the individual SV/HF yielded immortalized derivatives, which compares favourably with earlier studies using virus-transformed HF. The frequency of immortalized derivatives differed widely, ranging from 10^{-3} to 10^{-7} per cell for different independent transformants. Furthermore, 60% of the SV/HF clones did not yield any immortalized derivatives, even though all the cells from those colonies also expressed large T antigen and were indistinguishable in growth properties from those that produced immortal cells. This range amongst individual SV/HF clones suggests that the pre-immortal SV/HF may have undergone a mutation in the *SEN6* gene at different times in its passage history, resulting in variable numbers of heterozygotic cells when the second allele becomes inactivated. The results also suggest that uncontrolled factors may be influencing the outgrowth of immortalized derivatives.

Protocol 2 provides a method for isolation of immortal cell lines. It recommends the use of pooled SV/HF to maximize the likelihood that an SV/HF which has undergone a *SEN6* mutation is present in the population and, therefore, that an immortalized cell line will be obtained.

Protocol 2. Isolation of SV40 immortal human cells

This method is applicable for transformation by SV40 for many different cell types. Normal fetal HF characteristically senesce at 60–65 PD and life-span extension brings it up to approximately 90 PD; therefore, immortal derivatives are defined to exceed at least 100 population doublings.

Method

1. Subculture adherent cells in suitable serum-containing medium.

2. Plate 5×10^5 cells per 100 mm dish at 37 °C, 24 h prior to transfection of DNA.

3. Transfection of SV40 DNA can be done using several methods. This laboratory routinely uses the calcium phosphate (Ca-P) DNA co-precipitation method. SV40 DNA (ori-minus) (1–10 μg) in plasmid vectors is used, plus 10–20 μg calf thymus DNA per dish. Mock transfections utilize only calf thymus DNA as control. All subsequent steps and subcultures of cells are performed at 37 °C. In the case of ori-minus SVtsA58 DNA, serial passages of cells are done at 35 °C.

4. Add DNA Ca-P co-precipitates to each dish and incubate for 4–16 h.

5. Remove the precipitate and wash cells twice with PBS. Re-feed cells with the appropriate medium.

6. Allow cells to reach confluence. Change the medium twice weekly. Continue to incubate cultures until foci appear. Then calculate the transformation frequency.

7. Resuspend the cells once foci are clearly visible. Subculture at high density (1:3 to 1:5 ratios or $1–3 \times 10^6$ cells/100 mm dish). Culture all cells in order to maintain all transformants.

8. Allow cultures to attain confluence while monitoring the reappearance of foci. Freeze replicates cultures under conditions for viable cell storage.

9. Subculture cells repeatedly at high cell density until the normal cells in the culture become senescent. The culture becomes 100% large T antigen-positive at this time.

10. Eventually the cultures of the transformed cells exhibit a gradual decrease in the rate of growth, indicating entry into crisis. At this point, subculture at 1:3 rather than 1:5 ratios, without discarding any cells. In the absence of cell proliferation, maintain cultures without subcultivation.

11. Passage cells which survive crisis at high cell densities to ensure that immortal cells are actively growing.

12. Freeze immortal cells under conditions for viable cell storage at early passage for reference.

13. Verify that cells are 100% positive for large T antigen protein by immunohistochemical methods. It should be noted that the immortal cells might be derived from more than one SV40 transformant in steps 3–7.

In the event that immortal cell lines are not isolated following crisis, an earlier passage of cells (step 8) should be thawed and steps 9–13 repeated with larger cell numbers. In some cases it may be desirable to isolate immortal cells from individual transformants. Multiple foci should be picked at step 7 and passaged because some individual transformants may not yield immortalized cells at detectable frequencies.

Once the culture of immortal cells is obtained, the cells can be cloned. The cloned sublines can be assessed to determine whether they are derived from different original SV/HF based on Southern blot analysis of the integrated SV40 genome. Different pre-immortal SV/HF would be expected to have different patterns of SV40 sequences. We have shown that rearrangements of the SV40 genome are not required for immortalization. Hence, cell lineage can be verified under conditions where a few integrated copies are involved, such as in origin-minus transformants. In those cases where matched pre-immortal and immortal cell lines are required, multiple independent SV/HF should be tested in parallel; otherwise, the methods are similar.

The molecular basis for immortalization is unknown, although some intriguing information is available. Comprehensive analysis is required, and approaches assessing differential gene expression (e.g. cDNA libraries) in immortal versus pre-immortal SV/HF are in progress (26). Chromosome rearrangements involving 6q have been reported in multiple human tumours (see ref. 22 for examples) and introduction of chromosome 6 has been reported to suppress tumour (melanoma) metastasis in nude mice. We have introduced either an intact chromosome 6 or 6q into SV40 transformants and have found that both confer the loss of proliferation and induce a senescent morphology. Further mapping studies have localized the putative senescence gene *SEN6* to 6q27 (27).

5. Telomeres and telomerase

Earlier studies have shown that the rate of telomere loss on human chromosomes is approximately 40–50 bp per cell generation. This decrease correlated with the absence of detectable telomerase activity in somatic cells, whereas in germ cells and some stem cells, in which telomeres are comparably longer, telomerase is present. These observations spawned the hypothesis that telomere attrition may constitute a causal mechanism for replicative senescence and act as an important biological clock. This idea was further supported by numerous reports of telomerase activation in many human tumours. In any event, telomere stabilization, regardless of telomerase activation, appears to be a necessary component for continued cellular proliferation.

More recent and very exciting investigations have tested the telomere hypothesis of ageing. Both the catalytic subunit and RNA moiety of telomerase have been cloned and sequenced. Overexpression of the catalytic subunit of telomerase has been reported to significantly extend the life-span of human cells in culture (28), possibly indefinitely. Thus, telomere maintenance may play a dynamic role for both replicative senescence and immortalization. Analysis of telomerase activity provides a useful and sensitive tool to identify cultures of immortalized human cells. *Protocol 3* describes a method for determining telomerase in small numbers of cells (29).

It should be noted there is still controversy as to the universality of telomerase activation in immortalization. First, not all human tumours display detectable telomerase activity (30). Therefore, other telomere stabilization mechanisms must be operable, such as telomere recombination or telomere-binding proteins. Secondly, rodents have longer telomeres than humans but age faster. In mouse knock-out studies, in which the RNA component has been inactivated, mice are viable even in the presence of age-related decline of telomeric sequences. Thirdly, while rodent cells spontaneously immortalize in culture, this rarely occurs for human cells. One would predict that if a one-step mutational event could activate telomerase, it would lead to immortalization of human cells, and higher spontaneous immortalization rates would be operable for human cells. Yet this does not occur. Such paradoxes have elicited questions as to the exact nature of telomerase during immortalization. The answers to such issues should be extremely insightful for carcinogenesis.

Protocol 3. Telomerase assay [modification of Piatyszek *et al.*, (29)]

This assay is sensitive to detecting telomerase activity in extracts representing 10^3 cells.

Reagents

- 10 mM Hepes–KOH, pH7.5
- 1.5 mM MgCl$_2$
- 10 mM KCl
- 1 mM dithiothreitol

- lysis buffer: 10 mM Tris–HCl, pH 7.5, 1 mM MgCl$_2$, 1 mM EGTA, 0.1 mM phenylmethyl-sulfonyl fluoride, 5 mM β-mercaptoethanol, 0.5% CHAPS, and 10% glycerol

Method

1. Wash cells with PBS, remove from Petri dish with a rubber policeman in 1.5 ml, and then pellet in a microfuge tube at 10000 *g* for 1 min at 4°C.

---

2. Resuspend cells in cold 10 mM Hepes–KOH, pH7.5, 1.5 mM MgCl$_2$, 10 mM KCl, and 1 mM dithiothreitol. Pellet again at 10000 *g*.

3. Resuspend the pellet such that there are 10^4–10^6 cells in 20 μl of cold lysis buffer.[a]

4. Incubate for 30 min on ice, then centrifuge for 30 min at 12000 *g* at 4°C.

5. Remove supernatant, and assay as described below. Lysates can be quick frozen in acetone–dry ice and stored at –70°C and stored as multiple small aliquots. They should be thawed no more than 3–4 times.

[a] It is necessary to resuspend cells at a concentration of 10^6 cells per 20 μl to maintain stable activity. Extracts can be diluted at the time of assay to 10^3 cells per reaction.

Protocol 4. TRAP assay

Reagents

- 20 mM Tris–HCl (pH 8.3)
- 1.5 mM MgCl$_2$
- 63 mM KCl
- 0.005% Tween-20
- 1 mM EGTA
- 50 μM dNTPs
- TS[a] oligonucleotide

- T4g32 protein (optional)
- BSA
- 2 u Tag DNA polymerase
- CHAP cell extracts
- [α-^{32}P]dGTP (10 μCi/μl, 3000 Ci/mmol)
- [α-^{32}P]dCTP (10 μCi/μl, 3000 Ci/mmol)
- 0.5 × Tris–borate–EDTA

Method

1. 50 μl reactions containing 20 mM Tris–HCl (pH 8.3), 1.5 mM MgCl$_2$, 63 mM KCl, 0.005% Tween-20, 1 mM EGTA, 50 μM dNTPs, 0.1 μg TS[a] oligonucleotide, 1 μg of T4g32 protein (optional), 0.1 mg/ml BSA, 2 units of Tag DNA polymerase, and 1–2 μl CHAP cell extracts.

2. If using radiolabelling of products, add 0.2–04 μl of [α-^{32}P]dGTP or [α-^{32}P]dCTP (10 μCi/μl, 3000 Ci/mmol).

3. Incubate reactants at 30°C. This allows for extension of the TS oligo by telomerase.

4. Add the CX[b] primer, 0.1 μg DNA. This primer can be added prior to the extension reaction. If so, 7–10 μl of molten wax should be overlaid and allowed to cool before adding the remainder of the TRAP reactants.

5. PCR conditions now proceed at 27 rounds at 94°C for 30 seconds, 50°C for 30 seconds, and 72°C for 1–5 min.

6. Analyse 25 μl on 15% non-denaturing gels using 0.5 × Tris–borate–EDTA as the running buffer.

Protocol 4. *Continued*

Note. If initial activity is low, test serial dilutions of extracts because high protein concentrations can interfere with the assay. If diluted, the lysis buffer should be kept constant. It has also been recently reported that a combination of NP-40 and sodium deoxycholate is the most efficient for extracting telomerase activity (31). Protein concentrations should be used as a means for normalization of telomerase activity when cell counts are not used.

A TRAP kit (Trapeze™) can be obtained from commercial sources. Procedures are performed in accordance with the supplier (Oncor, Inc.)

[a] TS primer: 5'-AATCCGTCGAGCAGAGTT-3'
[b] CX primer: 5'-AATCCCATTCCCATTCCCATTCCC-3'

6. Conditional SV40 transformants

Temperature-sensitive SV40 constructs have been particularly useful in dissecting processes important for normal cell growth, as discussed briefly above. Furthermore, in many cases cells not only cease to grow when shifted to the non-permissive temperature, but also resume expression of normal differentiated properties. This aspect of conditional mutants of large T antigen makes it attractive for the study of differentiation pathways in diverse cell types, as the cell line at 35 °C serves as the control for the one shifted to 39 °C. The strategies outlined in this chapter for the preparation of SV40-transformed fibroblasts with extended life-span or immortalized cells (*Protocol 2*) can be exploited with other human cell types that are difficult to study in this regard owing to their limited ability to replicate these cells in culture. Human placental cells (infected with tsA mutant virus), thyroid epithelial cells (transfected with tsA DNA), and endometrial cells (transfected with origin-minus tsA DNA) are examples of cell lines that have been developed in this manner. The value of these cell lines is further underscored by the observation that transformation by SV40 may not necessarily lead to an undifferentiated phenotype. tsSV40-transformed cytotrophoblasts display properties of differentiation characteristic of early gestation regardless of the temperature of growth. For example, these cells expressed various cell adhesion molecules, metalloproteinases, and hormones indicative of cytotrophoblast differentiation at 35 °C. Additional properties, such as expression of chorionic gonadotrophin, were induced upon temperature shift to 39 °C. An even more extensive series of rodent cell lines has been established in culture after introduction of an SV40 tsA mutant genome, using virus vectors or DNA transfection. These include rat cerebellar and other central nervous system precursor cells, endometrial cells, hepatocytes, and macrophages.

Two additional classes of SV40-dependent transformants have been developed in the mouse. Cell lines can be isolated from transgenic mice which are generated using tissue-specific regulatory elements and the large T antigen coding region. These cell lines have been useful for the understanding of multistep carcinogenesis for a variety of tissues; for example, lens, liver, skin, pancreas, kidney, and brain. These studies employed a wild-type SV40 genome. Tumours and cell lines derived from them have proved useful in cases in which it was difficult to obtain the corresponding normal cell line for study. However, the use of temperature-sensitive mutants of large T antigen in transgenic mice is now providing insights not only into normal cell growth and senescence, but also into differentiation. In general, tumours do not occur in such transgenic animals. However, relevant cell lines isolated from these animals express a transformed phenotype, including becoming immortal at the permissive temperature in culture and reversion to a normal phenotype at the non-permissive temperature, similarly to what happens to cells trans-formed *in vitro*. In many cases, one or more phenotypic markers of the differentiated cell type also become expressed. Cell lines isolated from transgenic mice generated by introduction of SVtsA58 under the MHC (H-2kb) class I promoter behave in such a manner (32). This promoter is active in different tissues and can be induced for higher expression by exposure to interferon-γ. A variety of cell lines with different tissue specificities (fibroblast, thymic epithelial cells, crypt cells from colon and small intestines, etc.) have been established in culture from tissues of H-2K^b–tsA58 transgenic mice.

Exploitation of the ability to regulate large T antigen expression or function should therefore be useful for the study of the mechanisms required for cellular proliferation and senescence in cell types, which under normal circumstances prove difficult to study *in vitro*.

7. Other approaches to immortalization of human cells

Viral oncogenes other than SV40 have also been used to immortalize human cells. DNA viruses such as the papillomaviruses and adenoviruses appear to act through mechanisms of viral gene expression that mimic that of SV40. The E1A and E1B viral sequences of adenovirus transform and immortalize fibroblasts and epithelial cells. E1A binds to pRb-1 and immortalizes rodent cells; however, E1B (which binds p53) enhances cell survival and is necessary for full transformation. In the case of human cells, both E1A and E1B are required for immortalization. The gene products E7 (binds pRb-1) and E6 (binds p53) of human papillomavirus (HPV) are required to immortalize human fibroblasts and keratinocytes. Most recently it has been reported that inactivation of the Rb/p16 pathway by E7, in combination with telomerase activation, immortalized epithelial cells (but not fibroblasts) efficiently (33). Thus, it is clear that inactivation of the tumour suppressor genes p53 and pRb-l is often necessary at least as a first step in altering and extending the pattern of cell growth. Recently, DNA constructs encoding HPV genes have been increasingly used as an alternative to SV40 T antigen for the generation of immortal human cell lines. *Protocol 2* can be used with the exception that the appropriate sequences are used for DNA transformation in step 3. Constructs using a promoter derived from a retrovirus LTR (33) or other sources are often used instead of the HPV promoter to obtain higher levels of expression of the HPV E6 and E7 intracellularly.

The classes of viruses generally utilized for immortalization of lymphoid cells have been different from those discussed previously for fibroblasts and epithelial cells. Epstein–Barr virus (EBV), a member of the herpesviruses, has been extensively used to induce immortalization of B lymphocytes (34) for studies on human gene mapping and other purposes. This process is dependent on the expression of the EBV nuclear antigen 2 (*EBNA-2*) gene and the latent membrane protein 1 (LMP1). *EBNA-2* is a transcriptional regulator, which mediates EBV latency gene expression and cellular genes (*cfgr* and *CD23*), most likely accounting for its crucial role in immortalization. LMP1 is also necessary for immortalization (35) and can directly transform cell lines. LMP1-induced activation of cell surface markers may be mediated through signalling pathways, such as NF-κB and JNK kinase (36). In the case of T cells, infection with the human retrovirus T cell leukaemia virus (HTLV) has been used. Immortalization is dependent on expression of the viral *tax* gene. In each cell system, the isolation of the immortalized lymphoid cell requires methodologies appropriate to the cultivation of these cells in suspension and is therefore different from that described in *Protocol 2*.

It is recommended that all viruses and virus vectors be handled in accordance with safety guidelines recommended by the National Institutes of Health (US) or other agencies.

8. Summary

Immortalization of human cells by SV40 allows for the analysis of processes that affect cellular proliferation. Isolation of genetically matched sets of normal, SV40–pre-immortal, and SV40–immortal cells facilitates investigations of the mechanisms involved in senescence and immortalization. We have stressed in this chapter the usefulness of SV40-transformed human fibroblasts; however, many different cell types can be similarly immortalized by SV40. There are some instances where other viral agents may be more applicable, as in the case of B lymphocytes, which are readily immortalized by EBV.

Although some SV40 transformants do not yield immortal cell lines, such transformants with extended life-spans are nonetheless valuable for the development of model systems where replication *in vitro* was otherwise exceedingly difficult. Also, in this regard, the utilization of large T antigen function capable of differential regulation often permits the study of cellular differentiation. Thus, transformation by SV40 is applicable for the investigation of the multiple facets of normal cell growth and differentiation in human and other mammalian cell systems.

References

1. Hayflick L. (1965). *Exp. Cell Res.*, **37**, 614.
2. Radna, R., Caton, Y., Jha, K.K., Kaplan, P, Li, G., Traganos, F., and Ozer, H.L. (1989). *Mol. Cell. Biol.*, **9**, 3093.
3. Wright, W.E, Pereira-Smith, O.M., and Shay, J.W. (1989). *Mol. Cell. Biol.*, **9**, 3088.
4. Hubbard-Smith, K., Patsalis, P., Pardinas, J.R., Jha, K.K, Henderson, A.S., and Ozer, H.L. (1992). *Mol. Cell. Biol.*, **12**, 2273.
5. Wright, W.E. and Shay, J.W. (1992). *Trends Genet.*, **8**, 193.
6. Dimri, G.P., Lee, X., Basile, G., Acosta, M., Scott, G., Roskelley, C., Mendrano, E.E., Linskens, M., Rubelj, I., Pereira-Smith, O., Peacocke, M., and Campisi, J. (1995). *Proc. Natl. Acad. Sci. USA*, **92**, 9363.
7. Stein, G.H., Beeson, M., and Gordon, L. (1990). *Science*, **249**, 666.
8. Alcorta, D.A., Xiong, Y., Phelps, D., Hannon, G., Beach, D., and Barrett, J.C. (1996). *Proc. Natl. Acad. Sci. USA*, **93**, 13742.
9. Brown, J.P., Wei, W. and Sedivy, J.M. (1997). *Science*, **277**, 831.
10. Haber, D.A. (1997). *Cell*, **91**, 555.
11. Weinberg, R. (1995). *Cell*, **81**, 323.
12. Xiong, Y., Zhang, H., and Beach, D. (1993). *Genes Dev.*, **7**, 1572.
13. Ko, L. and Prives, C. (1996). *Genes Dev.*, **10**, 1054.
14. Ozer, H.L., Slater, M.L., Dermody, J.J., and Mandel, M. (1981). *J. Virol.*, **39**, 481.
15. Tooze, J. (ed.) (1981). *DNA tumor viruses*, 2nd edn. Cold Spring Harbor Laboratory Press, Cold Spring Harbor.
16. Girardi, A.J., Jensen, F.C., and Koprowski, H. (1965). *J. Cell. Comp. Physiol.*, **65**, 69.
17. Neufeld D., Ripley S., Henderson A., and Ozer H.L. (1987). *Mol. Cell. Biol.*, **7**, 2794.
18. Lin, J.-Y. and Simmons, D.T. (1991). *J. Virol.*, **65**, 6447.
19. Pereira-Smith, O.M. and Smith, J.R. (1988). *Proc. Natl. Acad. Sci. USA*, **85**, 6042.
20. Ray, F.A. and Kraemer, P.M. (1992). *Cancer Genet. Cytogenet.*, **59**, 39.
21. Sandhu, A.K., Hubbard, K., Kaur, G.P., Jha, K.K., Ozer, H.L., and Athwal, R.S. (1994). *Proc. Natl. Acad. Sci. USA*, **91**, 5498.
22. Ning, Y., Weber, J.L., Killary, A.M., Ledbetter, D.H., Smith, J.R., and Pereira-Smith, O.M., (1991). *Proc. Natl. Acad. Sci. USA*, **88**, 5635.
23. Hensler, P.J., Annab, L.A., Barrett, J.C., and Pereira-Smith, O.M. (1994). *Mol. Cell. Biol.*, **14**, 2291.
24. Ogata, T., Ayusawa, D., Namba, M., Takahashi, E., Oshimura, M., and Oishi, M. (1993). *Mol. Cell. Biol.*, **13**, 6036.
25. Bischoff, F., Yim, S.O., Pathak, S., Grant, G., Siciliano, M.J., Giovanella, B.S., Strong, L.C., and Tainsky, M.A. (1990). *Cancer Res.*, **50**, 7979.
26. Pardinas, J., Pang, Z., Houghton, J., Palejwala, V., Donnelly, R., Hubbard, K., Small, M.B., and Ozer, H.L. (1997). *J. Cell Phys.*, **171**, 325.

27. Banga, S.S., Kim, S., Hubbard., Dasgupta, T., Jha, K.K., Patsalis, P., Hauptschein, R., Gamberi, B., Dalla-Favera, P., Kraemer, P., and Ozer, H.L. (1997). *Oncogene*, **14**,313.

28. Bodnar, A.G., Ouellette, M., Frolkis, M., Holt, S.E., Chiu, C.P., Morin, G.B., Harley, C.B., Shay, J.W., Lichtsteiner, S., and Wright, W.E. (1998). *Science*, **279**, 349.

29. Piatyszek, M. A., Kim, N.W., Weinrich, S.L., Hiyama, K., Hiyama, E., Wright, W.E., and Shay, J.W. (1995). *Meth. Cell Sci.*, **17**, 1.

30. Bryan, T.M., Englezou, A., Gupta, J., Bacchetti, S., and Reddel, R.R. (1995). *EMBO J.*, **14**, 4240.

31. Norton, J.C., Holt, S.E., Wright, W.E., and Shay, J.W. (1998). *DNA Cell Biol.*, **17**, 217.

32. Jat, P.S., Noble, M.D., Atalcohis, P., Tanaka, Y., Yannoutsos, N., Larson, L., and Kioussis, D. (1991). *Proc. Natl. Acad. Sci. USA*, **88**, 5096.

33. Kiyono, T., Foster, S.A., Koop, J.I., McDougall, J.K., Galloway, D.A., and Klingelhutz, A.J. (1998). *Nature*, **396**, 84.

34. Middleton, T., Gahn, T.A., Martin, J.M., and Sugden, B. (1991). *Adv. Cancer Res.*, **40**, 19.

35. Kaye,K.M., Izumi, K.M., and Kieff, E. (1993). *Proc. Natl. Acad. Sci. USA*, **90**, 9150.

36. Liljeholm, S., Hughes, K., Grundstrom, T., and Brodin, P. (1998). *J. Gen. Virol.*, **78**, 7117.

A1

List of suppliers

Accumed International, Inc., Microbiology Division, 29299 Clemens Rd., Suite 1-K, Westlake, Ohio 44145, USA.

Accumed International, Inc., Microbiology Division, Imberhorne Lane, East Grinstead, West Sussex RH19 1QX, UK.

Accurate Chemical & Scientific Corp, Westbury, NY, USA

Amersham

Amersham International plc., Lincoln Place, Green End, Aylesbury, Buckinghamshire HP20 2TP, UK.

Amersham Corporation, 2636 South Clearbrook Drive, Arlington Heights, IL 60005, USA.

Amicon, 72 Cherry Hill Drive, Berverly, MA 01915.

Anderman

Anderman and Co. Ltd., 145 London Road, Kingston-Upon-Thames, Surrey KT17 7NH, UK.

Andreas Hettich GmbH & Co AG, Gartenstr. 100, D-78532, Germany.

Beckman Instruments

Beckman Instruments UK Ltd., Oakley Court, Kingsmead Business Park, London Road, High Wycombe, Bucks HP11 1J4, UK.

Beckman Instruments Inc., PO Box 3100, 2500 Harbor Boulevard, Fullerton, CA 92634, USA.

Becton Dickinson

Becton Dickinson and Co., Between Towns Road, Cowley, Oxford OX4 3LY, UK.

Becton Dickinson and Co., 2 Bridgewater Lane, Lincoln Park, NJ 07035, USA.

Bellco Biotechnology, Vineland, NJ, USA.

Bio

Bio 101 Inc., c/o Statech Scientific Ltd, 61–63 Dudley Street, Luton, Bedfordshire LU2 0HP, UK.

Bio 101 Inc., PO Box 2284, La Jolla, CA 92038–2284, USA.

Bio-Rad Laboratories

Bio-Rad Laboratories Ltd., Bio-Rad House, Maylands Avenue, Hemel Hempstead HP2 7TD, UK.

Bio-Rad Laboratories, Division Headquarters, 3300 Regatta Boulevard, Richmond, CA 94804, USA.

Boehringer Mannheim
Boehringer Mannheim UK (Diagnostics and Biochemicals) Ltd, Bell Lane, Lewes, East Sussex BN17 1LG, UK.
Boehringer Mannheim Corporation, Biochemical Products, 9115 Hague Road, P.O. Box 504 Indianapolis, IN 46250–0414, USA.
Boehringer Mannheim Biochemica, GmbH, Sandhofer Str. 116, Postfach 310120 D-6800 Ma 31, Germany.
British Drug Houses (BDH) Ltd, Poole, Dorset, UK.
Collaborative Biomedical Products, Through Becton Dichenson, 2 Oak, Park, Bedford, MA 01730.
DAKO A/S* Produktionsvej 42, DK-2600 Glostrup, Denmark.
Difco Laboratories
Difco Laboratories Ltd., P.O. Box 14B, Central Avenue, West Molesey, Surrey KT8 2SE, UK.
Difco Laboratories, P.O. Box 331058, Detroit, MI 48232–7058, USA.
Du Pont
Dupont (UK) Ltd., Industrial Products Division, Wedgwood Way, Stevenage, Herts, SG1 4Q, UK.
Du Pont Co. (Biotechnology Systems Division), P.O. Box 80024, Wilmington, DE 19880–002, USA.
European Collection of Animal Cell Culture, Division of Biologics, PHLS Centre for Applied Microbiology and Research, Porton Down, Salisbury, Wilts SP4 0JG, UK.
Falcon (Falcon is a registered trademark of Becton Dickinson and Co.).
Fisher Scientific Co., 711 Forbest Avenue, Pittsburgh, PA 15219–4785, USA.
Five Prime-Thre Prime, 5603 Arapahoe, Boulder, CO 80027.
Flow Laboratories, Woodcock Hill, Harefield Road, Rickmansworth, Herts. WD3 1PQ, UK.
Fluka
Fluka-Chemie AG, CH-9470, Buchs, Switzerland.
Fluka Chemicals Ltd., The Old Brickyard, New Road, Gillingham, Dorset SP8 4JL, UK.
Gibco BRL
Gibco BRL (Life Technologies Ltd.), Trident House, Renfrew Road, Paisley PA3 4EF, UK.
Gibco BRL (Life Technologies Inc.), 3175 Staler Road, Grand Island, NY 14072–0068, USA.
Arnold R. Horwell, 73 Maygrove Road, West Hampstead, London NW6 2BP, UK.
Hettich, *see* Andreas Hettich.
Hybaid
Hybaid Ltd., 111–113 Waldegrave Road, Teddington, Middlesex TW11 8LL, UK.

Hybaid, National Labnet Corporation, P.O. Box 841, Woodbridge, NJ. 07095, USA.

HyClone Laboratories 1725 South HyClone Road, Logan, UT 84321, USA.

International Biotechnologies Inc., 25 Science Park, New Haven, Connecticut 06535, USA.

Invitrogen Corporation

Invitrogen Corporation 3985 B Sorrenton Valley Building, San Diego, CA. 92121, USA.

Invitrogen Corporation c/o British Biotechnology Products Ltd., 4–10 The Quadrant, Barton Lane, Abingdon, OX14 3YS, UK.

Kodak: Eastman Fine Chemicals 343 State Street, Rochester, NY, USA.

Life Technologies Inc., 8451 Helgerman Court, Gaithersburg, MN 20877, USA.

Merck

Merck Industries Inc., 5 Skyline Drive, Nawthorne, NY 10532, USA.

Merck, Frankfurter Strasse, 250, Postfach 4119, D-64293, Germany.

Millipore

Millipore (UK) Ltd., The Boulevard, Blackmoor Lane, Watford, Herts WD1 8YW, UK.

Millipore Corp./Biosearch, P.O. Box 255, 80 Ashby Road, Bedford, MA 01730, USA.

New England Biolabs (NBL)

New England Biolabs (NBL), 32 Tozer Road, Beverley, MA 01915–5510, USA.

New England Biolabs (NBL), c/o CP Labs Ltd., P.O. Box 22, Bishops Stortford, Herts CM23 3DH, UK.

Nikon Corporation, Fuji Building, 2–3 Marunouchi 3-chome, Chiyoda-ku, Tokyo, Japan.

Perkin-Elmer

Perkin-Elmer Ltd., Maxwell Road, Beaconsfield, Bucks. HP9 1QA, UK.

Perkin-Elmer Ltd., Post Office Lane, Beaconsfield, Bucks, HP9 1QA, UK.

Perkin-Elmer-Cetus (The Perkin-Elmer Corporation), 761 Main Avenue, Norwalk, CT 0689, USA.

Pharmacia Biotech Europe Procordia EuroCentre, Rue de la Fuse-e 62, B-1130 Brussels, Belgium.

Pharmacia Biosystems

Pharmacia Biosystems Ltd. (Biotechnology Division), Davy Avenue, Knowlhill, Milton Keynes MK5 8PH, UK.

Pharmacia LKB Biotechnology AB, Björngatan 30, S-75182 Uppsala, Sweden.

Promega

Promega Ltd., Delta House, Enterprise Road, Chilworth Research Centre, Southampton, UK.

Promega Corporation, 2800 Woods Hollow Road, Madison, WI 53711–5399, USA.

Qiagen

Qiagen Inc., c/o Hybaid, 111–113 Waldegrave Road, Teddington, Middlesex, TW11 8LL, UK.

Qiagen Inc., 9259 Eton Avenue, Chatsworth, CA 91311, USA.

Schleicher and Schuell

Schleicher and Schuell Inc., Keene, NH 03431A, USA.

Schleicher and Schuell Inc., D-3354 Dassel, Germany. Schleicher and Schuell Inc., c/o Andermann and Company Ltd.

Shandon Scientific Ltd., Chadwick Road, Astmoor, Runcorn, Cheshire WA7 1PR, UK.

Sigma Chemical Company

Sigma Chemical Company (UK), Fancy Road, Poole, Dorset BH17 7NH, UK.

Sigma Chemical Company, 3050 Spruce Street, P.O. Box 14508, St. Louis, MO 63178–9916.

Sorvall DuPont Company, Biotechnology Division, P.O. Box 80022, Wilmington, DE 19880–0022, USA.

Stratagene

Stratagene Ltd., Unit 140, Cambridge Innovation Centre, Milton Road, Cambridge CB4 4FG, UK.

Strategene Inc., 11011 North Torrey Pines Road, La Jolla, CA 92037, USA.

United States Biochemical, P.O. Box 22400, Cleveland, OH 44122, USA.

Universal DAKI APAAP kit, Dako Corp., Santa Barbara, CA, USA.

Vectastain Elite ABC Kit, Vector Laboratories, Burlingame, CA, USA.

Vector Laboratories, Inc., 30 Ingold Rd. Burlingame, CA 94010.

Wellcome Reagents, Langley Court, Beckenham, Kent BR3 3BS, UK.

Worthington Biochemical Corp., 730 Vassar Ave. Lakewood, NH 08701.

ZYMED Laboratories, Inc., 458 Carlton Court, South San Francisco, CA 94080-2012, USA.

Index

Index